线性代数

(汉英双语版)

主编　牛大田　袁学刚　张　友

科学出版社

北　京

内 容 简 介

本书以线性方程组为主线，以行列式、矩阵和向量为工具，阐述线性代数的基本概念、基本理论和方法. 全书内容联系紧密，具有较强的逻辑性. 本书是根据教育部高等院校理工类专业以及经济和管理学科各专业线性代数教学大纲的要求编写而成的. 全书分为六章，各章内容分别是：行列式、矩阵、矩阵的初等变换、向量、方阵的特征值、相似与对角化、二次型. 在每一节都安排思考题的基础上，还为每章配备习题和补充题，习题是学生必做的题目，补充题是为考研学生和对线性代数有更高要求的同学而设计的. 本书采用汉英对照的方式编写，使学生在学习线性代数的同时，大大提高学生的英语读写能力.

本书可作为高等学校理工类学科各专业以及经济和管理学科各专业的教材或教学参考书.

图书在版编目（CIP）数据

线性代数(汉英双语版)/牛大田，袁学刚，张友主编. —北京：科学出版社，2016. 3

ISBN 978-7-03-047382-0

Ⅰ. ①线… Ⅱ. ①牛… ②袁… ③张… Ⅲ. ①线性代数–高等学校–教材–汉、英 Ⅳ. ①O151.2

中国版本图书馆 CIP 数据核字 (2016) 第 031973 号

责任编辑：张中兴／责任校对：张凤琴

责任印制：赵 博／封面设计：迷底书装

科学出版社 出版

北京东黄城根北街 16 号

邮政编码：100717

http://www.sciencep.com

三河市骏杰印刷有限公司印刷

科学出版社发行 各地新华书店经销

*

2016 年 3 月第 一 版 开本：787 × 1092 1/16

2024 年 8 月第七次印刷 印张：20 1/4

字数：480 000

定价：69.00 元

(如有印装质量问题，我社负责调换)

前　言

线性代数是理工和经管类专业的一门重要的基础课, 在自然科学、工程技术和管理科学等诸多领域有着广泛的应用. 线性代数的各个章节知识之间联系非常紧密, 其主要特点如下.

第一, 线性代数中概念抽象. 在刚开始的学习中, 学生的主要难点集中在对一些概念难于接受和理解, 例如, 行列式的定义、矩阵乘法的定义、矩阵的初等变换规则, 尤其是线性相关及线性无关的定义等.

第二, 线性代数中概念、结论、运算比较多, 而且这些概念、结论、运算联系紧密, 例如, 一个方阵是满秩的与方阵所对应的行列式值不为零、列向量组是线性无关的、齐次线性方程组只有零解等结论是等价的.

大部分教材一般是按逻辑顺序 —— 定义、公理、引理、定理、推论的模式来编写的. 但在实际教学中, 往往使学生抓不住知识的主干, “只见树木, 不见森林”, 不知道一开始学习的知识干什么, 只是被动地一步一步跟着走.

针对线性代数这门课程的特点和学生在学习中遇到的问题, 根据高等教育本科线性代数课程的教学基本要求, 编者在自编讲义的基础上, 结合多年来从事线性代数课程教学的体会编写本书. 其目的是为普通高等学校非数学专业学生提供一本适用面较宽的、易教易学的线性代数教材. 此外, 对于非英语专业的学生来说, 英语的主要功能应该是作为学生进行专业学习、阅读科技文献以及为未来工作服务的一种工具. 在编写过程中, 借鉴国内外许多优秀教材的思想和处理方法, 内容上突出精选够用, 表达上力求通俗易懂. 本教材具有以下特色:

(1) 以线性方程组为主线, 把行列式、矩阵和向量作为研究线性方程组的一种工具来学习. 这样有利于学生理解线性代数课程的基本概念和基本原理, 把线性代数中的 “抽象” 变具体、变简单. 使学生对线性代数有整体的把握, 目标明确.

(2) 将初等变换作为贯穿全书的计算工具, 强调它是矩阵的同秩变换; 是向量组的同线性关系变换, 是线性方程组的同解变换. 这样可以把线性代数各个章节的知识非常紧密的联系在一起, 把线性代数中比较多的概念、结论、运算变得易教易学.

(3) 在教材内容和习题的处理上都充分考虑到易教易学, 实用简明. 做到深入浅出, 努力做到每个抽象的定义和定理之前都给出简单、具体的引例, 通俗易懂, 讲清基本概念, 淡化理论证明, 没有令人费解的冗长证明, 努力做到每个抽象的定义和定理之前都给出简单、具体的引例.

(4) 在每一节都安排思考题的基础上, 还为每章配备习题和补充题, 习题是学生必做的题目, 补充题是为考研学生和对线性代数有更高要求的学生而设计的.

(5) 本书采用汉英对照的方式编写, 鉴于线性代数课程的特点, 为便于学生掌握, 英文翻译尽可能采用直译的方式, 使学生在学习线性代数的同时, 大大提高学生的英语读写能力.

本书可作为高等学校理工类学科各专业以及经济和管理学科各专业的教材或教学参考书. 适合 32~48 学时线性代数课程用教材. 对经济和管理学科各专业的学生, 打星号的内容和课后补充题不作为要求.

本书由大连民族大学理学院组织编写, 丛书主编为袁学刚、周文书. 本书主编为牛大田、袁学刚、张友. 参加编写的有王书臣、焦佳、吕娜、张文正、谢从波、赵巍、楚振艳、曲程远、张誉铎、田春艳.

限于编者的经验和水平, 书中难免会有不当与疏漏之处, 敬请专家与广大读者批评指正.

编　者

2015 年 11 月

目　录

第 1 章 Chapter 1

行列式 Determinants

在初等数学中, 我们学过用消元法求解未知数个数较少的**线性方程组**. 本章利用 n 阶行列式求解具有 n 个未知数、n 个方程构成的线性方程组. 行列式的相关理论是线性代数中的重要内容之一, 也是研究线性代数的一个重要工具. 它在数学的许多分支和工程技术中有着广泛的应用.

In elementary mathematics, we have learned how to solve a **system of linear equations** (for short, linear system) in few unknowns by elimination. In this chapter we will solve the systems of n linear equations in n unknowns by the n-th order determinants. The correlation theory of determinants is one of the important contents in Linear Algebra, and is also an important tool for studying Linear Algebra. It is widely used in many branches of mathematics and engineering.

行列式是一个数, 它是由一些数字按一定方式排成的数组所确定的. 这个思想分别早在 1683 年和 1693 就由日本数学家关孝和及德国数学家莱布尼茨提出. 在 1750 年, 瑞士数学家克拉默出版的《线性代数分析导言》一书中给出了行列式的定义, 并给出了利用 n 阶行列式求解 n 元线性方程组的方法, 即克拉默法则.

A determinant is a number and is determined by an array consisted of some numbers arranged in a certain way. This idea was proposed respectively by Seki Takakazu, a Japanese mathematician, in 1683 and by Gottfried Wilhelm Leibniz, a German mathematician, in 1693. Gabriel Cramer, a Swiss mathematician, published his book *Introduction à l'analyse des lignes courbes algériques* in 1750 and gave the definition of determinant and proposed how to solve systems of linear equations in n unknowns by the n-th order determinants, namely, Cramer's rule.

➢本章内容提要

Headline of this chapter

1. 行列式的定义、基本性质及其计算方法;

1. Definitions, basic properties and computational methods of determinants;

2. 利用行列式求解 n 元线性方程组, 即克拉默法则.

2. Solving linear systems in n unknowns by determinants, namely, Cramer's rule.

1.1 行列式的定义

1.1 Definitions of Determinants

1.1.1 二阶、三阶行列式

1.1.1 Second Order and Third Order Determinants

引例 用消元法解下面的方程组

Citing example Solve the following linear system by elimination

$$\begin{cases} 2x_1 + 3x_2 = 8 \\ x_1 \quad - 2x_2 = -3 \end{cases}$$

解 交换第 1 个和第 2 个方程得到

Solution Interchanging the first and the second equations, we have

$$\begin{cases} x_1 \quad - 2x_2 = -3 \\ 2x_1 + 3x_2 = 8 \end{cases}$$

将第 1 个方程两端乘以 −2 加到第 2 个方程得到

Adding (−2) times the first equation on both sides to the second equation, we get

$$\begin{cases} x_1 - 2x_2 = -3 \\ \qquad 7x_2 = 14 \end{cases}$$

将第 2 个方程两端乘以 $\frac{1}{7}$ 得到

Multiplying the second equation by $\frac{1}{7}$ on both sides, we obtain

$$\begin{cases} x_1 - 2x_2 = -3 \\ \qquad x_2 = 2 \end{cases}$$

将第 2 个方程两端乘以 2 加到第 1 个方程得到

Adding 2 times the second equation on both sides to the first equation yields

$$\begin{cases} x_1 = 1 \\ x_2 = 2 \end{cases}$$

问题 对于一般二元线性方程组是否用公式求解?

Question Are there exist formulas for solving general linear systems in two unknowns?

考虑如下二元线性方程组

Consider the following linear system in two unknowns

$$\begin{cases} a_{11}x_1 + a_{12}x_2 = b_1 \\ a_{21}x_1 + a_{22}x_2 = b_2 \end{cases} \tag{1.1}$$

利用消元法, 得

Using elimination method, we have

$$(a_{11}a_{22} - a_{12}a_{21})x_1 = b_1a_{22} - a_{12}b_2$$
$$(a_{11}a_{22} - a_{12}a_{21})x_2 = a_{11}b_2 - b_1a_{21}$$

当 $a_{11}a_{22} - a_{12}a_{21} \neq 0$ 时, 方程组 (1.1) 有唯一解

As $a_{11}a_{22} - a_{12}a_{21} \neq 0$, System (1.1) has a unique solution, given by

$$x_1 = \frac{b_1a_{22} - a_{12}b_2}{a_{11}a_{22} - a_{12}a_{21}}$$
$$x_2 = \frac{a_{11}b_2 - b_1a_{21}}{a_{11}a_{22} - a_{12}a_{21}}$$

为了进一步讨论方程组 (1.1) 的解与未知量的系数之间的关系, 引入记号

To further discuss the relation between the solutions and the coefficients of unknowns of (1.1), we introduce the following notation

$$\begin{vmatrix} a_{11} & a_{12} \\ a_{21} & a_{22} \end{vmatrix}$$

并称为**二阶行列式**, 它表示代数和 $a_{11}a_{22} - a_{12}a_{21}$, 即

and call it a **second order determinant**, which represents the algebraic sum $a_{11}a_{22} - a_{12}a_{21}$, i.e.,

$$\begin{vmatrix} a_{11} & a_{12} \\ a_{21} & a_{22} \end{vmatrix} = a_{11}a_{22} - a_{12}a_{21}$$

等于实线两个元素乘积与虚线两个元素乘积之差. 其中, 横排的称为行, 纵排的称为列, $a_{ij}(i,j=1,2)$ 称为行列式的元素, i 为行标, j 为列标.

which is equal to the difference between the product of two elements on the real line and the product of two elements on the imaginary line. In the notation, the horizontal array is called a row and the vertical array is called a column, $a_{ij}(i,j=1,2)$ are called the elements of the determinant, where i is the row index and j is the column index.

若令

If let

$$D = \begin{vmatrix} a_{11} & a_{12} \\ a_{21} & a_{22} \end{vmatrix}, \quad D_1 = \begin{vmatrix} b_1 & a_{12} \\ b_2 & a_{22} \end{vmatrix}, \quad D_2 = \begin{vmatrix} a_{11} & b_1 \\ a_{21} & b_2 \end{vmatrix}$$

则方程组 (1.1) 的解, 用二阶行列式可表示为

then the solution of System (1.1) can be expressed by the second order determinants, as follows,

$$x_1 = \frac{D_1}{D}, \quad x_2 = \frac{D_2}{D} \quad (D \neq 0)$$

上式即为二元线性方程组的求解公式 ($D \neq 0$).

The above formulas are those for solving the linear systems in two unknowns as $D \neq 0$.

例 1 解线性方程组

Example 1 Solve the linear system

$$\begin{cases} 2x_1 + 3x_2 = 8 \\ x_1 - 2x_2 = -3 \end{cases}$$

解

Solution

$$D = \begin{vmatrix} 2 & 3 \\ 1 & -2 \end{vmatrix} = 2 \times (-2) - 3 \times 1 = -7$$

$$D_1 = \begin{vmatrix} 8 & 3 \\ -3 & -2 \end{vmatrix} = 8 \times (-2) - 3 \times (-3) = -7$$

$$D_2 = \begin{vmatrix} 2 & 8 \\ 1 & -3 \end{vmatrix} = 2 \times (-3) - 8 \times 1 = -14$$

因为 $D=-7\neq 0$, 所以线性方程组有唯一解

Since $D=-7\neq 0$, the linear system has a unique solution, given by

$$x_1=\frac{D_1}{D}=\frac{-7}{-7}=1,\quad x_2=\frac{D_2}{D}=\frac{-14}{-7}=2 \qquad \blacksquare$$

类似地, 对于如下三元线性方程组

Similarly, for the following linear system in three unknowns

$$\begin{cases} a_{11}x_1+a_{12}x_2+a_{13}x_3=b_1\\ a_{21}x_1+a_{22}x_2+a_{23}x_3=b_2\\ a_{31}x_1+a_{32}x_2+a_{33}x_3=b_3 \end{cases} \tag{1.2}$$

在一定条件下该方程组有唯一解, 且其解也可以由相应的三阶行列式简化表示.

it has a unique solution under certain conditions and the solution can be simply expressed by the corresponding third order determinants.

记号

The notation

$$\begin{vmatrix} a_{11} & a_{12} & a_{13}\\ a_{21} & a_{22} & a_{23}\\ a_{31} & a_{32} & a_{33} \end{vmatrix}$$

称为**三阶行列式**, 它由三行三列共 9 个元素组成, 其值为

is called a **third order determinant** which is consisted of three rows and three columns and has totally 9 elements, moreover, its value is equal to

$$a_{11}a_{22}a_{33}+a_{13}a_{21}a_{32}+a_{12}a_{23}a_{31}-a_{13}a_{22}a_{31}-a_{11}a_{23}a_{32}-a_{12}a_{21}a_{33}$$

即

namely,

$$\begin{vmatrix} a_{11} & a_{12} & a_{13}\\ a_{21} & a_{22} & a_{23}\\ a_{31} & a_{32} & a_{33} \end{vmatrix}=a_{11}a_{22}a_{33}+a_{13}a_{21}a_{32}+a_{12}a_{23}a_{31}-a_{13}a_{22}a_{31}-a_{11}a_{23}a_{32}-a_{12}a_{21}a_{33} \tag{1.3}$$

可以借助下面图形来理解, 3 条实线上 3 个元素的乘积取正, 3 条虚线上 3 个元素的乘积取负.

It can be understood by the following graph. The products of the three elements on the three solid lines take positive signs and the products of the three elements on the three dashed lines take negative signs.

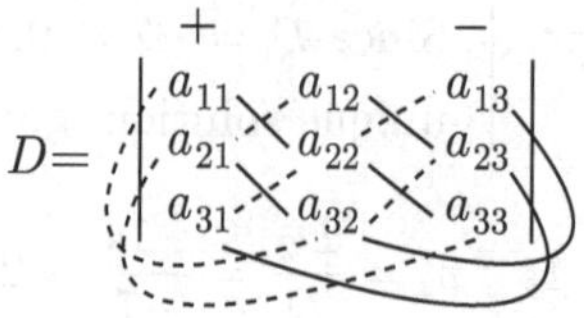

——三阶行列式

此方法也称为**对角线法则**.

This method is also called the **diagonal rule**.

若令

If let

$$D=\begin{vmatrix} a_{11} & a_{12} & a_{13} \\ a_{21} & a_{22} & a_{23} \\ a_{31} & a_{32} & a_{33} \end{vmatrix}, \quad D_1=\begin{vmatrix} b_1 & a_{12} & a_{13} \\ b_2 & a_{22} & a_{23} \\ b_3 & a_{32} & a_{33} \end{vmatrix}$$

$$D_2=\begin{vmatrix} a_{11} & b_1 & a_{13} \\ a_{21} & b_2 & a_{23} \\ a_{31} & b_3 & a_{33} \end{vmatrix}, \quad D_3=\begin{vmatrix} a_{11} & a_{12} & b_1 \\ a_{21} & a_{22} & b_2 \\ a_{31} & a_{32} & b_3 \end{vmatrix}$$

用消元法容易验证, 当 $D \neq 0$ 时, 方程组 (1.2) 的唯一解为

it can be easily verified by elimination that, as $D \neq 0$, System (1.2) has a unique solution, given by

$$x_1=\frac{D_1}{D}, \quad x_2=\frac{D_2}{D}, \quad x_3=\frac{D_3}{D}$$

上式为三元线性方程组的求解公式 ($D \neq 0$).

The above formulas are those for solving the linear systems in three unknowns as $D \neq 0$.

例 2 计算

Example 2 Calculate

$$\begin{vmatrix} 1 & 0 & -1 \\ 2 & 1 & -2 \\ 0 & 3 & 1 \end{vmatrix}$$

解 由对角线法则可得

Solution From the diagonal rule we have

$$\begin{aligned}\begin{vmatrix} 1 & 0 & -1 \\ 2 & 1 & -2 \\ 0 & 3 & 1 \end{vmatrix} &= 1\times 1\times 1+(-1)\times 2\times 3+0\times(-2)\times 0 \\ &\quad -(-1)\times 1\times 0-1\times(-2)\times 3-0\times 2\times 1 \\ &= 1\end{aligned}$$

■

例 3 求解线性方程组

Example 3 Solve the linear system

$$\begin{cases} 2x_1 - \ x_2 + \ x_3 = 0 \\ 3x_1 + 2x_2 - 5x_3 = 1 \\ x_1 + 3x_2 - 2x_3 = 4 \end{cases}$$

解 由前面的公式和对角线法则, 不难求得

Solution From the former formulas and the diagonal rule, it is not difficult to obtain

$$D = \begin{vmatrix} 2 & -1 & 1 \\ 3 & 2 & -5 \\ 1 & 3 & -2 \end{vmatrix} = 28, \quad D_1 = \begin{vmatrix} 0 & -1 & 1 \\ 1 & 2 & -5 \\ 4 & 3 & -2 \end{vmatrix} = 13$$

$$D_2 = \begin{vmatrix} 2 & 0 & 1 \\ 3 & 1 & -5 \\ 1 & 4 & -2 \end{vmatrix} = 47, \quad D_3 = \begin{vmatrix} 2 & -1 & 0 \\ 3 & 2 & 1 \\ 1 & 3 & 4 \end{vmatrix} = 21$$

因为 $D = 28 \neq 0$, 所以方程组有唯一解, 即

Since $D = 28 \neq 0$, the system has a unique solution, given by

$$x_1 = \frac{D_1}{D} = \frac{13}{28}, \quad x_2 = \frac{D_2}{D} = \frac{47}{28}, \quad x_3 = \frac{D_3}{D} = \frac{21}{28} = \frac{3}{4}$$ ■

1.1.2 n 阶行列式

1.1.2 n-th Order Determinants

引入二阶、三阶行列式概念之后, 二元、三元线性方程组的解可以很方便地由二阶、三阶行列式表示出来.

After introducing the second order and the third order determinants, the solutions of the linear systems in two and three unknowns can be expressed conveniently by the second order and the third order determinants.

问题 对于 n 元线性方程组

Question For the following linear system in n unknowns

$$\begin{cases} a_{11}x_1 + a_{12}x_2 + \cdots + a_{1n}x_n = b_1 \\ a_{21}x_1 + a_{22}x_2 + \cdots + a_{2n}x_n = b_2 \\ \qquad \cdots\cdots \\ a_{n1}x_1 + a_{n2}x_2 + \cdots + a_{nn}x_n = b_n \end{cases} \tag{1.4}$$

在一定条件下, 它的解是否有类似的结论?

under certain conditions, whether there has a similar conclusion for its solution?

回答是肯定的, 可见 1.4 节. 为此, 引入 n 阶行列式的定义. 根据前面的讨论, 我们暂且将下面的记号

The answer is yes, see Section 1.4 for details. Therefore, we introduce the definition of n-th order determinants. From the above discussions we call temporarily the following notation

$$\begin{vmatrix} a_{11} & a_{12} & \cdots & a_{1n} \\ a_{21} & a_{22} & \cdots & a_{2n} \\ \vdots & \vdots & & \vdots \\ a_{n1} & a_{n2} & \cdots & a_{nn} \end{vmatrix} \tag{1.5}$$

称为 n**阶行列式**, 记为 D. 它是由 n 行 n 列共 n^2 个元素组成的. 在明确式 (1.5) 的具体含义之前, 先给出余子式、代数余子式的定义.

an **n-th order determinant**, written as D, which is consisted of n rows and n columns and has totally n^2 elements. Before giving the concrete meaning of Equation (1.5), we introduce the definitions of cofactors and algebraic cofactors.

定义 1 在式 (1.5) 中, 划去元素 $a_{ij}(i,j=1,2,\cdots,n)$ 所在的第 i 行和第 j 列后, 剩下的元素按原来的顺序构成的 $n-1$ 阶行列式, 称为元素 a_{ij} 的**余子式**, 记作 M_{ij}, 即

Definition 1 The $(n-1)$-th order determinant obtained by deleting the i-th row and the j-th column in Equation (1.5) is called the **cofactor** of $a_{ij}(i,j=1,2,\cdots,n)$, written as M_{ij}, namely,

$$M_{ij}=\begin{vmatrix} a_{11} & \cdots & a_{1,j-1} & a_{1,j+1} & \cdots & a_{1n} \\ \vdots & & \vdots & \vdots & & \vdots \\ a_{i-11} & \cdots & a_{i-1,j-1} & a_{i-1,j+1} & \cdots & a_{i-1,n} \\ a_{i+11} & \cdots & a_{i+1,j-1} & a_{i+1,j+1} & \cdots & a_{i+1,n} \\ \vdots & & \vdots & \vdots & & \vdots \\ a_{n1} & \cdots & a_{n,j-1} & a_{n,j+1} & \cdots & a_{nn} \end{vmatrix}$$

并且

and call

$$A_{ij}=(-1)^{i+j}M_{ij}$$

称为元素 a_{ij} 的**代数余子式**.

is called the **algebraic cofactor** of a_{ij}.

例如, 在如下的三阶行列式中

For example, in the following third order determinant,

$$\begin{vmatrix} a_{11} & a_{12} & a_{13} \\ a_{21} & a_{22} & a_{23} \\ a_{31} & a_{32} & a_{33} \end{vmatrix}$$

第 1 行元素 a_{11}, a_{12}, a_{13} 的代数余子式分别为

the algebraic cofactors of the elements a_{11}, a_{12}, a_{13} on the first row are respectively given by

$$A_{11}=(-1)^{1+1}\begin{vmatrix} a_{22} & a_{23} \\ a_{32} & a_{33} \end{vmatrix}, \quad A_{12}=(-1)^{1+2}\begin{vmatrix} a_{21} & a_{23} \\ a_{31} & a_{33} \end{vmatrix}, \quad A_{13}=(-1)^{1+3}\begin{vmatrix} a_{21} & a_{22} \\ a_{31} & a_{32} \end{vmatrix}$$

利用以上结果, 可将式 (1.3) 重新记为

Using the above results, we rewrite Equation (1.3) as

$$\begin{vmatrix} a_{11} & a_{12} & a_{13} \\ a_{21} & a_{22} & a_{23} \\ a_{31} & a_{32} & a_{33} \end{vmatrix} = a_{11}A_{11} + a_{12}A_{12} + a_{13}A_{13} \tag{1.6}$$

表明三阶行列式等于它的第一行元素 a_{11}, a_{12}, a_{13} 与所对应的代数余子式 A_{11}, A_{12}, A_{13} 乘积之和. 这种表示方法具有一般性. 按照这样的思想方法, 我们给出 n 阶行列式 (1.5) 的递归定义.

This means that a third order determinant is equal to the sum of the multiplications of the elements a_{11}, a_{12}, a_{13} on first row by the corresponding cofactors A_{11}, A_{12}, A_{13}. This kind of representation methods has universality. Following this idea, we give the recursive definition of the n-th order determinant (1.5).

定义 2 由 n^2 个元素 $a_{ij}(i,j = 1,2,\cdots,n)$ 组成 n 阶行列式可以定义为

Definition 2 The n-th order determinant consisted of n^2 elements $a_{ij}(i,j = 1,2,\cdots,n)$ can be defined as

$$\begin{vmatrix} a_{11} & a_{12} & \cdots & a_{1n} \\ a_{21} & a_{22} & \cdots & a_{2n} \\ \vdots & \vdots & & \vdots \\ a_{n1} & a_{n2} & \cdots & a_{nn} \end{vmatrix} = a_{11}A_{11} + \cdots + a_{1n}A_{1n}$$

其中 $A_{11},\cdots,A_{1n}$ 分别为第 1 行的元素 $a_{11},\cdots,a_{1n}$ 对应的代数余子式.

where $A_{11},\cdots,A_{1n}$ are algebraic cofactors associated with the elements $a_{11},\cdots,a_{1n}$ on the first row, respectively.

注 (1) 上式也称为 n 阶行列式按第 1 行的展开式.

Note (1) The above formula is also called the expansion of an n-th order determinant in accordance with the first row.

(2) 行列式的递归定义表明, n 阶行列式可以由 n 个 $n-1$ 阶行列式表示, 而每一个 $n-1$ 阶行列式又可由 $n-1$ 个 $n-2$ 阶行列式来表示 …… 如此进行下去, 可将 n 阶行列式用 $(n-1),\cdots,3,2$ 阶行列式表示, 最后表示成 $n!$ 项的代数和.

(2) The above recursive definition of determinants shows that, an n-th order determinant can be expressed by n determinants of order $n-1$, while an $(n-1)$-th order determinant can also be expressed by $n-1$ determinants of order $n-2$, and so on. Performing the above procedure, an n-th order determinant can be expressed by determinants of orders $(n-1),\cdots,3,2$, respectively, and finally by the algebraic sum of $n!$ terms.

(3) 由行列式定义知, 如果行列式中第 1 行元素除 a_{1j} 外都为零, 则行列式等于 a_{1j} 与其对应的代数余子式的乘积, 即

(3) From the definition of determinants we know that, if the elements except for a_{1j} in the first row of a determinant are all zeros, then the determinant is equal to a_{1j} times its algebraic cofactor, i.e.,

$$D=a_{1j}A_{1j}$$

此时, n 阶行列式被约化为一个 $n-1$ 阶行列式.

In this case, the n-th order determinant is reduced to an $(n-1)$-th order determinant.

(4) 当 $n=1$ 时, $|a_{11}|=a_{11}$. 它不能与数的绝对值相混淆, 如一阶行列式, $|-3|=-3$.

(4) $|a_{11}|=a_{11}$ as $n=1$. Do not confuse it with the absolute value of a number. For example, the first order determinant, $|-3|=-3$.

例 4 计算

Example 4 Calculate

$$D=\begin{vmatrix} a_{11} & 0 & \cdots & 0 \\ a_{21} & a_{22} & \cdots & 0 \\ \vdots & \vdots & & \vdots \\ a_{n1} & a_{n2} & \cdots & a_{nn} \end{vmatrix}$$

这个行列式称为**下三角形行列式**, 它的特点是当 $i<j$ 时 $a_{ij}=0\ (i,j=1,2,\cdots,n)$.

This determinant is called a **lower triangular determinant**, whose feature is that $a_{ij}=0\ (i,j=1,2,\cdots,n)$ as $i<j$.

解 在行列式第 1 行中, $a_{12}=a_{13}=\cdots=a_{1n}=0$, 由定义 2, 得

Solution In the first row of the determinant, $a_{12}=a_{13}=\cdots=a_{1n}=0$. From Definition 2 we have

$$D=a_{11}A_{11}$$

显然, A_{11} 是 $n-1$ 阶下三角形行列式, 则

Obviously, A_{11} is a lower triangular determinant of order $n-1$, then

$$A_{11}=a_{22}\begin{vmatrix} a_{33} & 0 & \cdots & 0 \\ a_{43} & a_{44} & \cdots & 0 \\ \vdots & \vdots & & \vdots \\ a_{n3} & a_{n4} & \cdots & a_{nn} \end{vmatrix}$$

依此类推, 不难求出

and so on, it is easy to calculate that

$$D=a_{11}a_{22}\cdots a_{nn}$$

即下三角形行列式等于主对角线上诸元素的乘积. ■

that is to say, the lower triangular determinant is equal to the products of its main diagonal elements. ■

特别地, 主对角行列式

In particular, for the main diagonal determinant, we have

$$\begin{vmatrix} \lambda_1 & 0 & \cdots & 0 \\ 0 & \lambda_2 & \cdots & 0 \\ \vdots & \vdots & & \vdots \\ 0 & 0 & \cdots & \lambda_n \end{vmatrix}=\lambda_1\lambda_2\cdots\lambda_n$$

例 5 证明

Example 5 Prove that

$$D=\begin{vmatrix} 0 & 0 & \cdots & 0 & a_{1n} \\ 0 & 0 & \cdots & a_{2,n-1} & a_{2n} \\ \vdots & \vdots & & \vdots & \vdots \\ a_{n1} & a_{n2} & \cdots & a_{n,n-1} & a_{nn} \end{vmatrix}=(-1)^{\frac{n(n-1)}{2}}a_{1n}a_{2,n-1}\cdots a_{n1}$$

证 在行列式的第 1 行元素中, $a_{11} = a_{12} = \cdots = a_{1,n-1} = 0$, 由定义 2, 得

Proof In the first row of the determinant, $a_{11} = a_{12} = \cdots = a_{1,n-1} = 0$. From Definition 2 we have

$$\begin{aligned}
D =& a_{1n}A_{1n} = (-1)^{1+n}a_{1n}\begin{vmatrix} 0 & \cdots & 0 & a_{2,n-1} \\ 0 & \cdots & a_{3,n-2} & a_{3,n-1} \\ \vdots & & \vdots & \vdots \\ a_{n1} & \cdots & a_{n,n-2} & a_{n,n-1} \end{vmatrix} \\
=&(-1)^{1+n}a_{1n}\cdot(-1)^{1+(n-1)}a_{2,n-1}\begin{vmatrix} 0 & \cdots & 0 & a_{3,n-2} \\ 0 & \cdots & a_{4,n-3} & a_{4,n-2} \\ \vdots & & \vdots & \vdots \\ a_{n1} & \cdots & a_{n,n-3} & a_{n,n-2} \end{vmatrix} \\
=&\cdots = (-1)^{1+n}\cdot(-1)^{1+(n-1)}\cdot\cdots\cdot(-1)^{1+2}a_{1n}a_{2,n-1}\cdot\cdots\cdot a_{n1} \\
=&(-1)^{\frac{(n+4)(n-1)}{2}}a_{1n}a_{2,n-1}\cdots a_{n1} \\
=&(-1)^{\frac{n(n-1)}{2}}a_{1n}a_{2,n-1}\cdots a_{n1}
\end{aligned}$$

特别地, 次对角行列式

In particular, for the minor diagonal determinant, we have

$$\begin{vmatrix} 0 & \cdots & 0 & \lambda_1 \\ 0 & \cdots & \lambda_2 & 0 \\ \vdots & & \vdots & \vdots \\ \lambda_n & \cdots & 0 & 0 \end{vmatrix} = (-1)^{\frac{n(n-1)}{2}}\lambda_1\lambda_2\cdots\lambda_n$$

■

例 6 计算

Example 6 Calculate

$$D = \begin{vmatrix} 1 & 2 & -1 \\ 3 & 1 & 0 \\ -1 & 0 & -2 \end{vmatrix}$$

解 按第 1 行展开, 得

Solution Expanding the determinant in accordance with the first row, we have

$$A_{11} = (-1)^{1+1}\begin{vmatrix} 1 & 0 \\ 0 & -2 \end{vmatrix} = -2$$

$$A_{12} = (-1)^{1+2}\begin{vmatrix} 3 & 0 \\ -1 & -2 \end{vmatrix} = 6$$

$$A_{13} = (-1)^{1+3} \begin{vmatrix} 3 & 1 \\ -1 & 0 \end{vmatrix} = 1$$

$$D = a_{11}A_{11} + a_{12}A_{12} + a_{13}A_{13} = 1 \times (-2) + 2 \times 6 + (-1) \times 1 = 9 \quad \blacksquare$$

当然, 此题也可以用对角线法则计算.

Of course, this example can also be calculated by the diagonal rule.

思 考 题

Questions

1. 余子式 M_{ij} 和元素 a_{ij} 有关系吗?

1. Are there relations between the cofactor M_{ij} and the element a_{ij}?

2. 余子式 M_{ij} 和代数余子式 A_{ij} 之间有什么关系?

2. What is the relation between the cofactor M_{ij} and the algebraic cofactor A_{ij}?

1.2 行列式的性质

1.2 Properties of Determinants

利用行列式的定义直接计算行列式一般很困难, 行列式的阶数越高, 困难越大. 为了简化相应的计算, 下面讨论行列式的性质. 为了方便起见, 我们略去了部分性质的理论证明.

Generally, it is very difficult to calculate determinants directly by the definition. The higher order a determinant has, the more difficult the calculation is. To simplify the corresponding calculation, in the following part we discuss the properties of determinants. For convenience, we omit the theoretical proofs for some properties.

根据 n 阶行列式的定义, n 阶行列式等于它的第 1 行的各元素与其对应的代数余子式的乘积之和. 事实上, 可以证明如下定理.

In accordance with the definition of determinats, an n-th order determinant is equal to the sum of the multiplications of the elements on the first row by the corresponding algebraic cofactors. In fact, we can also prove the following theorem.

定理 1 n 阶行式等于它的任意一行 (列) 的各元素与其对应代数余子式的乘积之和, 即

Theorem 1 An n-th order determinant is equal to the sum of the multiplications of the elements in an arbitrary row (or column) by the corresponding algebraic cofactors, namely,

$$D = a_{i1}A_{i1} + a_{i2}A_{i2} + \cdots + a_{in}A_{in} \quad (i = 1, 2, \cdots, n)$$

或

or

$$D = a_{1j}A_{1j} + a_{2j}A_{2j} + \cdots + a_{nj}A_{nj} \quad (j = 1, 2, \cdots, n)$$

该定理又称为**行列式的按行 (列) 展开定理**. 特别地, 有如下推论.

This theorem is also called the **expansion theorem in accordance with a row (or column) of a determinant**. In particular, we have the following corollary.

推论 1 如果 n 阶行列式中第 i 行 (j 列) 元素除 a_{ij} 外都为零, 那么行列式等于 a_{ij} 与其对应的代数余子式的乘积, 即

Corollary 1 If the elements in the i-th row (or the j-th column) except for a_{ij} of an n-th order determinant are all zero, then the determinant is equal to the multiplication of a_{ij} by its corresponding algebraic cofactor, namely,

$$D = a_{ij}A_{ij}$$

对于如下的 n 阶行列式

For the following n-th order determinant

$$D = \begin{vmatrix} a_{11} & a_{12} & \cdots & a_{1n} \\ a_{21} & a_{22} & \cdots & a_{2n} \\ \vdots & \vdots & & \vdots \\ a_{n1} & a_{n2} & \cdots & a_{nn} \end{vmatrix}$$

若将 D 中行与列互换, 得到新的行列式

Interchanging the rows and the columns of D, we have a new determinant

$$D^{\mathrm{T}} = \begin{vmatrix} a_{11} & a_{21} & \cdots & a_{n1} \\ a_{12} & a_{22} & \cdots & a_{n2} \\ \vdots & \vdots & & \vdots \\ a_{1n} & a_{2n} & \cdots & a_{nn} \end{vmatrix}$$

则称 D^{T}(或记为 D') 为 D 的**转置行列式**.

D^{T}(or written as D') is called the **transpose determinant** of D.

性质 1 (转置性) 行列式与它的转置行列式相等.

Property 1 (Transpose) A determinant is equal to its transpose determinant.

注 性质 1 表明, 行列式中行与列具有同等的地位, 即行列式的性质凡是对行成立的, 对列也同样成立, 反之亦然.

Note Property 1 shows that the rows and the columns of a determinant have the same status, that is to say, the properties of a determinant holding for rows are also valid for columns, and vice versa.

例如, 对于如下的上三角形行列式

For example, for the following upper triangular determinant

$$D = \begin{vmatrix} a_{11} & a_{12} & \cdots & a_{1n} \\ 0 & a_{22} & \cdots & a_{2n} \\ \vdots & \vdots & & \vdots \\ 0 & 0 & \cdots & a_{nn} \end{vmatrix}$$

由性质 1 得

from Poperty 1 we have

$$D = D^{\mathrm{T}} = \begin{vmatrix} a_{11} & 0 & \cdots & 0 \\ a_{12} & a_{22} & \cdots & 0 \\ \vdots & \vdots & & \vdots \\ a_{1n} & a_{2n} & \cdots & a_{nn} \end{vmatrix} = a_{11}a_{22}\cdots a_{nn}$$

性质 2 (反号性) 互换行列式任意两行 (列), 行列式改变符号.

Property 2 (Opposite-sign) Interchanging arbitrary two rows (or columns) of a determinant changes the sign of the determinant.

通常以 r_i 表示行列式的第 i 行, 以 c_i 表示第 i 列, 交换 i, j 两行, 记作 $r_i \leftrightarrow r_j$, 而交换 i, j 两列, 记为 $c_i \leftrightarrow c_j$.

In general, denoted by r_i the i-th row of a determinant, by c_i the i-th column, by $r_i \leftrightarrow r_j$ interchanging the i-th row and the j-th row, by $c_i \leftrightarrow c_j$ interchanging the i-th column and the j-th column.

例如,

For example,

$$\begin{vmatrix} 2 & 3 \\ 1 & -2 \end{vmatrix} = -7, \quad \begin{vmatrix} 1 & -2 \\ 2 & 3 \end{vmatrix} = 7$$

推论 1 行列式中有两行 (列) 完全相同, 行列式等于零.

Corollary 1 A determinant is equal to zero if there exist two identical rows (or columns) in the determinant.

证 互换行列式 D 中对应元素相等的两行 (列), 则由性质 2 可知 $D=-D$, 故 $D=0$. ▲

Proof Interchange the two identical rows (or columns). From Property 2 we know that $D=-D$, then $D=0$. ▲

性质 3 (倍乘性) 行列式中某一行 (列) 的所有元素都乘以数 k, 等于用数 k 乘此行列式, 即

Property 3 (Multiple-multiplication) Multiplying all elements on a row (or column) of a determinant by k is equal to multiplying this determinant by k, namely,

$$\begin{vmatrix} a_{11} & a_{12} & \cdots & a_{1n} \\ \vdots & \vdots & & \vdots \\ ka_{i1} & ka_{i2} & \cdots & ka_{in} \\ \vdots & \vdots & & \vdots \\ a_{n1} & a_{n2} & \cdots & a_{nn} \end{vmatrix} = k \begin{vmatrix} a_{11} & a_{12} & \cdots & a_{1n} \\ \vdots & \vdots & & \vdots \\ a_{i1} & a_{i2} & \cdots & a_{in} \\ \vdots & \vdots & & \vdots \\ a_{n1} & a_{n2} & \cdots & a_{nn} \end{vmatrix}$$

证 将左边行列式按第 i 行展开, 再将 k 提出来即得. ▲

Proof Expanding the left determinant in accordance with the i-th row and then extracting k from each term, the proof is complete. ▲

第 i 行 (列) 乘以 k, 记为 $kr_i(kc_i)$.

Denoted by $kr_i(kc_i)$ multiplying the i-th row (or column) by k.

例如

For example,

$$\begin{vmatrix} 4 & 6 \\ 1 & -2 \end{vmatrix} = 2\begin{vmatrix} 2 & 3 \\ 1 & -2 \end{vmatrix} = 2\times(-7) = -14$$

推论 1 行列式中某一行 (列) 的所有元素的公因子可以提到行列式符号外面.

Corollary 1 The common factor of all elements on a certain row (or column) of a determinant can be extracted out of the determinant.

推论 2 若行列式中有一行 (列) 的元素全为零, 则行列式等于零.

Corollary 2 If the elements on a row (or column) of a determinant are all zeros, then the determinant is equal to zero.

推论 3 若行列式中有两行 (列) 对应元素成比例, 则行列式等于零.

Corollary 3 If the elements on two rows (or columns) of a determinant are proportional correspondingly, then the determinant is equal to zero.

性质 4 (可加性) 若行列式中某一行 (列) 的各元素 a_{ij} 都是两个元素 b_{ij} 与 c_{ij} 之和, 即 $a_{ij} = b_{ij} + c_{ij}(j = 1, 2, \cdots, n, 1 \leqslant i \leqslant n)$, 则该行列式可分解为两个相应的行列式之和, 即

Property 4 (Additivity) If each element a_{ij} on a certain row (or column) of a determinant is the sum of two numbers b_{ij} and c_{ij}, i.e., $a_{ij} = b_{ij} + c_{ij}(j = 1, 2, \cdots, n, 1 \leqslant i \leqslant n)$, then the determinant can be decomposed into the sum of two corresponding determinants, namely,

$$D = \begin{vmatrix} a_{11} & a_{12} & \cdots & a_{1n} \\ \vdots & \vdots & & \vdots \\ b_{i1} + c_{i1} & b_{i2} + c_{i2} & \cdots & b_{in} + c_{in} \\ \vdots & \vdots & & \vdots \\ a_{n1} & a_{n2} & \cdots & a_{nn} \end{vmatrix}$$

$$= \begin{vmatrix} a_{11} & a_{12} & \cdots & a_{1n} \\ \vdots & \vdots & & \vdots \\ b_{i1} & b_{i2} & \cdots & b_{in} \\ \vdots & \vdots & & \vdots \\ a_{n1} & a_{n2} & \cdots & a_{nn} \end{vmatrix} + \begin{vmatrix} a_{11} & a_{12} & \cdots & a_{1n} \\ \vdots & \vdots & & \vdots \\ c_{i1} & c_{i2} & \cdots & c_{in} \\ \vdots & \vdots & & \vdots \\ a_{n1} & a_{n2} & \cdots & a_{nn} \end{vmatrix}$$

证 将 D 按第 i 行展开

Proof Expanding D in accordance with the i-th row, we have

$$D = \sum_{j=1}^{n} (b_{ij} + c_{ij}) A_{ij} = \sum_{j=1}^{n} b_{ij} A_{ij} + \sum_{j=1}^{n} c_{ij} A_{ij} = D_1 + D_2$$

其中 D_1, D_2 分别是上式右边第一、二个行列式. ▲

where D_1 and D_2 are the first and the second determinants on the right hand of the above equation, respectively. ▲

例如

For example,

$$\begin{vmatrix} 4 & 6 \\ 1 & -2 \end{vmatrix} = \begin{vmatrix} 1+3 & 1+5 \\ 1 & -2 \end{vmatrix} = \begin{vmatrix} 1 & 1 \\ 1 & -2 \end{vmatrix} + \begin{vmatrix} 3 & 5 \\ 1 & -2 \end{vmatrix} = (-3) + (-11) = -14$$

性质 5 (倍加性) 行列式任一行 (列) 的各元素乘以一个常数 k 后加到另一行 (列) 对应的元素上, 行列式不变, 即

Property 5 (Multiple-additivity) Adding a constant k times the elements on an arbitrary row (or column) to the corresponding elements on another row (or column), the determinant is invariant, namely,

$$D=\begin{vmatrix} a_{11} & a_{12} & \cdots & a_{1n} \\ \vdots & \vdots & & \vdots \\ a_{i1} & a_{i2} & \cdots & a_{in} \\ \vdots & \vdots & & \vdots \\ a_{j1} & a_{j2} & \cdots & a_{jn} \\ \vdots & \vdots & & \vdots \\ a_{n1} & a_{n2} & \cdots & a_{nn} \end{vmatrix} = \begin{vmatrix} a_{11} & a_{12} & \cdots & a_{1n} \\ \vdots & \vdots & & \vdots \\ a_{i1}+ka_{j1} & a_{i2}+ka_{j2} & \cdots & a_{in}+ka_{jn} \\ \vdots & \vdots & & \vdots \\ a_{j1} & a_{j2} & \cdots & a_{jn} \\ \vdots & \vdots & & \vdots \\ a_{n1} & a_{n2} & \cdots & a_{nn} \end{vmatrix}$$

证 由性质 4 得

Proof From Property 4 we have

$$\begin{vmatrix} a_{11} & a_{12} & \cdots & a_{1n} \\ \vdots & \vdots & & \vdots \\ a_{i1}+ka_{j1} & a_{i2}+ka_{j2} & \cdots & a_{in}+ka_{jn} \\ \vdots & \vdots & & \vdots \\ a_{j1} & a_{j2} & \cdots & a_{jn} \\ \vdots & \vdots & & \vdots \\ a_{n1} & a_{n2} & \cdots & a_{nn} \end{vmatrix}$$

$$= \begin{vmatrix} a_{11} & a_{12} & \cdots & a_{1n} \\ \vdots & \vdots & & \vdots \\ a_{i1} & a_{i2} & \cdots & a_{in} \\ \vdots & \vdots & & \vdots \\ a_{j1} & a_{j2} & \cdots & a_{jn} \\ \vdots & \vdots & & \vdots \\ a_{n1} & a_{n2} & \cdots & a_{nn} \end{vmatrix} + \begin{vmatrix} a_{11} & a_{12} & \cdots & a_{1n} \\ \vdots & \vdots & & \vdots \\ ka_{j1} & ka_{j2} & \cdots & ka_{jn} \\ \vdots & \vdots & & \vdots \\ a_{j1} & a_{j2} & \cdots & a_{jn} \\ \vdots & \vdots & & \vdots \\ a_{n1} & a_{n2} & \cdots & a_{nn} \end{vmatrix}$$

上面等号右边第一个行列式为 D, 第二个行列式两行成比例, 由推论 3 知行列式为零. 因此上面等号右边等于 D. ▲

The first determinant on the right side of the above equation is D, while the second determinant, whose two rows are proportional, is zero by Corollary 3. Hence, the right side of the above equation is equal to D. ▲

性质 5 是简化行列式的一种基本方法. 若用数 k 乘第 j 行 (列) 加到第 i 行 (列) 上, 简记为 $r_i + kr_j(c_i + kc_j)$.

Property 5 is a basic method for simplifying determinants. Denoted by $r_i+kr_j(c_i+kc_j)$ adding k times the j-th row (or column) to the i-th row (or column).

例如

For example,

$$\begin{vmatrix} 4 & 6 \\ 1 & -2 \end{vmatrix} \xlongequal{r_1+(-4)r_2} \begin{vmatrix} 0 & 14 \\ 1 & -2 \end{vmatrix} = 0 - 14 = -14$$

由定理 1 和上述性质, 可推出下面的定理.

From Theorem 1 and the above properties we have the following theorem.

定理 2 行列式中某一行 (列) 的元素与另一行 (列) 的对应元素的代数余子式的乘积之和等于零, 即

Theorem 2 The sum of the multiplications of the elements on a certain row (or column) by the algebraic cofactors associated with the elements on another row (or column) is equal to zero, namely,

$$a_{i1}A_{j1} + a_{i2}A_{j2} + \cdots + a_{in}A_{jn} = 0 \quad (i \neq j)$$

或

or

$$a_{1i}A_{1j} + a_{2i}A_{2j} + \cdots + a_{ni}A_{nj} = 0 \quad (i \neq j)$$

证 设

Proof Let

$$D = \begin{vmatrix} a_{11} & a_{12} & \cdots & a_{1n} \\ \vdots & \vdots & & \vdots \\ a_{i1} & a_{i2} & \cdots & a_{in} \\ \vdots & \vdots & & \vdots \\ a_{j1} & a_{j2} & \cdots & a_{jn} \\ \vdots & \vdots & & \vdots \\ a_{n1} & a_{n2} & \cdots & a_{nn} \end{vmatrix}$$

构造行列式

Construct the following determinant

$$D_1 = \begin{vmatrix} a_{11} & a_{12} & \cdots & a_{1n} \\ \vdots & \vdots & & \vdots \\ a_{i1} & a_{i2} & \cdots & a_{in} \\ \vdots & \vdots & & \vdots \\ a_{i1} & a_{i2} & \cdots & a_{in} \\ \vdots & \vdots & & \vdots \\ a_{n1} & a_{n2} & \cdots & a_{nn} \end{vmatrix} \begin{matrix} \\ \\ \leftarrow i \\ \\ \leftarrow j \\ \\ \\ \end{matrix}$$

显然, $D_1 = 0$, 且 D_1 与 D 的第 j 行各元素的代数余子式对应相等, 由定理 1, 将 D_1 按第 j 行展开, 得

Obviously, $D_1 = 0$, and the algebraic cofactors of the corresponding elements on the j-th rows of D_1 and D are equal. Expanding D_1 in accordance with the j-th row, from Theorem 1 we have

$$a_{i1}A_{j1} + a_{i2}A_{j2} + \cdots + a_{in}A_{jn} = \sum_{k=1}^{n} a_{ik}A_{jk} = 0 \quad (i \neq j)$$

同理可证

Similarly, we have

$$a_{1i}A_{1j} + a_{2i}A_{2j} + \cdots + a_{ni}A_{nj} = \sum_{k=1}^{n} a_{ki}A_{kj} = 0 \quad (i \neq j)$$ ▲

综合定理 1 和定理 2, 对于代数余子式有如下重要结论:

Summarizing Theorem 1 and Theorem 2, we have the following important conclusions for algebraic cofactors,

$$\sum_{k=1}^{n} a_{ik}A_{jk} = \begin{cases} D, & i = j \\ 0, & i \neq j \end{cases}$$

和

and

$$\sum_{k=1}^{n} a_{ki}A_{kj} = \begin{cases} D, & i = j \\ 0, & i \neq j \end{cases}$$

例 1 对于如下的行列式

Example 1 For the following determinant

$$D = \begin{vmatrix} 3 & 6 & 9 & 12 \\ 2 & 4 & 6 & 8 \\ 1 & 2 & 0 & 3 \\ 5 & 6 & 4 & 3 \end{vmatrix}$$

求 $A_{41}+2A_{42}+3A_{44}$.

Find $A_{41}+2A_{42}+3A_{44}$.

解 根据定理 1

Solution From Theorem 1, we have

$$A_{41}+2A_{42}+3A_{44}=A_{41}+2A_{42}+0A_{43}+3A_{44}=\begin{vmatrix}3&6&9&12\\2&4&6&8\\1&2&0&3\\1&2&0&3\end{vmatrix}=0$$

■

例 2 计算

Example 2 Calculate

$$\begin{vmatrix}246&427&327\\1014&543&443\\-342&721&621\end{vmatrix}$$

解 由行列式的性质及按行 (列) 展开定理, 得

Solution From the properties and the expansion theorem in accordance with a row (or column) of determinants, we have

$$\begin{vmatrix}246&427&327\\1014&543&443\\-342&721&621\end{vmatrix}=\begin{vmatrix}1000&100&327\\2000&100&443\\1000&100&621\end{vmatrix}=10^5\begin{vmatrix}1&1&327\\2&1&443\\1&1&621\end{vmatrix}=10^5\begin{vmatrix}0&1&327\\1&1&443\\0&1&621\end{vmatrix}$$

$$=-10^5\begin{vmatrix}1&327\\1&621\end{vmatrix}=-294\times10^5$$

■

例 3 计算

Example 3 Calculate

$$D=\begin{vmatrix}1&1&-1&2\\-1&-1&-4&1\\2&4&-6&1\\1&2&4&2\end{vmatrix}$$

解 由行列式的性质及行列式按行 (列) 展开定理, 得

Solution From the properties and the expansion theorem in accordance with a row (or column) of determinants, we obtain

$$D=\begin{vmatrix}1&1&-1&2\\-1&-1&-4&1\\2&4&-6&1\\1&2&4&2\end{vmatrix}=\begin{vmatrix}1&1&-1&2\\0&0&-5&3\\0&2&-4&-3\\0&1&5&0\end{vmatrix}=\begin{vmatrix}0&-5&3\\2&-4&-3\\1&5&0\end{vmatrix}$$

$$=\begin{vmatrix}0&-5&3\\0&-14&-3\\1&5&0\end{vmatrix}=\begin{vmatrix}-5&3\\-14&-3\end{vmatrix}=57$$ ■

以上两例的求解过程为：首先利用性质 5 将行列式某一行 (列) 只保留一个非零元素；再按这一行 (列) 展开；通过降阶计算行列式的值. 这是计算行列式常用方法之一.

The above two examples proceed as follows. Retain only a nonzero element on a certain row (or column) by Property 5; and then expand the determinant in accordance with this row (or column); subsequently, calculate the determinant by reducing its order. It is one of the common methods for calculating determinants.

例 4 计算

Example 4 Calculate

$$D=\begin{vmatrix}0&1&-1&2\\-1&-1&2&1\\-1&0&-1&1\\2&2&0&0\end{vmatrix}$$

解 根据行列式的性质, 将 D 化为三角形行列式

Solution In accordance with the properties of determinants, we reduce D to a triangular form,

$$D=\begin{vmatrix}0&1&-1&2\\-1&-1&2&1\\-1&0&-1&1\\2&2&0&0\end{vmatrix}\overset{r_2\leftrightarrow r_1}{=\!=\!=}\begin{vmatrix}-1&-1&2&1\\0&1&-1&2\\-1&0&-1&1\\2&2&0&0\end{vmatrix}\overset{r_3-r_1,r_4+2r_1}{=\!=\!=\!=\!=}-\begin{vmatrix}-1&-1&2&1\\0&1&-1&2\\0&1&-3&0\\0&0&4&2\end{vmatrix}$$

$$\overset{r_3-r_2}{=\!=\!=}-\begin{vmatrix}-1&-1&2&1\\0&1&-1&2\\0&0&-2&-2\\0&0&4&2\end{vmatrix}\overset{r_4+2r_3}{=\!=\!=}-\begin{vmatrix}-1&-1&2&1\\0&1&-1&2\\0&0&-2&-2\\0&0&0&-2\end{vmatrix}$$

$$=-(-1)\times 1\times(-2)\times(-2)=4$$ ■

例 5 对于给定的多项式

Example 5 For the given polynomial

$$f(x)=\begin{vmatrix}1 & 1 & 2 & 3\\ 1 & 2-x^2 & 2 & 3\\ 2 & 3 & 1 & 5\\ 2 & 3 & 1 & 9-x^2\end{vmatrix}$$

求 $f(x)=0$ 的根.

Find the roots of $f(x)=0$.

解 根据行列式的性质, 将 $f(x)$ 化为三角形行列式

Solution From the properties of determinants, $f(x)$ can be reduced to a triangular determinant,

$$f(x)=\begin{vmatrix}1 & 1 & 2 & 3\\ 1 & 2-x^2 & 2 & 3\\ 2 & 3 & 1 & 5\\ 2 & 3 & 1 & 9-x^2\end{vmatrix}=\begin{vmatrix}1 & 0 & 0 & 0\\ 1 & 1-x^2 & 0 & 0\\ 2 & 1 & -3 & -1\\ 2 & 1 & -3 & 3-x^2\end{vmatrix}$$

$$=-3\begin{vmatrix}1 & 0 & 0 & 0\\ 1 & 1-x^2 & 0 & 0\\ 2 & 1 & 1 & 0\\ 2 & 1 & 1 & 4-x^2\end{vmatrix}=-3(1-x^2)(4-x^2)$$

由 $f(x)=0$, 即 $-3(1-x^2)(4-x^2)=0$, 得 $f(x)=0$ 的根为

Since $f(x)=0$, i.e., $-3(1-x^2)(4-x^2)=0$, the roots of $f(x)=0$ are

$$x_1=-1,\quad x_2=1,\quad x_3=-2,\quad x_4=2$$ ■

例 6 令

Example 6 Let

$$D=\begin{vmatrix}a_{11} & \cdots & a_{1k} & 0 & \cdots & 0\\ \vdots & & \vdots & \vdots & & \vdots\\ a_{k1} & \cdots & a_{kk} & 0 & \cdots & 0\\ c_{11} & \cdots & c_{1k} & b_{11} & \cdots & b_{1n}\\ \vdots & & \vdots & \vdots & & \vdots\\ c_{n1} & \cdots & c_{nk} & b_{n1} & \cdots & b_{nn}\end{vmatrix}$$

$$D_1=\begin{vmatrix}a_{11} & \cdots & a_{1k}\\ \vdots & & \vdots\\ a_{k1} & \cdots & a_{kk}\end{vmatrix},\quad D_2=\begin{vmatrix}b_{11} & \cdots & b_{1n}\\ \vdots & & \vdots\\ b_{n1} & \cdots & b_{nn}\end{vmatrix}$$

证明: $D = D_1D_2$.

Prove that $D = D_1D_2$.

证 利用本节的性质及推论, 对 D_1 实施行运算, 如 $r_i \leftrightarrow r_j, kr_j$, $r_i + kr_j$, 对 D_2 实施列运算, 如 $c_i \leftrightarrow c_j, kc_j$, $c_i + kc_j$. 最终可分别将 D_1 和 D_2 化为下三角形行列式.

Proof Utilizing the properties and the corollaries in this section, we perform the row operations on D_1, such as kr_i, $r_i + kr_j$, and the column operations on D_2, such as kc_j, $c_i + kc_j$. Finally, we can reduce D_1 and D_2 to the following lower triangular determinants, respectively,

$$D_1 = \begin{vmatrix} p_{11} & & \\ \vdots & \ddots & \\ p_{k1} & \cdots & p_{kk} \end{vmatrix} = p_{11} \cdots p_{kk}, \quad D_2 = \begin{vmatrix} q_{11} & & \\ \vdots & \ddots & \\ q_{n1} & \cdots & q_{nn} \end{vmatrix} = q_{11} \cdots q_{nn}$$

对 D 的前 k 行作与对 D_1 相同的行运算, 再对后 n 列作与对 D_2 相同的列运算, 最终可将 D 化为下三角形行列式

Performing the row operations, the same as those on D_1, on the first k rows of D, and then performing the column operations, the same as those on D_2, on the last n columns of D, finally, we can reduce D to the following lower triangular determinant

$$\begin{vmatrix} p_{11} & & & & & \\ \vdots & \ddots & & & & \\ p_{k1} & \cdots & p_{kk} & & & \\ c_{11} & \cdots & c_{1k} & q_{11} & & \\ \vdots & & \vdots & \vdots & \ddots & \\ c_{n1} & \cdots & c_{nk} & q_{n1} & \cdots & q_{nn} \end{vmatrix}$$

故

Hence,

$$D = p_{11} \cdots p_{kk} \cdot q_{11} \cdots q_{nn} = D_1D_2$$ ■

思 考 题

Question

判断下列等式是否成立?

Decide whether or not the following equation is valid?

$$\begin{vmatrix} a_{11}+b_{11} & a_{12}+b_{12} \\ a_{21}+b_{21} & a_{22}+b_{22} \end{vmatrix} = \begin{vmatrix} a_{11} & a_{12} \\ a_{21} & a_{22} \end{vmatrix} + \begin{vmatrix} b_{11} & b_{12} \\ b_{21} & b_{22} \end{vmatrix}$$

1.3 行列式的计算

行列式的计算是本章的主要任务. 本节介绍计算行列式几种常用的方法. 计算行列式时, 首先要观察行列式的结构上有什么特点, 然后针对这些特点对行列式进行化简、计算. 计算行列式的常见方法有:

(1) 用定义计算行列式;

(2) 用行列式性质化简、计算行列式;

(3) 用递推公式或数学归纳法计算行列式.

例 1 证明

1.3 Calculation of Determinants

Calculation of determinants is the main task of this chapter. In this section, we introduce some common methods for calculating determinants. When we calculate a determinant, it needs to observe the features about the determinant's structure, and then simplify the determinant according to these features, finally calculate the determinant. The common methods for calculating determinants are as follows:

(1) Use the definition to calculate determinants;

(2) Use the properties to simplify and then calculate determinants;

(3) Use recursion formulas or mathematical induction to calculate determinants.

Example 1 Prove that

$$\begin{vmatrix} b+c & c+a & a+b \\ b_1+c_1 & c_1+a_1 & a_1+b_1 \\ b_2+c_2 & c_2+a_2 & a_2+b_2 \end{vmatrix} = 2\begin{vmatrix} a & b & c \\ a_1 & b_1 & c_1 \\ a_2 & b_2 & c_2 \end{vmatrix}$$

证 先用行列式的性质 4(可加性) 拆第 1 列, 再用性质 5(倍加性) 化简得

Proof Splitting the first column by Property 4 (Additivity) of the determinant and then reducing it by Property 5 (Multiple-Additivity), we have

$$
左=\begin{vmatrix} b & c+a & a+b \\ b_1 & c_1+a_1 & a_1+b_1 \\ b_2 & c_2+a_2 & a_2+b_2 \end{vmatrix}+\begin{vmatrix} c & c+a & a+b \\ c_1 & c_1+a_1 & a_1+b_1 \\ c_2 & c_2+a_2 & a_2+b_2 \end{vmatrix}
$$

$$
=\begin{vmatrix} b & c+a & a \\ b_1 & c_1+a_1 & a_1 \\ b_2 & c_2+a_2 & a_2 \end{vmatrix}+\begin{vmatrix} c & a & a+b \\ c_1 & a_1 & a_1+b_1 \\ c_2 & a_2 & a_2+b_2 \end{vmatrix}=\begin{vmatrix} b & c & a \\ b_1 & c_1 & a_1 \\ b_2 & c_2 & a_2 \end{vmatrix}+\begin{vmatrix} c & a & b \\ c_1 & a_1 & b_1 \\ c_2 & a_2 & b_2 \end{vmatrix}
$$

$$
=\begin{vmatrix} a & b & c \\ a_1 & b_1 & c_1 \\ a_2 & b_2 & c_2 \end{vmatrix}+\begin{vmatrix} a & b & c \\ a_1 & b_1 & c_1 \\ a_2 & b_2 & c_2 \end{vmatrix}=2\begin{vmatrix} a & b & c \\ a_1 & b_1 & c_1 \\ a_2 & b_2 & c_2 \end{vmatrix}=右
$$

■

例 2 计算 n 阶行列式

Example 2 Calculate the n-th order determinant

$$
D_n=\begin{vmatrix} a_1+b_1 & a_1+b_2 & \cdots & a_1+b_n \\ a_2+b_1 & a_2+b_2 & \cdots & a_2+b_n \\ \vdots & \vdots & & \vdots \\ a_n+b_1 & a_n+b_2 & \cdots & a_n+b_n \end{vmatrix}
$$

解 显然, 当 $n=1$ 时, $D_1=a_1+b_1$.

Solution Obviously, as $n=1$, $D_1=a_1+b_1$.

当 $n=2$ 时, 容易求得

As $n=2$, we easily get

$$
D_2=(a_1+b_1)(a_2+b_2)-(a_1+b_2)(a_2+b_1)=(a_1-a_2)(b_2-b_1)
$$

当 $n\geqslant 3$ 时, 将第 1 行乘 -1 加到其余各行后, 这些行除第 1 行外其余各行对应成比例, 即

As $n\geqslant 3$, adding (-1) times the first row to the other rows, then each row is proportional to the other rows except for the first row, namely,

$$
D_n=\begin{vmatrix} a_1+b_1 & a_1+b_2 & \cdots & a_1+b_n \\ a_2-a_1 & a_2-a_1 & \cdots & a_2-a_1 \\ a_3-a_1 & a_3-a_1 & \cdots & a_3-a_1 \\ \vdots & \vdots & & \vdots \\ a_n-a_1 & a_n-a_1 & \cdots & a_n-a_1 \end{vmatrix}=0
$$

■

例 3 已知

Example 3 Known that

$$D=\begin{vmatrix}1&2&3&4&5\\5&5&5&3&3\\3&2&5&4&2\\2&2&2&1&1\\4&6&5&2&3\end{vmatrix}$$

(1) 求 $A_{51}+2A_{52}+3A_{53}+4A_{54}+5A_{55}$;

(2) 求 $A_{31}+A_{32}+A_{33}$ 及 $A_{34}+A_{35}$.

(1) Find $A_{51}+2A_{52}+3A_{53}+4A_{54}+5A_{55}$;

(2) Find $A_{31}+A_{32}+A_{33}$ and $A_{34}+A_{35}$.

解 由 1.2 节定理 1 和行列式的性质可知

Solution From Theorem 1 in Section 1.2 and the properties of determinants, we have

(1)

$$A_{51}+2A_{52}+3A_{53}+4A_{54}+5A_{55}=\begin{vmatrix}1&2&3&4&5\\5&5&5&3&3\\3&2&5&4&2\\2&2&2&1&1\\1&2&3&4&5\end{vmatrix}=0$$

(2)

$$5A_{31}+5A_{32}+5A_{33}+3A_{34}+3A_{35}=\begin{vmatrix}1&2&3&4&5\\5&5&5&3&3\\5&5&5&3&3\\2&2&2&1&1\\4&6&5&2&3\end{vmatrix}=0$$

$$2A_{31}+2A_{32}+2A_{33}+A_{34}+A_{35}=\begin{vmatrix}1&2&3&4&5\\5&5&5&3&3\\2&2&2&1&1\\2&2&2&1&1\\4&6&5&2&3\end{vmatrix}=0$$

求解上面两式可得

Solving the above two equations, we have

$$A_{31}+A_{32}+A_{33}=0,\quad A_{34}+A_{35}=0$$

■

此题也可以通过直接求代数余子式得到结果, 但这样做比较麻烦. 也可以直接用定理 1 计算下面行列式.

This example can also be solved by calculating the algebraic cofactors directly, but this approach is more troublesome. We can also calculate them directly from Theorem 1, as follows,

$$A_{31}+A_{32}+A_{33}=\begin{vmatrix}1&2&3&4&5\\5&5&5&3&3\\3&2&5&4&2\\1&1&1&0&0\\4&6&5&2&3\end{vmatrix}=0,\quad A_{34}+A_{35}=\begin{vmatrix}1&2&3&4&5\\5&5&5&3&3\\3&2&5&4&2\\0&0&0&1&1\\4&6&5&2&3\end{vmatrix}=0$$

例 4 计算 $n+1$ 阶 (爪形) 行列式

Example 4 Calculate the $(n+1)$-th order determinant (of a claw form)

$$D_{n+1}=\begin{vmatrix}a_0&1&1&1&\cdots&1\\1&a_1&0&0&\cdots&0\\1&0&a_2&0&\cdots&0\\\vdots&\vdots&\vdots&\vdots&&\vdots\\1&0&0&0&\cdots&a_n\end{vmatrix}$$

其中 $a_1a_2\cdots a_n\neq 0$.

where $a_1a_2\cdots a_n\neq 0$.

解 将行列式 D_{n+1} 的第 2 列, 第 3 列, $\cdots$, 第 n 列分别乘以 $-\frac{1}{a_1},-\frac{1}{a_2},\cdots,-\frac{1}{a_n}$ 都加到第 1 列上得

Solution Adding $\left(-\frac{1}{a_1}\right),\left(-\frac{1}{a_2}\right),\cdots,\left(-\frac{1}{a_n}\right)$ time the second, the third, $\cdots$, and the n-th columns to the first column, respectively, we have

$$D_{n+1}=\begin{vmatrix}a_0-\sum\limits_{i=1}^{n}\frac{1}{a_i}&1&1&1&\cdots&1\\0&a_1&0&0&\cdots&0\\0&0&a_2&0&\cdots&0\\\vdots&\vdots&\vdots&\vdots&&\vdots\\0&0&0&0&\cdots&a_n\end{vmatrix}=a_1a_2\cdots a_n\left(a_0-\sum_{i=1}^{n}\frac{1}{a_i}\right)\quad\blacksquare$$

例 5 计算 n 阶行列式

Example 5 Calculate the n-th order determinant

$$D_n=\begin{vmatrix} x & a & a & \cdots & a \\ a & x & a & \cdots & a \\ \vdots & \vdots & \vdots & & \vdots \\ a & a & a & \cdots & x \end{vmatrix}$$

解 将行列式 D_n 的第 2 列, 第 3 列, $\cdots$, 第 n 列都加到第 1 列上, 再从第 1 列中提取公因子 $x+(n-1)a$, 最后利用行列式的性质将其约化为上三角形行列式, 得

Solution Adding the second, the third, $\cdots$, the n-th columns of D_n to the first column and then extracting the common factor $x+(n-1)a$ from the first column, finally reducing it to an upper triangular determinant by the properties of determinants, we have

$$D_n=[x+(n-1)a]\begin{vmatrix} 1 & a & a & \cdots & a \\ 1 & x & a & \cdots & a \\ \vdots & \vdots & \vdots & & \vdots \\ 1 & a & a & \cdots & x \end{vmatrix}$$

$$=[x+(n-1)a]\begin{vmatrix} 1 & a & \cdots & a & a \\ 0 & x-a & \cdots & 0 & 0 \\ \vdots & \vdots & & \vdots & \vdots \\ 0 & 0 & \cdots & x-a & 0 \\ 0 & 0 & \cdots & 0 & x-a \end{vmatrix}=[x+(n-1)a](x-a)^{n-1} \quad \blacksquare$$

例 6* 计算 $n+1$ 阶行列式

Example 6* Calculate the $(n+1)$-th order determinant

$$D_{n+1}=\begin{vmatrix} x & a_1 & a_2 & \cdots & a_n \\ a_1 & x & a_2 & \cdots & a_n \\ a_1 & a_2 & x & \cdots & a_n \\ \vdots & \vdots & \vdots & & \vdots \\ a_1 & a_2 & a_3 & \cdots & x \end{vmatrix}$$

解 将 D_{n+1} 的第 2 列、第 3 列、$\cdots$、第 $n+1$ 列全加到第 1 列上, 然后从第 1 列提取公因子 $x+\sum\limits_{i=1}^{n}a_i$, 最后利用行列式的性质将其约化为下三角行列式, 得

Solution Adding the second, the third, $\cdots$, the $(n+1)$-th columns of D_{n+1} to the first column and then extracting the common factor $x+\sum\limits_{i=1}^{n}a_i$ from the first column, finally reducing it to a lower triangular determinant by the properties of determinants, we have

$$D_{n+1}=\left(x+\sum_{i=1}^{n}a_i\right)\begin{vmatrix}1 & a_1 & a_2 & \cdots & a_n\\ 1 & x & a_2 & \cdots & a_n\\ 1 & a_2 & x & \cdots & a_n\\ \vdots & \vdots & \vdots & & \vdots\\ 1 & a_2 & a_3 & \cdots & x\end{vmatrix}$$

$$=\left(x+\sum_{i=1}^{n}a_i\right)\begin{vmatrix}1 & 0 & 0 & \cdots & 0\\ 1 & x-a_1 & 0 & \cdots & 0\\ 1 & a_2-a_1 & x-a_2 & \cdots & 0\\ \vdots & \vdots & \vdots & & \vdots\\ 1 & a_2-a_1 & a_3-a_2 & \cdots & x-a_n\end{vmatrix}$$

$$=\left(x+\sum_{i=1}^{n}a_i\right)(x-a_1)(x-a_2)\cdots(x-a_n)$$ ■

例 7* 计算 n 阶行列式

Example 7* Calculate the n-th order determinant

$$D_n=\begin{vmatrix}x+a_1 & a_2 & \cdots & a_n\\ a_1 & x+a_2 & \cdots & a_n\\ \vdots & \vdots & & \vdots\\ a_1 & a_2 & \cdots & x+a_n\end{vmatrix}$$

解 (1) 当 $x=0$ 时, 若 $n=1$, 则 $D_1=a_1$; 若 $n\geqslant 2$, 则 $D_n=0$.

Solution (1) As $x=0$, if $n=1$, then $D_1=a_1$; while if $n\geqslant 2$ then $D_n=0$.

(2) 当 $x\neq 0$ 时, 对 D_n 增加 1 行 1 列 (此方法也称为**加边法**), 得到

(2) As $x\neq 0$, appending a row and a column on D_n (this method is also called the **bordering method**), we obtain

$$D_n=\begin{vmatrix}1 & a_1 & a_2 & \cdots & a_n\\ 0 & x+a_1 & a_2 & \cdots & a_n\\ \vdots & \vdots & \vdots & & \vdots\\ 0 & a_1 & a_2 & \cdots & x+a_n\end{vmatrix}$$

将第 1 行的 (-1) 倍加到其余各行, 再将第 $2,3,\cdots,n+1$ 列分别乘上 $\frac{1}{x}$ 加到第 1 列, 得

Adding (-1) times the first row to the other rows and then adding $\frac{1}{x}$ times the second, the third, $\cdots$, the $(n+1)$-th columns to the first column, we obtain

$$D_n=\begin{vmatrix}1&a_1&a_2&\cdots&a_n\\-1&x&0&\cdots&0\\-1&0&x&\cdots&0\\\vdots&\vdots&\vdots&&\vdots\\-1&0&0&\cdots&x\end{vmatrix}=\begin{vmatrix}1+\sum\limits_{i=1}^{n}\frac{a_i}{x}&a_1&a_2&\cdots&a_n\\0&x&0&\cdots&0\\0&0&x&\cdots&0\\\vdots&\vdots&\vdots&&\vdots\\0&0&0&\cdots&x\end{vmatrix}$$

$$=x^n\left(1+\frac{1}{x}\sum_{i=1}^{n}a_i\right)=x^{n-1}\left(x+\sum_{i=1}^{n}a_i\right)$$ ■

例 8 计算 $2n$ 阶行列式

Example 8 Calculate the $2n$-th order determinant

$$D_{2n}=\begin{vmatrix}a&0&\cdots&0&0&\cdots&0&b\\0&a&\cdots&0&0&\cdots&b&0\\\vdots&\vdots&&\vdots&\vdots&&\vdots&\vdots\\0&0&\cdots&a&b&\cdots&0&0\\0&0&\cdots&c&d&\cdots&0&0\\\vdots&\vdots&&\vdots&\vdots&&\vdots&\vdots\\0&c&\cdots&0&0&\cdots&d&0\\c&0&\cdots&0&0&\cdots&0&d\end{vmatrix}$$

解 将 D_{2n} 按第 1 行展开, 得

Solution Expanding D_{2n} in accordance with the first row, we have

$$D_{2n}=a\begin{vmatrix}a&0&\cdots&0&0&\cdots&b&0\\\vdots&\vdots&&\vdots&\vdots&&\vdots&\vdots\\0&0&\cdots&a&b&\cdots&0&0\\0&0&\cdots&c&d&\cdots&0&0\\\vdots&\vdots&&\vdots&\vdots&&\vdots&\vdots\\c&0&\cdots&0&0&\cdots&d&0\\0&0&\cdots&0&0&\cdots&0&d\end{vmatrix}$$

$$
+b(-1)^{2n+1}\begin{vmatrix} 0 & a & \cdots & 0 & 0 & \cdots & 0 & b \\ \vdots & \vdots & & \vdots & \vdots & & \vdots & \vdots \\ 0 & 0 & \cdots & a & b & \cdots & 0 & 0 \\ 0 & 0 & \cdots & c & d & \cdots & 0 & 0 \\ \vdots & \vdots & & \vdots & \vdots & & \vdots & \vdots \\ 0 & c & \cdots & 0 & 0 & \cdots & 0 & d \\ c & 0 & \cdots & 0 & 0 & \cdots & 0 & 0 \end{vmatrix}
$$

$$
\begin{aligned}
&= ad(-1)^{2n-1+2n-1}D_{2n-2} + bc(-1)^{2n+1}(-1)^{1+2n-1}D_{2n-2} \\
&= (ad-bc)D_{2n-2}
\end{aligned}
$$

即有递推公式

namely, we have the following recursive formula,

$$
D_{2n} = (ad-bc)D_{2n-2}
$$

进一步, 可得

Further, we obtain

$$
\begin{aligned}
D_{2n} &= (ad-bc)D_{2(n-1)} = (ad-bc)^2 D_{2(n-2)} = \cdots \\
&= (ad-bc)^{n-1}D_2 = (ad-bc)^{n-1}\begin{vmatrix} a & b \\ c & d \end{vmatrix} = (ad-bc)^n
\end{aligned}
$$

■

例 9 证明范德蒙德行列式

Example 9 Prove the Vandermonde determinant

$$
D_n = \begin{vmatrix} 1 & 1 & \cdots & 1 \\ x_1 & x_2 & \cdots & x_n \\ x_1^2 & x_2^2 & \cdots & x_n^2 \\ \vdots & \vdots & & \vdots \\ x_1^{n-1} & x_2^{n-1} & \cdots & x_n^{n-1} \end{vmatrix} = \prod_{n\geqslant i>j\geqslant 1}(x_i - x_j)
$$

其中 "$\prod$" 是连乘记号, 它表示 $(x_i - x_j)$ 全体同类因子的乘积.

where "$\prod$" is the notation of continued product, which denotes the product of all factors with the same form $(x_i - x_j)$.

证 用数学归纳法证明.

Proof Prove it by mathematical induction.

当 $n=2$ 时,

As $n=2$,

$$D_2=\begin{vmatrix} 1 & 1 \\ x_1 & x_2 \end{vmatrix}=x_2-x_1$$

结论成立.

this means that the result holds.

假设对 $n-1$ 阶范德蒙德行列式结论成立, 下面证明对 n 阶范德蒙德行列式结论也成立.

Assume that the result holds for the $(n-1)$-th order Vandermonde determinant, then we prove that the result also holds for the n-th order Vandermonde determinant.

将 D_n 从第 n 行开始, 各行依次加上前一行的 $-x_1$ 倍, 然后将得到的行列式按第 1 列展开 (只剩下一个 $n-1$ 阶行列式), 最后提取 $n-1$ 阶行列式每一列的公因子, 即 $x_2-x_1, x_3-x_1, \cdots, x_n-x_1$, 得

From the n-th row of D_n, adding $(-x_1)$ times the previous row to each row in turn, and then expanding the obtained determinant in accordance with the first column (only remain an $(n-1)$-th order determinant), finally extracting the factors $x_2-x_1, x_3-x_1, \cdots, x_n-x_1$ of the $(n-1)$-th order determinant from each column, we obtain

$$\begin{aligned}
D_n&=\begin{vmatrix} 1 & 1 & \cdots & 1 \\ 0 & x_2-x_1 & \cdots & x_n-x_1 \\ 0 & x_2(x_2-x_1) & \cdots & x_n(x_n-x_1) \\ \vdots & \vdots & & \vdots \\ 0 & x_2^{n-2}(x_2-x_1) & \cdots & x_n^{n-2}(x_n-x_1) \end{vmatrix} \\
&=\begin{vmatrix} x_2-x_1 & x_2-x_1 & \cdots & x_n-x_1 \\ x_2(x_2-x_1) & x_3(x_3-x_1) & \cdots & x_n(x_n-x_1) \\ \vdots & \vdots & & \vdots \\ x_2^{n-2}(x_2-x_1) & x_3^{n-2}(x_3-x_1) & \cdots & x_n^{n-2}(x_n-x_1) \end{vmatrix} \\
&=(x_2-x_1)(x_3-x_1)\cdots(x_n-x_1)\begin{vmatrix} 1 & 1 & \cdots & 1 \\ x_2 & x_3 & \cdots & x_n \\ \vdots & \vdots & & \vdots \\ x_2^{n-2} & x_3^{n-2} & \cdots & x_n^{n-2} \end{vmatrix}
\end{aligned}$$

显然, 最后一个表达式中的行列式为 $n-1$ 阶范德蒙德行列式. 由归纳假设得

Obviously, the determinant in the last expression is the $(n-1)$-th order Vandermonde determinant. From the inductive assumption, we have

$$D_n = (x_2 - x_1)(x_3 - x_1)\cdots(x_n - x_1)\prod_{n\geqslant i>j\geqslant 2}(x_i - x_j) = \prod_{n\geqslant i>j\geqslant 1}(x_i - x_j) \qquad \blacksquare$$

思 考 题

范德蒙德行列式有什么特点?

Question

What is the feature of the Vandermonde determinant?

1.4 克拉默法则

1.4 Cramer's Rule

本节利用 n 阶行列式来求解 n 元线性方程组, 以此回答 1.1.2 小节提出的问题.

In this section, we solve the systems of linear equations in n unknowns by n-th order determinants, which can answer the question proposed in Subsection 1.1.2.

对于如下的 n 元线性方程组

For the following linear system in n unknowns

$$\begin{cases} a_{11}x_1 + a_{12}x_2 + \cdots + a_{1n}x_n = b_1 \\ a_{21}x_1 + a_{22}x_2 + \cdots + a_{2n}x_n = b_2 \\ \qquad\cdots\cdots \\ a_{n1}x_1 + a_{n2}x_2 + \cdots + a_{nn}x_n = b_n \end{cases} \tag{1.7}$$

其未知量的系数构成的行列式为

the determinant formed by the coefficients of unknowns is as follows

$$D = \begin{vmatrix} a_{11} & a_{12} & \cdots & a_{1n} \\ a_{21} & a_{22} & \cdots & a_{2n} \\ \vdots & \vdots & & \vdots \\ a_{n1} & a_{n2} & \cdots & a_{nn} \end{vmatrix}$$

称为方程组 (1.7) 的**系数行列式**.

It is called the **coefficient determinant** associated with System (1.7).

定理 1 (克拉默法则) 如果方程组 (1.7) 的系数行列式 $D \neq 0$, 则方程组 (1.7) 有唯一解, 且

Theorem 1 (Cramer's rule) If the coefficient determinant of System (1.7) satisfies $D \neq 0$, then System (1.7) has a unique solution, given by

$$x_1 = \frac{D_1}{D}, \ x_2 = \frac{D_2}{D}, \ \cdots, \ x_n = \frac{D_n}{D} \tag{1.8}$$

其中 D_j 是将 D 中第 j 列的元素用常数项 $b_1, b_2, \cdots, b_n$ 代替后所得到的 n 阶行列式，即

where D_j is an n-th order determinant obtained by replacing the elements on the j-th row of D by the constant terms $b_1, b_2, \cdots, b_n$, namely,

$$D_j = \begin{vmatrix} a_{11} & \cdots & a_{1j-1} & b_1 & a_{1j+1} & \cdots & a_{1n} \\ a_{21} & \cdots & a_{2j-1} & b_2 & a_{2j+1} & \cdots & a_{2n} \\ \vdots & & \vdots & \vdots & \vdots & & \vdots \\ a_{n1} & \cdots & a_{nj-1} & b_n & a_{nj+1} & \cdots & a_{nn} \end{vmatrix} \quad (j = 1, 2, \cdots, n)$$

证 用 D 的第 j 列的元素的代数余子式 $A_{1j}, A_{2j}, \cdots, A_{nj}$ 分别乘方程组 (1.7) 的第 1, 2, $\cdots$, n 个方程两端, 然后竖式相加, 由 1.2 节的定理 1 和定理 2 得到

Proof Multiplying both sides of the first, the second, $\cdots$ and the n-th equations of System (1.7) by the algebraic cofactors $A_{1j}, A_{2j}, \cdots, A_{nj}$ associated with the elements on the j-th column of D, respectively, and then adding them vertically, from Theorem 1 and Theorem 2 in Section 1.2 we have

$$x_j D = D_j \quad (j = 1, 2, \cdots, n) \tag{1.9}$$

由于 $D \neq 0$, 得方程组 (1.7) 的唯一解

Since $D \neq 0$, we obtain the unique solution of System (1.7), given by

$$x_j = \frac{D_j}{D} \quad (j = 1, 2, \cdots, n)$$

下面验证式 (1.8) 一定是方程组 (1.7) 的解. 将式 (1.8) 代入方程组 (1.7) 中第 i 个方程的左边, 也就是要证明

In the following part we verify that Equation(1.8) must be a solution of (1.7). Substituting (1.8) into the left hand of the i-th equation in System (1.7), that is to say, we need to prove that

$$a_{i1}x_1 + a_{i2}x_2 + \cdots + a_{in}x_{in} = a_{i1}\frac{D_1}{D} + a_{i2}\frac{D_2}{D} + \cdots + a_{in}\frac{D_n}{D} = b_i \quad (i = 1, 2, \cdots, n)$$

为此, 构造有两行相同的 $n+1$ 阶行列式

For this purpose, construct the following $(n+1)$-th order determinants with two equal rows,

$$\begin{vmatrix} b_i & a_{i1} & \cdots & a_{in} \\ b_1 & a_{11} & \cdots & a_{1n} \\ \vdots & \vdots & & \vdots \\ b_n & a_{n1} & \cdots & a_{nn} \end{vmatrix} \quad (i = 1, 2, \cdots, n) \tag{1.10}$$

显然, 这些行列式等于零. 由于第 1 行的 a_{ij} 的代数余子式

Obviously, these determinants are all equal to zero. Since the algebraic cofactor of a_{ij} on the first row is

$$\begin{aligned} A_{ij} =& (-1)^{1+j+1} \begin{vmatrix} b_1 & a_{11} & \cdots & a_{1j-1} & a_{1j+1} & \cdots & a_{1n} \\ \vdots & \vdots & & \vdots & \vdots & & \vdots \\ b_n & a_{n1} & \cdots & a_{nj-1} & a_{nj+1} & & a_{nn} \end{vmatrix} \\ =& (-1)^{j+2}(-1)^{j-1} D_j = -D_j \end{aligned}$$

将式 (1.10) 按第 1 行展开有

expanding Equation (1.10) in accordance with the first row, we have

$$0 = b_i D - a_{i1} D_1 - \cdots - a_{in} D_n$$

即当 $D \neq 0$, 则

that is, if $D \neq 0$, then

$$a_{i1}x_1 + a_{i2}x_2 + \cdots + a_{in}x_{in} = a_{i1}\frac{D_1}{D} + a_{i2}\frac{D_2}{D} + \cdots + a_{in}\frac{D_n}{D} = b_i \qquad \blacktriangle$$

注 克拉默法则在理论上有一定的应用价值. 由于它要求的条件较高 (方程个数与未知数个数必须相等, 且系数行列式不为零), 且计算量大 (需计算 $n+1$ 个 n 阶行列式), 故不适合用来求解线性方程组.

Note Theoretically, Cramer's rule has certain application value. Since it requires strict conditions (the equation number must be equal to the unknown number and the coefficient determinant must be nonzero), and the computational cost is very large (it needs to calculate $n+1$ determinants of order n), it is not suitable for solving linear systems.

定理 1 的逆否定理为如下定理.

The converse-negative theorem of Theorem 1 is as follows.

定理 1′ 若方程组 (1.7) 无解或有至少两个不同的解, 则它的系数行列式必等于零, 即 $D = 0$.

Theorem 1′ If System (1.7) has no solution or has at least two distinct solutions, then its coefficient determinant must be equal to zero, namely, $D = 0$.

例 1 解线性方程组

Example 1 Solve the linear system

$$\begin{cases} x_1 + 2x_2 - \ x_3 + 3x_4 = 2 \\ 2x_1 - \ x_2 + 3x_3 - 2x_4 = 7 \\ 3x_2 - \ x_3 + \ x_4 = 6 \\ x_1 - \ x_2 + \ x_3 + 4x_4 = -4 \end{cases}$$

解 由于

Solution Since

$$D = \begin{vmatrix} 1 & 2 & -1 & 3 \\ 2 & -1 & 3 & -2 \\ 0 & 3 & -1 & 1 \\ 1 & -1 & 1 & 4 \end{vmatrix} = \begin{vmatrix} 1 & 2 & -1 & 3 \\ 0 & -5 & 5 & -8 \\ 0 & 3 & -1 & 1 \\ 0 & -3 & 2 & 1 \end{vmatrix}$$

$$= \begin{vmatrix} -5 & 5 & -8 \\ 3 & -1 & 1 \\ -3 & 2 & 1 \end{vmatrix} = \begin{vmatrix} 19 & -3 & -8 \\ 0 & 0 & 1 \\ -6 & 3 & 1 \end{vmatrix}$$

$$= - \begin{vmatrix} 19 & -3 \\ -6 & 3 \end{vmatrix} = -39 \neq 0$$

可知方程组有唯一解；此外, 有

we know that the system has a unique solution. Furthermore, we have

$$D_1 = \begin{vmatrix} 2 & 2 & -1 & 3 \\ 7 & -1 & 3 & -2 \\ 6 & 3 & -1 & 1 \\ -4 & -1 & 1 & 4 \end{vmatrix} = -39$$

$$D_2 = \begin{vmatrix} 1 & 2 & -1 & 3 \\ 2 & 7 & 3 & -2 \\ 0 & 6 & -1 & 1 \\ 1 & -4 & 1 & 4 \end{vmatrix} = -117$$

$$D_3 = \begin{vmatrix} 1 & 2 & 2 & 3 \\ 2 & -1 & 7 & -2 \\ 0 & 3 & 6 & 1 \\ 1 & -1 & -4 & 4 \end{vmatrix} = -78$$

$$D_4 = \begin{vmatrix} 1 & 2 & -1 & 2 \\ 2 & -1 & 3 & 7 \\ 0 & 3 & -1 & 6 \\ 1 & -1 & 1 & -4 \end{vmatrix} = 39$$

所以方程组的唯一解为

Hence, the unique solution of the system is given by

$$x_1 = \frac{D_1}{D} = 1, \quad x_2 = \frac{D_2}{D} = 3, \quad x_3 = \frac{D_3}{D} = 2, \quad x_4 = \frac{D_4}{D} = -1 \qquad \blacksquare$$

当方程组(1.7)的右端常数项$b_1, \cdots, b_n$不全为零时, 称 (1.7) 为**非齐次线性方程组**. 当 $b_1, b_2, \cdots, b_n$ 全为零时, 方程组

System (1.7) is called a **system of non-homogeneous linear equations** if the constant terms $b_1, \cdots, b_n$ of the right hand of (1.7) are not all zeros. The linear system

$$\begin{cases} a_{11}x_1 + a_{12}x_2 + \cdots + a_{1n}x_n = 0 \\ a_{21}x_1 + a_{22}x_2 + \cdots + a_{2n}x_n = 0 \\ \qquad \cdots\cdots \\ a_{n1}x_1 + a_{n2}x_2 + \cdots + a_{nn}x_n = 0 \end{cases} \tag{1.11}$$

称为**齐次线性方程组**. 显然, $x_1 = x_2 = \cdots = x_n = 0$ 是方程组 (1.11) 的解, 这个解称为 (1.11) 的零解.

is called a **system of homogeneous linear equations** if $b_1, \cdots, b_n$ are all zeros. Obviously, the set, $x_1 = x_2 = \cdots = x_n = 0$, is a solution of System (1.11), and is called the zero solution of (1.11).

当 $D \neq 0$ 时, 对齐次线性方程组应用克拉默法则, 得到 $D_j = 0\ (j = 1, 2, \cdots, n)$, 所以它有唯一零解. 于是有如下定理.

As $D \neq 0$, applying Cramer's rule to a system of homogeneous linear equations, we have that $D_j = 0\ (j = 1, 2, \cdots, n)$ and hence the system has a unique zero solution. Then we have the following theorems.

定理 2 如果方程组 (1.11) 的系数行列式满足 $D \neq 0$, 则 (1.11) 有唯一零解.

Theorem 2 If the coefficient determinant of System (1.11) satisfies $D \neq 0$, then (1.11) has a unique zero solution.

其逆否定理为如下.

The converse-negative theorem is as follows.

定理 2′ 如果方程组 (1.11) 有非零解, 则其系数行列式必为零, 即 $D=0$.

Theorem 2′ If System (1.11) has nonzero solutions, then the coefficient determinant must be zero, i.e., $D=0$.

反之亦然, 结论将在第 3 章中给予证明. 于是有下面重要的定理.

Vice versa, this will be proved in Chapter 3. Thus we have the following important theorem.

定理 3 方程组 (1.11) 有非零解的充分必要条件是 (1.11) 的系数行列式为零, 即 $D=0$.

Theorem 3 System (1.11) has nonzero solutions if and only if its coefficient determinant is equal to zero, i.e., $D=0$.

例 2 问 λ 取何值时, 齐次线性方程组

Example 2 What values do λ take on such that the following system of homogeneous linear equations

$$\begin{cases}(\lambda-3)x_1+x_2-x_3=0\\ x_1+(\lambda-5)x_2+x_3=0\\ -x_1+x_2+(\lambda-3)x_3=0\end{cases}$$

有非零解?

has nonzero solutions?

解 将系数行列式的第 2、3 列都加到第 1 列, 然后从得到的行列式中提取公因子 $\lambda-3$, 最后利用行列式的性质对其进行化简, 得到

Solution Adding the second and the third columns of the coefficient determinant to the first column, and then extracting the common factor $\lambda-3$ from the obtained determinant, finally reducing it by the properties of determinants, we have

$$D=\begin{vmatrix}\lambda-3 & 1 & -1\\ 1 & \lambda-5 & 1\\ -1 & 1 & \lambda-3\end{vmatrix}=\begin{vmatrix}\lambda-3 & 1 & -1\\ \lambda-3 & \lambda-5 & 1\\ \lambda-3 & 1 & \lambda-3\end{vmatrix}=(\lambda-2)(\lambda-3)(\lambda-6)$$

根据定理 3, 当 $D=0$ 时, 即 $\lambda=2$ 或 3 或 6 时, 齐次线性方程组有非零解.

In accordance with Theorem 3, as $D=0$, i.e., $\lambda=2$, 3 or 6, the system of homogeneous linear equations has nonzero solutions. ■

例 3 设 $f(x)=c_0+c_1x+c_2x^2+\cdots+c_nx^n$, 用克拉默法则证明: 若 $f(x)$ 有 $n+1$ 个不同的零点, 则 $f(x)$ 是一个零多项式.

Example 3 Suppose that $f(x)=c_0+c_1x+c_2x^2+\cdots+c_nx^n$. Use Cramer's rule to prove that $f(x)$ is a zero polynomial if it has $(n+1)$ distinct zero points.

证 设 $a_1, a_2, \cdots, a_n, a_{n+1}$ 是 $f(x)$ 的 $n+1$ 个不同的零点, 则有

Proof Suppose that $a_1, a_2, \cdots, a_n, a_{n+1}$ are $n+1$ distinct zero points of $f(x)$, it means that

$$\begin{cases} c_0 + c_1 a_1 + c_2 a_1^2 + \cdots + c_n a_1^n = 0 \\ c_0 + c_1 a_2 + c_2 a_2^2 + \cdots + c_n a_2^n = 0 \\ \cdots\cdots \\ c_0 + c_1 a_{n+1} + c_2 a_{n+1}^2 + \cdots + c_n a_{n+1}^n = 0 \end{cases}$$

这是以 $c_0, c_1, c_2, \cdots, c_n$ 为未知数的齐次线性方程组, 其系数行列式为

It is a system of homogeneous linear equations in unknowns $c_0, c_1, c_2, \cdots, c_n$, and the corresponding coefficient determinant is

$$D = \begin{vmatrix} 1 & a_1 & a_1^2 & \cdots & a_1^n \\ 1 & a_2 & a_2^2 & \cdots & a_2^n \\ 1 & a_3 & a_3^2 & \cdots & a_3^n \\ \vdots & \vdots & \vdots & & \vdots \\ 1 & a_{n+1} & a_{n+1}^2 & \cdots & a_{n+1}^n \end{vmatrix} = \begin{vmatrix} 1 & 1 & \cdots & 1 \\ a_1 & a_2 & \cdots & a_{n+1} \\ a_1^2 & a_2^2 & \cdots & a_{n+1}^2 \\ \vdots & \vdots & & \vdots \\ a_1^n & a_2^n & \cdots & a_{n+1}^n \end{vmatrix} = D^{\mathrm{T}}$$

显然, 此行列式是范德蒙德行列式. 由于 $a_i \neq a_j (i \neq j)$, 所以

Obviously, it is the Vandermonde determinant. Since $a_i \neq a_j (i \neq j)$, we have

$$D = \prod_{1 \leqslant j < i \leqslant n+1} (a_i - a_j) \neq 0$$

根据定理 2 知, 方程组有唯一零解, 即

From Theorem 2 we know that the system has a unique zero solution, i.e.,

$$c_0 = c_1 = c_2 = \cdots = c_n = 0$$

故 $f(x)$ 是一个零多项式. ■

Thus, $f(x)$ is a zero polynomial. ■

思 考 题

Question

用克拉默法则求解线性方程组时, 对未知数的个数和方程个数有什么要求?

As we solve a linear system by Cramer's rule, what is the requirement for the unknown number and the equation number?

习 题 1

Exercise 1

1. 计算下列行列式:

1. Calculate the following determinants:

(1) $\begin{vmatrix} \sqrt{a} & 1 \\ -1 & -\sqrt{a} \end{vmatrix}$; (2) $\begin{vmatrix} 1 & -1 & 2 \\ 0 & 3 & -1 \\ -2 & 2 & -4 \end{vmatrix}$;

(3) $\begin{vmatrix} 1 & 1 & 0 & 0 \\ 2 & -1 & 1 & 0 \\ 3 & 0 & 0 & -1 \\ -1 & 2 & 1 & 2 \end{vmatrix}$; (4) $\begin{vmatrix} a_{11} & a_{12} & 0 & 0 & 0 \\ a_{21} & a_{22} & 0 & 0 & 0 \\ a_{31} & a_{32} & 1 & 0 & 0 \\ a_{41} & a_{42} & 0 & 1 & 0 \\ a_{51} & a_{52} & 0 & 0 & 1 \end{vmatrix}$;

(5) $\begin{vmatrix} 0 & 1 & 1 & 1 \\ 1 & 0 & 1 & 1 \\ 1 & 1 & 0 & 1 \\ 1 & 1 & 1 & 0 \end{vmatrix}$; (6) $\begin{vmatrix} 3 & 1 & 1 & 1 \\ 1 & 3 & 1 & 1 \\ 1 & 1 & 3 & 1 \\ 1 & 1 & 1 & 3 \end{vmatrix}$;

(7) $\begin{vmatrix} 1 & 2 & -1 & 2 \\ 3 & 0 & 1 & 5 \\ 1 & -2 & 0 & 3 \\ -2 & -4 & 1 & 6 \end{vmatrix}$; (8) $\begin{vmatrix} a & b & c & d \\ a & a+b & a+b+c & a+b+c+d \\ 0 & a & a+b & a+b+c \\ 0 & 0 & a & a+b \end{vmatrix}$.

2. 计算下列 n 阶行列式:

2. Calculate the following n-th order determinants:

(1) $D_n = \begin{vmatrix} 0 & 0 & \cdots & 0 & 1 & 0 \\ 0 & 0 & \cdots & 2 & 0 & 0 \\ \vdots & \vdots & & \vdots & \vdots & \vdots \\ n-1 & 0 & \cdots & 0 & 0 & 0 \\ 0 & 0 & \cdots & 0 & 0 & n \end{vmatrix}$;

(2) $D_n = \begin{vmatrix} 1+a_1 & 1 & \cdots & 1 \\ 1 & 1+a_2 & \cdots & 1 \\ \vdots & \vdots & & \vdots \\ 1 & 1 & \cdots & 1+a_n \end{vmatrix}$ $(a_1, a_2, \cdots, a_n \neq 0)$;

(3) $D_n = \begin{vmatrix} x_1 - m & x_2 & \cdots & x_n \\ x_1 & x_2 - m & \cdots & x_n \\ \vdots & \vdots & & \vdots \\ x_1 & x_2 & \cdots & x_n - m \end{vmatrix}$;

(4) $D_{n+1}=\begin{vmatrix} a & -1 & 0 & \cdots & 0 & 0 \\ ax & a & -1 & \cdots & 0 & 0 \\ ax^2 & ax & a & \cdots & 0 & 0 \\ \vdots & \vdots & \vdots & & \vdots & \vdots \\ ax^n & ax^{n-1} & ax^{n-2} & \cdots & ax & a \end{vmatrix}$.

3. 证明:

3. Prove that

(1) $\begin{vmatrix} a_1+b_1x & a_1x+b_1 & c_1 \\ a_2+b_2x & a_2x+b_2 & c_2 \\ a_3+b_3x & a_3x+b_3 & c_3 \end{vmatrix}=(1-x^2)\begin{vmatrix} a_1 & b_1 & c_1 \\ a_2 & b_2 & c_2 \\ a_3 & b_3 & c_3 \end{vmatrix}$;

(2) $\begin{vmatrix} 1 & 1 & 1 \\ a & b & c \\ a^3 & b^3 & c^3 \end{vmatrix}=(a+b+c)(a-b)(a-c)(c-b)$;

(3) $\begin{vmatrix} 1 & a & b & c+d \\ 1 & b & c & a+d \\ 1 & c & d & a+b \\ 1 & d & a & b+c \end{vmatrix}=0$;

(4) $\begin{vmatrix} x & 0 & 0 & \cdots & 0 & a_0 \\ -1 & x & 0 & \cdots & 0 & a_1 \\ 0 & -1 & x & \cdots & 0 & a_2 \\ \vdots & \vdots & \vdots & & \vdots & \vdots \\ 0 & 0 & 0 & \cdots & x & a_{n-2} \\ 0 & 0 & 0 & \cdots & -1 & x+a_{n-1} \end{vmatrix}=x^n+a_{n-1}x^{n-1}+\cdots+a_1x+a_0$.

4. 已知行列式

4. For the known determinant

$$D=\begin{vmatrix} 3 & -5 & 2 & 1 \\ 1 & 1 & 0 & -5 \\ -1 & 3 & 1 & 3 \\ 2 & -4 & -1 & -3 \end{vmatrix}$$

求 $A_{11}+A_{12}+A_{13}+A_{14}$ 和 $M_{11}+M_{21}+M_{31}+M_{41}$.

find $A_{11}+A_{12}+A_{13}+A_{14}$ and $M_{11}+M_{21}+M_{31}+M_{41}$.

5. 解方程

5. Solve the equation

$$\begin{vmatrix} 1 & 1 & 1 & \cdots & 1 \\ 1 & 1-x & 1 & \cdots & 1 \\ 1 & 1 & 2-x & \cdots & 1 \\ \vdots & \vdots & \vdots & & \vdots \\ 1 & 1 & 1 & \cdots & n-x \end{vmatrix} = 0$$

6. 设

6. Suppose that

$$\begin{vmatrix} a_{11} & a_{12} & a_{13} \\ a_{21} & a_{22} & a_{23} \\ a_{31} & a_{32} & a_{33} \end{vmatrix} = 1$$

求

Find

$$\begin{vmatrix} 4a_{11} & 2a_{11}-3a_{12} & a_{13} \\ 4a_{21} & 2a_{21}-3a_{22} & a_{23} \\ 4a_{31} & 2a_{31}-3a_{32} & a_{33} \end{vmatrix}$$

7. 用克拉默法则求解下列线性方程组:

7. Use Cramer's rule to solve the following linear systems:

(1) $$\begin{cases} x+y+z=0; \\ 2x-5y-3z=10; \\ 4x+8y+2z=4; \end{cases}$$

(2) $$\begin{cases} 2x_1+x_2-5x_3+x_4=8; \\ x_1-3x_2-6x_4=9; \\ 2x_2-x_3+2x_4=-5; \\ x_1+4x_2-7x_3+6x_4=0. \end{cases}$$

8. λ 取何值时, 齐次线性方程组

8. What values do λ take on such that the following system of homogeneous linear equations

$$\begin{cases} (5-\lambda)x_1+2x_2+2x_3=0 \\ 2x_1+(6-\lambda)x_2=0 \\ 2x_1+(4-\lambda)x_3=0 \end{cases}$$

有非零解?

has nonzero solutions?

9. a,b 取何值时, 使得方程组

9. What values do a,b take on such that the system

$$\begin{cases} ax_1 + \quad x_2 + x_3 = 0 \\ x_1 + \quad bx_2 + x_3 = 0 \\ x_1 + 2bx_2 + x_3 = 0 \end{cases}$$

只有零解?

has only a zero solution?

补充题 1

Supplement Exercise 1

1. 证明: 若 n 为奇数, 有

1. Prove that if n is an odd number, we have

$$D_n = \begin{vmatrix} 0 & a_{12} & \cdots & a_{1n} \\ -a_{12} & 0 & \cdots & a_{2n} \\ \vdots & \vdots & & \vdots \\ -a_{1n} & -a_{2n} & \cdots & 0 \end{vmatrix} = 0$$

2. 证明

2. Prove that

$$\begin{vmatrix} 1+x & 1 & 1 & 1 \\ 1 & 1-x & 1 & 1 \\ 1 & 1 & 1+y & 1 \\ 1 & 1 & 1 & 1-y \end{vmatrix} = x^2y^2$$

3. 证明

3. Prove that

$$D_n = \begin{vmatrix} a_1 & -a_2 & 0 & 0 & 0 & \cdots & 0 & 0 \\ 0 & a_2 & -a_3 & 0 & 0 & \cdots & 0 & 0 \\ 0 & 0 & a_3 & -a_4 & 0 & \cdots & 0 & 0 \\ \vdots & \vdots & \vdots & \vdots & \vdots & & \vdots & \vdots \\ 0 & 0 & 0 & 0 & 0 & \cdots & a_{n-1} & -a_n \\ 1 & 1 & 1 & 1 & 1 & \cdots & 1 & 1+a_n \end{vmatrix}$$

$$= a_1a_2\cdots a_n\left(1+\sum_{i=1}^{n}\frac{1}{a_i}\right) \quad (a_1, a_2, \cdots, a_n \neq 0)$$

4. 证明

4. Prove that

$$D_n = \begin{vmatrix} n & n-1 & n-2 & \cdots & 2 & 1 \\ -1 & x & 0 & \cdots & 0 & 0 \\ 0 & -1 & x & \cdots & 0 & 0 \\ \vdots & \vdots & \vdots & & \vdots & \vdots \\ 0 & 0 & 0 & \cdots & x & 0 \\ 0 & 0 & 0 & \cdots & -1 & x \end{vmatrix} = nx^{n-1} + (n-1)x^{n-2} + \cdots + 2x + 1$$

5. 求方程 $f(x) = 0$ 的根, 其中

5. Find the roots of the equation $f(x) = 0$, where

$$f(x) = \begin{vmatrix} x-1 & x-2 & x-1 & x \\ x-2 & x-4 & x-2 & x \\ x-3 & x-6 & x-4 & x-1 \\ x-4 & x-8 & 2x-5 & x-2 \end{vmatrix}$$

6. 求三次多项式 $f(x) = a_0 + a_1x + a_2x^2 + a_3x^3$, 使得

6. Find a polynomial of degree 3, i.e., $f(x) = a_0 + a_1x + a_2x^2 + a_3x^3$, such that

$$f(-1) = 0, f(1) = 4, f(2) = 3, f(3) = 16$$

第 2 章
Chapter 2

矩 阵
Matrices

矩阵是线性代数中的一个重要概念，它贯穿于线性代数的各部分内容. 矩阵不仅用于求解线性方程组，同时也是学习管理科学及其他学科不可缺少的基本工具. 早在 1858 年，英国数学家凯莱发表了论文《矩阵论的研究报告》，首先定义了矩阵的某些运算，因而他被认为是矩阵论的创始人.

Matrix is used as an important concept in Linear Algebra and runs through each part of the contents. Moreover, matrix not only is used for solving systems of linear equations, but also is an indispensable basic tool for studying management science and other subjects. Earlier in 1858, A. Cayley, a British mathematician, published a paper named "A Memoir on the Theory of Matrices", in which he firstly defined some operations on matrices, and thus he is viewed as the founder of matrix theory.

第 1 章我们利用行列式求解线性方程组. 因为线性方程组的一些重要性质都反映在对应的系数矩阵和增广矩阵上，所以本章将利用矩阵的相关理论求解线性方程组.

In Chapter 1, we solved linear systems by determinants. Since some important properties of a linear system are reflected in the corresponding coefficient matrix and augmented matrix, this chapter will solve the linear system by the correlation theory of matrices.

➢本章内容提要

1. 矩阵的运算及其运算律;

2. 利用矩阵求解线性方程组.

Headline of this chapter

1. Operations and operation rules of matrices;

2. Solving linear systems by matrices.

2.1 矩阵及其运算

2.1 Matrices and Operations

2.1.1 矩阵的概念

2.1.1 Concepts of Matrices

考虑产品调运问题. 设某产品有 3 个产地 A_1, A_2, A_3, 4 个销地 B_1, B_2, B_3, B_4. 从产地 A_i 到销地 B_j 调运该产品的数量为 $a_{ij}(i=1,2,3; j=1,2,3,4)$, 则该产品的调运方案可用数表表示, 即

Consider the problem of product transportation. Suppose that a certain product has 3 producing areas A_1, A_2, A_3 and 4 sale places B_1, B_2, B_3, B_4. If the quantity of the product transported from the producing area A_i to the sale place B_j is $a_{ij}(i=1,2,3; j=1,2,3,4)$, then the transportation scheme can be expressed by the following array, namely,

$$\begin{pmatrix} a_{11} & a_{12} & a_{13} & a_{14} \\ a_{21} & a_{22} & a_{23} & a_{24} \\ a_{31} & a_{32} & a_{33} & a_{34} \end{pmatrix}$$

又如, 考虑如下的线性方程组

For another example, consider the following linear system

$$\begin{cases} x_1 + x_2 + 3x_3 = 2 \\ 2x_1 + 2x_2 + x_3 = 3 \\ 3x_1 + 4x_2 + 3x_3 = 4 \end{cases}$$

它的解由未知量 x_1, x_2, x_3 的系数及常数项确定. 因此, 可将该方程组用如下的数表简单表示, 即

its solution is determined by the coefficients of the unknowns x_1, x_2, x_3 and the constant terms. Therefore, this system can be expressed simply by the following array, i.e.,

$$\begin{pmatrix} 1 & 1 & 3 & 2 \\ 2 & 2 & 1 & 3 \\ 3 & 4 & 3 & 4 \end{pmatrix}$$

以上两个数表称为矩阵. 下面给出一般定义.

The above two arrays are said to be matrices. Next we give a general definition.

定义 1 由 $m \times n$ 个数 $a_{ij}\,(i=1,2,\cdots,m;j=1,2,\cdots,n)$ 排成 m 行 n 列的数表

Definition 1 An array of $m \times n$ scalars $a_{ij}\,(i=1,2,\cdots,m;j=1,2,\cdots,n)$ arranged by m rows and n columns, given by

$$\boldsymbol{A}=\begin{pmatrix} a_{11} & a_{12} & \cdots & a_{1n} \\ a_{21} & a_{22} & \cdots & a_{2n} \\ \vdots & \vdots & & \vdots \\ a_{m1} & a_{m2} & \cdots & a_{mn} \end{pmatrix} \tag{2.1}$$

称为 m 行 n 列**矩阵**, 或称为 $m \times n$ 矩阵 $\boldsymbol{A}$. 数 a_{ij} 称为矩阵 $\boldsymbol{A}$ 的第 i 行、第 j 列的元素. 元素是实数的矩阵称为实矩阵, 元素是复数的矩阵称为复矩阵. 除特别说明, 本书中讨论的矩阵都是实矩阵. 式 (2.1) 也简记为 $\boldsymbol{A}=(a_{ij})_{m\times n}$ 或 $\boldsymbol{A}_{m\times n}$.

is called a **matrix** of m rows and n columns, or is called an $m \times n$ matrix $\boldsymbol{A}$. The scalar a_{ij} is called an element located at the i-th row and the j-th column of the matrix $\boldsymbol{A}$. The matrix is said to be a real matrix if the elements are all real numbers, and is said to be a complex matrix if the elements are all complex numbers. Except special instruction, the matrices discussed in this textbook are all real matrices. Equation (2.1) can also be written as $\boldsymbol{A}=(a_{ij})_{m\times n}$ or $\boldsymbol{A}_{m\times n}$.

注 (1) 两个行数、列数分别相等的矩阵, 称为同型矩阵.

Note (1) Two matrices are said to be matrices of the same size if their row number and column number are equal, respectively.

(2) 两个同型矩阵 $\boldsymbol{A}=(a_{ij})_{m\times n}$ 与 $\boldsymbol{B}=(b_{ij})_{m\times n}$ 中, 若

(2) In the two matrices of the same size, $\boldsymbol{A}=(a_{ij})_{m\times n}$ and $\boldsymbol{B}=(b_{ij})_{m\times n}$, if

$$a_{ij}=b_{ij} \quad (i=1,2,\cdots,m;j=1,2,\cdots,n)$$

则称矩阵 $\boldsymbol{A}$ 与 $\boldsymbol{B}$ **相等**, 记为 $\boldsymbol{A}=\boldsymbol{B}$.

then $\boldsymbol{A}$ is said to be **equal** to $\boldsymbol{B}$, written as $\boldsymbol{A}=\boldsymbol{B}$.

(3) 元素全为零的矩阵称为**零矩阵**. 注意不同型的零矩阵不相等.

(3) The matrix is said to be a **zero matrix** if its elements are all zeros. Note that zero matrices of different size are not equal.

(4) 行数和列数都等于 n 的矩阵称为 n **阶方阵**, 有时也称为 n **阶矩阵**.

(4) The matrix is said to be a **square matrix of order** n if the row number and the column number are all equal to n, sometimes it is also said to be a **matrix of order** n.

(5) 当 $m = 1$ 时,

(5) As $m = 1$,

$$\boldsymbol{A} = (a_{ij})_{1\times n} = (a_{11}, a_{12}, \cdots, a_{1n})$$

称为**行矩阵** (也称为**行向量**).

is said to be a **row matrix** (is also said to be a **row vector**).

(6) 当 $n = 1$ 时,

(6) As $n = 1$,

$$\boldsymbol{A} = (a_{ij})_{m\times 1} = \begin{pmatrix} a_{11} \\ a_{21} \\ \vdots \\ a_{m1} \end{pmatrix}$$

称为**列矩阵** (也称为**列向量**).

it is said to be a **column matrix** (is also said to be a **column vector**).

下面介绍几类特殊的方阵.

Next we introduce several special square matrices.

如下的两个矩阵分别称为**上三角形矩阵**和**下三角形矩阵**, 统称为**三角形矩阵**.

The following two matrices are respectively called an **upper triangular matrix** and a **lower triangular matrix**, collectively called **triangular matrices**.

$$\begin{pmatrix} a_{11} & a_{12} & \cdots & a_{1n} \\ 0 & a_{22} & \cdots & a_{2n} \\ \vdots & \vdots & & \vdots \\ 0 & 0 & \cdots & a_{nn} \end{pmatrix}, \quad \begin{pmatrix} a_{11} & 0 & \cdots & 0 \\ a_{21} & a_{22} & \cdots & 0 \\ \vdots & \vdots & & \vdots \\ a_{n1} & a_{n2} & \cdots & a_{nn} \end{pmatrix}$$

特别地, 矩阵

Particularly, the matrix

$$\begin{pmatrix} \lambda_1 & 0 & \cdots & 0 \\ 0 & \lambda_2 & \cdots & 0 \\ \vdots & \vdots & & \vdots \\ 0 & 0 & \cdots & \lambda_n \end{pmatrix}$$

称为**对角矩阵**, 其特点是不在主对角线上的元素全为零, 简记为

is said to be a **diagonal matrix**, the feature is that the off-diagonal elements are all zeros, for short, written as

$$\boldsymbol{\Lambda} = \mathrm{diag}(\lambda_1, \lambda_2, \cdots, \lambda_n)$$

对角线上的元素均相等的对角矩阵, 称为**数量矩阵**, 即

The diagonal matrix is said to be a **scalar matrix** if the elements on the diagonal are all equal, namely,

$$\begin{pmatrix} \lambda & 0 & \cdots & 0 \\ 0 & \lambda & \cdots & 0 \\ \vdots & \vdots & & \vdots \\ 0 & 0 & \cdots & \lambda \end{pmatrix}$$

对角线上的元素均为 1 的 n 阶对角矩阵称为 n **阶单位矩阵**, 即

The diagonal matrix of order n is said to be the **unit matrix of order** n if its elements on the diagonal are all equal to 1, namely,

$$\begin{pmatrix} 1 & 0 & \cdots & 0 \\ 0 & 1 & \cdots & 0 \\ \vdots & \vdots & & \vdots \\ 0 & 0 & \cdots & 1 \end{pmatrix}$$

记为 $\boldsymbol{E}_{n\times n}$ 或 $\boldsymbol{I}_{n\times n}$.

written as $\boldsymbol{E}_{n\times n}$ or $\boldsymbol{I}_{n\times n}$.

2.1.2 矩阵的运算

2.1.2 Operations of Matrices

1. 矩阵的加法

1. Addition of matrices

定义 2 令

Definition 2 Let

$$\boldsymbol{A} = (a_{ij})_{m\times n}, \boldsymbol{B} = (b_{ij})_{m\times n}$$

称矩阵 $\boldsymbol{C}$ 为 $\boldsymbol{A}$ 与 $\boldsymbol{B}$ 的和, 记为 $\boldsymbol{C} = \boldsymbol{A}+\boldsymbol{B}$, 即

the matrix $\boldsymbol{C}$ is called the sum of $\boldsymbol{A}$ and $\boldsymbol{B}$, written as $\boldsymbol{C} = \boldsymbol{A}+\boldsymbol{B}$, namely,

$$\boldsymbol{A}+\boldsymbol{B} = (a_{ij})_{m\times n} + (b_{ij})_{m\times n} = (a_{ij}+b_{ij})_{m\times n}$$

$$= \begin{pmatrix} a_{11}+b_{11} & a_{12}+b_{12} & \cdots & a_{1n}+b_{1n} \\ a_{21}+b_{21} & a_{22}+b_{22} & \cdots & a_{2n}+b_{2n} \\ \vdots & \vdots & & \vdots \\ a_{m1}+b_{m1} & a_{m2}+b_{m2} & \cdots & a_{mn}+b_{mn} \end{pmatrix}_{m\times n}$$

注 两个矩阵 $\boldsymbol{A}$ 与 $\boldsymbol{B}$ 必须是同型矩阵才能作加法运算.

Note The two matrices $\boldsymbol{A}$ and $\boldsymbol{B}$ can perform the addition operation only when they are of the same size.

因为矩阵的加法运算是利用数的加法而定义的, 所以数的加法的某些运算规律会传递给矩阵的加法. 设 $\boldsymbol{A},\boldsymbol{B},\boldsymbol{C},\boldsymbol{O}$ 均为 $m\times n$ 矩阵, 不难证明:

Since the addition operation of matrices is defined by using the addition of numbers, some operation rules about the addition of numbers can be transmitted to the addition operation of matrices. Suppose that $\boldsymbol{A},\boldsymbol{B},\boldsymbol{C},\boldsymbol{O}$ are all $m\times n$ matrices, it is not difficult to show that:

(1) 交换律

(1) Commutative law

$$\boldsymbol{A}+\boldsymbol{B}=\boldsymbol{B}+\boldsymbol{A}$$

(2) 结合律

(2) Associative law

$$(\boldsymbol{A}+\boldsymbol{B})+\boldsymbol{C}=\boldsymbol{A}+(\boldsymbol{B}+\boldsymbol{C})$$

(3) 零矩阵

(3) Zero matrix

$$\boldsymbol{A}+\boldsymbol{O}=\boldsymbol{A}$$

(4) 负矩阵

(4) Negative matrix

$$-\boldsymbol{A}=(-a_{ij})_{m\times n}=\begin{pmatrix} -a_{11} & -a_{12} & \cdots & -a_{1n} \\ -a_{21} & -a_{22} & \cdots & -a_{2n} \\ \vdots & \vdots & & \vdots \\ -a_{m1} & -a_{m2} & \cdots & -a_{mn} \end{pmatrix}_{m\times n}$$

显然,

Obviously,

$$\boldsymbol{A}+(-\boldsymbol{A})=\boldsymbol{O}$$

因此, 矩阵的减法可定义为

Therefore, we can define the subtraction of matrices as follows

$$\boldsymbol{A}-\boldsymbol{B}=\boldsymbol{A}+(-\boldsymbol{B})=(a_{ij}-b_{ij})_{m\times n}$$

2. 矩阵的数乘

定义 3 矩阵 $(\lambda a_{ij})_{m\times n}$ 称为数 λ 与矩阵 $\boldsymbol{A}=(a_{ij})_{m\times n}$ 的乘积 (简称数乘), 记为 $\lambda\boldsymbol{A}$ 或 $\boldsymbol{A}\lambda$, 即

2. Scalar multiplication of matrices

Definition 3 The matrix $(\lambda a_{ij})_{m\times n}$ is called a multiplication of the matrix $\boldsymbol{A}=(a_{ij})_{m\times n}$ by the scalar λ (for short, scalar multiplication), written as $\lambda\boldsymbol{A}$ or $\boldsymbol{A}\lambda$, namely,

$$\lambda\boldsymbol{A}=(\lambda a_{ij})_{m\times n}=\begin{pmatrix}\lambda a_{11} & \lambda a_{12} & \cdots & \lambda a_{1n}\\ \lambda a_{21} & \lambda a_{22} & \cdots & \lambda a_{2n}\\ \vdots & \vdots & & \vdots\\ \lambda a_{m1} & \lambda a_{m2} & \cdots & \lambda a_{mn}\end{pmatrix}_{m\times n}$$

设 $\boldsymbol{A}$ 和 $\boldsymbol{B}$ 为 $m\times n$ 矩阵, λ 和 μ 为常数. 根据数乘运算的定义, 容易验证它满足下列运算规律:

Suppose that $\boldsymbol{A}$ and $\boldsymbol{B}$ are $m\times n$ matrices, λ and μ are constants. From the definition of scalar multiplication, we can easily show that it satisfies the following operation rules:

(1) $\lambda(\mu\boldsymbol{A})=(\lambda\mu)\boldsymbol{A}$;
(2) $(\lambda+\mu)\boldsymbol{A}=\lambda\boldsymbol{A}+\mu\boldsymbol{A}$;
(3) $\lambda(\boldsymbol{A}+\boldsymbol{B})=\lambda\boldsymbol{A}+\lambda\boldsymbol{B}$.

(1) $\lambda(\mu\boldsymbol{A})=(\lambda\mu)\boldsymbol{A}$;
(2) $(\lambda+\mu)\boldsymbol{A}=\lambda\boldsymbol{A}+\mu\boldsymbol{A}$;
(3) $\lambda(\boldsymbol{A}+\boldsymbol{B})=\lambda\boldsymbol{A}+\lambda\boldsymbol{B}$.

注 (1)

Note (1)

$$\begin{pmatrix}6 & 0 & -14\\ -2 & 2 & 4\end{pmatrix}=2\begin{pmatrix}3 & 0 & -7\\ -1 & 1 & 2\end{pmatrix}$$

但是,

however,

$$\begin{pmatrix}6 & 0 & -14\\ -2 & 2 & 4\end{pmatrix}\neq 2\begin{pmatrix}6 & 0 & -14\\ -1 & 1 & 2\end{pmatrix}$$

即不要把矩阵的数乘运算与行列式的性质 3 的推论 1 相混淆.

that is to say, do not confuse the scalar multiplication of matrices with Corollary 1 of Property 3 of determinants.

(2) 矩阵的加法和矩阵与数的乘法统称为矩阵的**线性运算**.

(2) The addition and the scalar multiplication of matrices are said to be **linear operations** of matrices collectively.

例 1 设

Example 1 Suppose that

$$A=\begin{pmatrix}-1 & 4 & -5\\ 2 & 0 & 1\end{pmatrix},\quad B=\begin{pmatrix}3 & 0 & -7\\ -1 & 1 & 2\end{pmatrix}$$

求 $2A-3B$.

Find $2A-3B$.

解 利用矩阵的线性运算, 得

Solution Using the linear operations of matrices, we obtain

$$\begin{aligned}2A-3B&=2\begin{pmatrix}-1 & 4 & -5\\ 2 & 0 & 1\end{pmatrix}-3\begin{pmatrix}3 & 0 & -7\\ -1 & 1 & 2\end{pmatrix}\\ &=\begin{pmatrix}-2 & 8 & -10\\ 4 & 0 & 2\end{pmatrix}-\begin{pmatrix}9 & 0 & -21\\ -3 & 3 & 6\end{pmatrix}\\ &=\begin{pmatrix}-11 & 8 & 11\\ 7 & -3 & -4\end{pmatrix}\end{aligned}$$

■

3. 矩阵乘法

3. Matrix multiplication

定义 4 令

Definition 4 Let

$$A=(a_{ij})_{m\times s}, B=(b_{ij})_{s\times n}, C=(c_{ij})_{m\times n}$$

称为 A 与 B 的乘积, 其中

is called the multiplication of A and B, where

$$c_{ij}=a_{i1}b_{1j}+a_{i2}b_{2j}+\cdots+a_{is}b_{sj}=\sum_{k=1}^{s}a_{ik}b_{kj}\quad(i=1,2,\cdots,m;j=1,2,\cdots,n)$$

记为 $C=AB$, 即

written as $C=AB$, i.e.,

$$\begin{aligned}&\begin{pmatrix}a_{11} & a_{12} & \cdots & a_{1s}\\ a_{21} & a_{22} & \cdots & a_{2s}\\ \vdots & \vdots & & \vdots\\ a_{m1} & a_{m2} & \cdots & a_{ms}\end{pmatrix}\begin{pmatrix}b_{11} & b_{12} & \cdots & b_{1n}\\ b_{21} & b_{22} & \cdots & b_{2n}\\ \vdots & \vdots & & \vdots\\ b_{s1} & b_{s2} & \cdots & b_{sn}\end{pmatrix}\\ =&\begin{pmatrix}a_{11}b_{11}+\cdots+a_{1s}b_{s1} & \cdots & a_{11}b_{1n}+\cdots+a_{1s}b_{sn}\\ a_{21}b_{11}+\cdots+a_{2s}b_{s1} & \cdots & a_{21}b_{1n}+\cdots+a_{2s}b_{sn}\\ \vdots & & \vdots\\ a_{m1}b_{11}+\cdots+a_{ms}b_{s1} & \cdots & a_{m1}b_{1n}+\cdots+a_{ms}b_{sn}\end{pmatrix}_{m\times n}\end{aligned}$$

注 (1) $\boldsymbol{C}$ 中的元素 c_{ij} 等于第一个矩阵 $\boldsymbol{A}$ 中的第 i 行与第二个矩阵 $\boldsymbol{B}$ 中的第 j 列对应元素乘积之和.

Note (1) The element c_{ij} in $\boldsymbol{C}$ is obtained by adding the multiplications of the corresponding elements on the i-th row in the first matrix $\boldsymbol{A}$ by those on the j-th column in the second matrix $\boldsymbol{B}$.

(2) 只有当左边矩阵 $\boldsymbol{A}$ 的列数等于右边矩阵 $\boldsymbol{B}$ 的行数时, 乘积 $\boldsymbol{AB}$ 才有意义, 并且 $\boldsymbol{AB}$ 的行数等于矩阵 $\boldsymbol{A}$ 的行数, $\boldsymbol{AB}$ 的列数等于矩阵 $\boldsymbol{B}$ 的列数.

(2) The product $\boldsymbol{AB}$ makes sense only when the column number of the left matrix $\boldsymbol{A}$ is equal to the row number of the right matrix $\boldsymbol{B}$, moreover, the row number of $\boldsymbol{AB}$ is equal to that of $\boldsymbol{A}$ and the column number of $\boldsymbol{AB}$ is equal to that of $\boldsymbol{B}$.

(3) 记号 $\boldsymbol{AB}$ 常读为 "$\boldsymbol{A}$**左乘**$\boldsymbol{B}$" 或 "$\boldsymbol{B}$**右乘**$\boldsymbol{A}$".

(3) The notation $\boldsymbol{AB}$ always reads "$\boldsymbol{A}$ premultiplys $\boldsymbol{B}$" or "$\boldsymbol{B}$ postmultiplys $\boldsymbol{A}$".

例 2 设

Example 2 Suppose that

$$\boldsymbol{A}=\begin{pmatrix} 1 & 0 & 2 & -1 \\ 0 & 1 & -1 & 3 \\ -1 & 2 & 0 & 1 \end{pmatrix}, \quad \boldsymbol{B}=\begin{pmatrix} 1 & 2 \\ 1 & 1 \\ 1 & 3 \\ 1 & 4 \end{pmatrix}$$

求 $\boldsymbol{AB}$.

Find $\boldsymbol{AB}$.

解 由矩阵乘法, 得

Solution From matrix multiplication we have

$$\begin{aligned}\boldsymbol{AB}&=\begin{pmatrix} 1\times1+0\times1+2\times1+(-1)\times1 & 1\times2+0\times1+2\times3+(-1)\times4 \\ 0\times1+1\times1+(-1)\times1+3\times1 & 0\times2+1\times1+(-1)\times3+3\times4 \\ (-1)\times1+2\times1+0\times1+1\times1 & (-1)\times2+2\times1+0\times3+1\times4 \end{pmatrix}\\ &=\begin{pmatrix} 2 & 4 \\ 3 & 10 \\ 2 & 4 \end{pmatrix}\end{aligned}$$

■

注 例 2 中 $\boldsymbol{B}$ 的列数为 2, $\boldsymbol{A}$ 的行数为 3, 所以乘积 $\boldsymbol{BA}$ 无意义.

Note In Example 2, the column number of $\boldsymbol{B}$ is 2 and the row number of $\boldsymbol{A}$ is 3, this means that the product $\boldsymbol{BA}$ makes no sense.

例 3 设

Example 3 Suppose that

$$A = \begin{pmatrix} a_1 \\ a_2 \\ \vdots \\ a_n \end{pmatrix}, \quad B = (b_1, b_2, \cdots, b_n)$$

求 $\boldsymbol{AB}$ 与 $\boldsymbol{BA}$.

Find $\boldsymbol{AB}$ and $\boldsymbol{BA}$.

解 由矩阵乘法, 得

Solution From matrix multiplication we get

$$AB = \begin{pmatrix} a_1 \\ a_2 \\ \vdots \\ a_n \end{pmatrix} (b_1, b_2, \cdots, b_n) = \begin{pmatrix} a_1b_1 & a_1b_2 & \cdots & a_1b_n \\ a_2b_1 & a_2b_2 & \cdots & a_2b_n \\ \vdots & \vdots & & \vdots \\ a_nb_1 & a_nb_2 & \cdots & a_nb_n \end{pmatrix}$$

$$BA = (b_1, b_2, \cdots b_n) \begin{pmatrix} a_1 \\ a_2 \\ \vdots \\ a_n \end{pmatrix} = b_1a_1 + b_2a_2 + \cdots + b_na_n$$ ■

此例说明, 即使 $\boldsymbol{AB}$ 与 $\boldsymbol{BA}$ 都有意义, 但却未必是同型矩阵, 当然也未必相等.

This example illustrates that even if $\boldsymbol{AB}$ and $\boldsymbol{BA}$ all make sense, they may not be matrices of the same size, of course, they may be not identical.

例 4 设线性方程组为

Example 4 Suppose that the linear system is given by

$$\begin{cases} a_{11}x_1 + a_{12}x_2 + \cdots + a_{1n}x_n = b_1 \\ a_{21}x_1 + a_{22}x_2 + \cdots + a_{2n}x_n = b_2 \\ \cdots\cdots \\ a_{m1}x_1 + a_{m2}x_2 + \cdots + a_{mn}x_n = b_m \end{cases} \tag{2.2}$$

令

Let

$$A=\begin{pmatrix} a_{11} & a_{12} & \cdots & a_{1n} \\ a_{21} & a_{22} & \cdots & a_{2n} \\ \vdots & \vdots & & \vdots \\ a_{m1} & a_{m2} & \cdots & a_{mn} \end{pmatrix}, \quad \overline{A}=\begin{pmatrix} a_{11} & a_{12} & \cdots & a_{1n} & b_1 \\ a_{21} & a_{22} & \cdots & a_{2n} & b_2 \\ \vdots & \vdots & & \vdots & \vdots \\ a_{m1} & a_{m2} & \cdots & a_{mn} & b_m \end{pmatrix},$$

$$x=\begin{pmatrix} x_1 \\ x_2 \\ \vdots \\ x_n \end{pmatrix}, \quad b=\begin{pmatrix} b_1 \\ b_2 \\ \vdots \\ b_m \end{pmatrix}$$

利用矩阵乘法, 得到线性方程组的矩阵形式

Using matrix multiplication, we obtain the matrix form of the linear system

$$Ax = b \tag{2.3}$$

其中, 矩阵 A 称为 (2.2) 的**系数矩阵**, 矩阵 $\bar{A}=(A,b)$ 称为线性方程组 (2.2) 的**增广矩阵**. ■

where the matrix A is called the **coefficient matrix** of (2.2) and the matrix $\bar{A}=(A,b)$ is called the **augmented matrix** of (2.2). ■

例 5 设

Example 5 Suppose that

$$A=\begin{pmatrix} 1 & 1 \\ -1 & -1 \end{pmatrix}, \quad B=\begin{pmatrix} -2 & 1 \\ 2 & -1 \end{pmatrix}, \quad C=\begin{pmatrix} 2 & 3 \\ 1 & -3 \end{pmatrix}, \quad D=\begin{pmatrix} 1 & -1 \\ 2 & 1 \end{pmatrix}$$

计算 AB, BA, AC, AD.

Calculate AB, BA, AC, AD.

解 由矩阵乘法, 得

Solution From matrix multiplication we obtain

$$AB=\begin{pmatrix} 1 & 1 \\ -1 & -1 \end{pmatrix}\begin{pmatrix} -2 & 1 \\ 2 & -1 \end{pmatrix}=\begin{pmatrix} 0 & 0 \\ 0 & 0 \end{pmatrix}$$

$$BA=\begin{pmatrix} -2 & 1 \\ 2 & -1 \end{pmatrix}\begin{pmatrix} 1 & 1 \\ -1 & -1 \end{pmatrix}=\begin{pmatrix} -3 & -3 \\ 3 & 3 \end{pmatrix}$$

$$AC=\begin{pmatrix} 1 & 1 \\ -1 & -1 \end{pmatrix}\begin{pmatrix} 2 & 3 \\ 1 & -3 \end{pmatrix}=\begin{pmatrix} 3 & 0 \\ -3 & 0 \end{pmatrix}$$

$$AD=\begin{pmatrix} 1 & 1 \\ -1 & -1 \end{pmatrix}\begin{pmatrix} 1 & -1 \\ 2 & 1 \end{pmatrix}=\begin{pmatrix} 3 & 0 \\ -3 & 0 \end{pmatrix}$$

■

由例 5 可知:

From Example 5 we know that:

(1) $\boldsymbol{AB}$, $\boldsymbol{BA}$ 都有意义且同型, 但 $\boldsymbol{AB} \neq \boldsymbol{BA}$, 即矩阵乘法不满足交换律.

(1) $\boldsymbol{AB}$ and $\boldsymbol{BA}$ all make sense and have the same size, but $\boldsymbol{AB} \neq \boldsymbol{BA}$, that is to say, matrix multiplication does not satisfy the commutative law.

(2) 两个非零矩阵的乘积可能为零矩阵, 也就是, 由等式 $\boldsymbol{AB} = \boldsymbol{O}$, 不能推出 $\boldsymbol{A} = \boldsymbol{O}$ 或 $\boldsymbol{B} = \boldsymbol{O}$.

(2) The product of two nonzero matrices may be a zero matrix, in other words, from the equation $\boldsymbol{AB} = \boldsymbol{O}$ we can not conclude that $\boldsymbol{A} = \boldsymbol{O}$ or $\boldsymbol{B} = \boldsymbol{O}$.

(3) 若等式 $\boldsymbol{AC} = \boldsymbol{AD}$ 成立, 不一定有 $\boldsymbol{C} = \boldsymbol{D}$, 即矩阵乘法不满足消去律.

(3) If the equation $\boldsymbol{AC} = \boldsymbol{AD}$ is valid, there may not exist $\boldsymbol{C} = \boldsymbol{D}$, namely, matrix multiplication does not satisfy the cancellation law.

矩阵乘法虽然不满足交换律和消去律, 但可以证明, 满足下列运算规律 (假设运算都是可行的):

Although matrix multiplication does not satisfy the commutative law and the cancellation law, it can be shown that if the operations are feasible, they satisfy the following operation rules:

(1) 结合律

(1) Associative law

$$(\boldsymbol{AB})\boldsymbol{C} = \boldsymbol{A}(\boldsymbol{BC})$$

(2) 左、右分配律

(2) Left and right distributive laws

$$\boldsymbol{A}(\boldsymbol{B}+\boldsymbol{C}) = \boldsymbol{AB}+\boldsymbol{AC}, \quad (\boldsymbol{B}+\boldsymbol{C})\boldsymbol{A} = \boldsymbol{BA}+\boldsymbol{CA}$$

(3)

(3)

$$\lambda(\boldsymbol{AB}) = (\lambda\boldsymbol{A})\boldsymbol{B} = \boldsymbol{A}(\lambda\boldsymbol{B})$$

其中 λ 为常数.

where λ is a constant.

定义 5 令 $\boldsymbol{A}, \boldsymbol{B}$ 为 n 阶方阵, 如果 $\boldsymbol{AB} = \boldsymbol{BA}$, 则称矩阵 $\boldsymbol{A}$ 与矩阵 $\boldsymbol{B}$ **可交换**.

Definition 5 Let $\boldsymbol{A}, \boldsymbol{B}$ be square matrices of order n. If $\boldsymbol{AB} = \boldsymbol{BA}$, then it is called that the matrices $\boldsymbol{A}$ and $\boldsymbol{B}$ are **commutable**.

注 对于单位矩阵 $\boldsymbol{E}$, 容易证明

Note For the unit matrix $\boldsymbol{E}$, it is easy to show that

$$\boldsymbol{E}_{m\times m}\boldsymbol{A}_{m\times n}=\boldsymbol{A}_{m\times n},\quad \boldsymbol{A}_{m\times n}\boldsymbol{E}_{n\times n}=\boldsymbol{A}_{m\times n}$$

或简写成

or briefly written as

$$\boldsymbol{EA}=\boldsymbol{AE}=\boldsymbol{A}$$

由此可见, 若矩阵乘法可行, 单位矩阵 $\boldsymbol{E}$ 可与任何矩阵可换.

It follows that the unit matrix $\boldsymbol{E}$ is commutable with any matrix if matrix multiplication is feasible.

由于矩阵乘法满足结合律, 因此可定义矩阵的方幂如下.

Since matrix multiplication satisfies the associative law, we can define the power of a matrix, as follows.

设 $\boldsymbol{A}$ 为 n 阶方阵, 用 $\boldsymbol{A}^k$ 表示 k 个方阵 $\boldsymbol{A}$ 的连乘积, 则称 $\boldsymbol{A}^k$ 为 $\boldsymbol{A}$ 的 k **次方幂**. 容易看出

Let $\boldsymbol{A}$ be a square matrix of order n, and let $\boldsymbol{A}^k$ be a product of k square matrices, then $\boldsymbol{A}^k$ is said to be a k-th **power** of $\boldsymbol{A}$. It is easy to see that

$$\boldsymbol{A}^k\boldsymbol{A}^l=\boldsymbol{A}^{k+l},\quad (\boldsymbol{A}^k)^l=\boldsymbol{A}^{kl}$$

其中 k,l 都是正整数.

where k,l are all positive integers.

设 $f(x)$ 为 x 的 m 次多项式, 即 $f(x)=a_0+a_1x+\cdots+a_mx^m$, $\boldsymbol{A}$ 为 n 阶方阵. 令

Suppose that $f(x)$ is a polynomial of degree m with respect to x, given by $f(x)=a_0+a_1x+\cdots+a_mx^m$, and that $\boldsymbol{A}$ is a square matrix of order n. Let

$$f(\boldsymbol{A})=a_0\boldsymbol{E}+a_1\boldsymbol{A}+\cdots+a_m\boldsymbol{A}^m$$

$f(\boldsymbol{A})$ 称为 $\boldsymbol{A}$ 的 m **次多项式**.

$f(\boldsymbol{A})$ is said to be a **polynomial of degree** m with respect to $\boldsymbol{A}$.

例如, 若 $f(x)=2+3x+x^2$, 则有 $f(\boldsymbol{A})=2\boldsymbol{E}+3\boldsymbol{A}+\boldsymbol{A}^2$. 注意不能把 $f(\boldsymbol{A})=2\boldsymbol{E}+3\boldsymbol{A}+\boldsymbol{A}^2$ 写成 $f(\boldsymbol{A})=2+3\boldsymbol{A}+\boldsymbol{A}^2$.

For example, if $f(x)=2+3x+x^2$, then we have $f(\boldsymbol{A})=2\boldsymbol{E}+3\boldsymbol{A}+\boldsymbol{A}^2$. Note that $f(\boldsymbol{A})=2\boldsymbol{E}+3\boldsymbol{A}+\boldsymbol{A}^2$ can not be written as $f(\boldsymbol{A})=2+3\boldsymbol{A}+\boldsymbol{A}^2$.

因为矩阵 $\boldsymbol{A}^k, \boldsymbol{A}^l$ 和 $\boldsymbol{E}$ 都是可交换的, 所以 $\boldsymbol{A}$ 的两个多项式 $g(\boldsymbol{A})$ 和 $f(\boldsymbol{A})$ 总是可交换的, 即

Since the matrices $\boldsymbol{A}^k, \boldsymbol{A}^l$ and $\boldsymbol{E}$ are all commutable, the two polynomials $g(\boldsymbol{A})$ and $f(\boldsymbol{A})$ with respect to $\boldsymbol{A}$ are always commutable, namely,

$$g(\boldsymbol{A})f(\boldsymbol{A}) = f(\boldsymbol{A})g(\boldsymbol{A})$$

例 6 设 $\boldsymbol{A}, \boldsymbol{B}$ 均为 n 阶方阵, 计算 $(\boldsymbol{A}+\boldsymbol{B})^2$.

Example 6 Let $\boldsymbol{A}, \boldsymbol{B}$ be square matrices of order n, calculate $(\boldsymbol{A}+\boldsymbol{B})^2$.

解 由矩阵乘法, 得

Solution From matrix multiplication we have

$$(\boldsymbol{A}+\boldsymbol{B})^2 = (\boldsymbol{A}+\boldsymbol{B})(\boldsymbol{A}+\boldsymbol{B}) = \boldsymbol{A}^2 + \boldsymbol{AB} + \boldsymbol{BA} + \boldsymbol{B}^2 \qquad \blacksquare$$

注 只有当 $\boldsymbol{A}, \boldsymbol{B}$ 可交换时, 才有

Note Only when $\boldsymbol{A}, \boldsymbol{B}$ are commutable, we have

$$(\boldsymbol{A}+\boldsymbol{B})^2 = (\boldsymbol{A}+\boldsymbol{B})(\boldsymbol{A}+\boldsymbol{B}) = \boldsymbol{A}^2 + 2\boldsymbol{AB} + \boldsymbol{A}^2$$
$$(\boldsymbol{A}+\boldsymbol{B})(\boldsymbol{A}-\boldsymbol{B}) = \boldsymbol{A}^2 - \boldsymbol{B}^2$$

例 7 已知

Example 7 Known that

$$\boldsymbol{A} = \begin{pmatrix} 2 & -1 & 2 \\ 4 & -2 & 4 \\ 2 & -1 & 2 \end{pmatrix}$$

求 $\boldsymbol{A}^n$.

Find $\boldsymbol{A}^n$.

解 由矩阵乘法的定义和矩阵的结合律知

Solution From the definition of matrix multiplication and the associative law of matrices, we see that

$$\boldsymbol{A} = \begin{pmatrix} 1 \\ 2 \\ 1 \end{pmatrix} (\ 2, \quad -1, \quad 2\ )$$

$$\boldsymbol{A}^2 = \begin{pmatrix} 1 \\ 2 \\ 1 \end{pmatrix} (\ 2, \quad -1, \quad 2\ ) \begin{pmatrix} 1 \\ 2 \\ 1 \end{pmatrix} (\ 2, \quad -1, \quad 2\ ) = 2\boldsymbol{A}$$

$$\cdots\cdots$$

$$\boldsymbol{A}^n = 2^{n-1}\boldsymbol{A} = \begin{pmatrix} 2^n & -2^{n-1} & 2^n \\ 2^{n+1} & -2^n & 2^{n+1} \\ 2^n & -2^{n-1} & 2^n \end{pmatrix} \qquad \blacksquare$$

4. 矩阵的转置

4. Transpose of matrices

定义 6 将 $m \times n$ 矩阵 $\boldsymbol{A}$ 的行换成同序数的列得到的 $n \times m$ 矩阵, 称为 $\boldsymbol{A}$ 的**转置矩阵**, 记为 $\boldsymbol{A}^{\mathrm{T}}$ 或 $\boldsymbol{A}'$.

Definition 6 Interchanging the rows of an $m \times n$ matrix $\boldsymbol{A}$ by the columns of the same ordinal, we obtain a new $n \times m$ matrix, which is called the **transpose matrix** of $\boldsymbol{A}$, written as $\boldsymbol{A}^{\mathrm{T}}$ or $\boldsymbol{A}'$.

例如, 矩阵 $\boldsymbol{A} = \begin{pmatrix} 1 & 2 & 0 \\ 3 & 1 & -1 \end{pmatrix}$ 的转置矩阵为

For example, the transpose matrix of the original matrix $\boldsymbol{A} = \begin{pmatrix} 1 & 2 & 0 \\ 3 & 1 & -1 \end{pmatrix}$ is

$$\boldsymbol{A}^{\mathrm{T}} = \begin{pmatrix} 1 & 3 \\ 2 & 1 \\ 0 & -1 \end{pmatrix}$$

矩阵的转置也是一种运算, 不难证明其满足下列运算规律 (假设运算都是可行的):

Transpose of a matrix is also a kind of operations, it is not difficult to prove that it satisfies the following operation rules if the operations are all feasible:

(1) $(\boldsymbol{A}^{\mathrm{T}})^{\mathrm{T}} = \boldsymbol{A}$;

(2) $(\boldsymbol{A}+\boldsymbol{B})^{\mathrm{T}} = \boldsymbol{A}^{\mathrm{T}} + \boldsymbol{B}^{\mathrm{T}}$;

(3) $(\lambda \boldsymbol{A})^{\mathrm{T}} = \lambda \boldsymbol{A}^{\mathrm{T}}$;

(4) $(\boldsymbol{A}\boldsymbol{B})^{\mathrm{T}} = \boldsymbol{B}^{\mathrm{T}}\boldsymbol{A}^{\mathrm{T}}$.

注 (2) 和 (4) 可推广到多个矩阵的情形.

Note (2) and (4) can be generalized to the cases of many matrices.

这里只证明 (4). 设

Here we only prove (4). Suppose that

$$\boldsymbol{A} = \begin{pmatrix} a_{11} & a_{12} & \cdots & a_{1s} \\ a_{21} & a_{22} & \cdots & a_{2s} \\ \vdots & \vdots & & \vdots \\ a_{m1} & a_{m2} & \cdots & a_{ms} \end{pmatrix}, \quad \boldsymbol{B} = \begin{pmatrix} b_{11} & b_{12} & \cdots & b_{1n} \\ b_{21} & b_{22} & \cdots & b_{2n} \\ \vdots & \vdots & & \vdots \\ b_{s1} & b_{s2} & \cdots & b_{sn} \end{pmatrix}$$

容易看出, $(\boldsymbol{A}\boldsymbol{B})^{\mathrm{T}}$ 和 $\boldsymbol{B}^{\mathrm{T}}\boldsymbol{A}^{\mathrm{T}}$ 都是 $n \times m$ 矩阵. 其次, 位于 $(\boldsymbol{A}\boldsymbol{B})^{\mathrm{T}}$ 的第 i 行第 j 列的元素就是位于 $\boldsymbol{A}\boldsymbol{B}$ 的第 j 行第 i 列的元素, 且等于

Obviously, $(\boldsymbol{A}\boldsymbol{B})^{\mathrm{T}}$ and $\boldsymbol{B}^{\mathrm{T}}\boldsymbol{A}^{\mathrm{T}}$ are all $n \times m$ matrices. Secondly, the element locating at the i-th row and the j-th column of $(\boldsymbol{A}\boldsymbol{B})^{\mathrm{T}}$ is just the element locating at the j-th row and the i-th column of $\boldsymbol{A}\boldsymbol{B}$, and it is equal to

$$a_{j1}b_{1i}+a_{j2}b_{2i}+\cdots+a_{js}b_{si}=\sum_{k=1}^{s}a_{jk}b_{ki}$$

利用矩阵乘法, 位于 $\boldsymbol{B}^{\mathrm{T}}\boldsymbol{A}^{\mathrm{T}}$ 的第 i 行第 j 列的元素为

Utilizing matrix multiplication, we obtain the element at the i-th row and the j-th column of $\boldsymbol{B}^{\mathrm{T}}\boldsymbol{A}^{\mathrm{T}}$, given by

$$b_{1i}a_{j1}+b_{2i}a_{j2}+\cdots+b_{si}a_{js}=\sum_{k=1}^{s}b_{ki}a_{jk}$$

上面两个式子显然相等, 所以

Obviously, the above two equations are equal, so we have

$$(\boldsymbol{AB})^{\mathrm{T}}=\boldsymbol{B}^{\mathrm{T}}\boldsymbol{A}^{\mathrm{T}}$$

一般地, $(\boldsymbol{AB})^{\mathrm{T}}\neq\boldsymbol{A}^{\mathrm{T}}\boldsymbol{B}^{\mathrm{T}}$.

In general, $(\boldsymbol{AB})^{\mathrm{T}}\neq\boldsymbol{A}^{\mathrm{T}}\boldsymbol{B}^{\mathrm{T}}$.

例 8 设

Example 8 Suppose that

$$\boldsymbol{A}=\begin{pmatrix}-1&1&2\\0&1&1\end{pmatrix},\quad \boldsymbol{B}=\begin{pmatrix}-1&0\\1&3\\2&1\end{pmatrix}$$

求 $(\boldsymbol{AB})^{\mathrm{T}}$ 和 $\boldsymbol{A}^{\mathrm{T}}\boldsymbol{B}^{\mathrm{T}}$.

Find $(\boldsymbol{AB})^{\mathrm{T}}$ and $\boldsymbol{A}^{\mathrm{T}}\boldsymbol{B}^{\mathrm{T}}$.

解 因为

Solution Since

$$\boldsymbol{A}^{\mathrm{T}}=\begin{pmatrix}-1&0\\1&1\\2&1\end{pmatrix},\quad \boldsymbol{B}^{\mathrm{T}}=\begin{pmatrix}-1&1&2\\0&3&1\end{pmatrix}$$

有

we have

$$(\boldsymbol{AB})^{\mathrm{T}}=\boldsymbol{B}^{\mathrm{T}}\boldsymbol{A}^{\mathrm{T}}=\begin{pmatrix}-1&1&2\\0&3&1\end{pmatrix}\begin{pmatrix}-1&0\\1&1\\2&1\end{pmatrix}=\begin{pmatrix}6&3\\5&4\end{pmatrix}$$

$$\boldsymbol{A}^{\mathrm{T}}\boldsymbol{B}^{\mathrm{T}}=\begin{pmatrix}-1&0\\1&1\\2&1\end{pmatrix}\begin{pmatrix}-1&1&2\\0&3&1\end{pmatrix}=\begin{pmatrix}1&-1&-2\\-1&4&3\\-2&5&5\end{pmatrix}$$ ■

设 $\boldsymbol{A}$ 为 n 阶矩阵, 如果满足 $\boldsymbol{A}^{\mathrm{T}}=\boldsymbol{A}$, 即 $a_{ij}=a_{ji}(i,j=1,2,\cdots,n)$, 则称 $\boldsymbol{A}$ 为**对称矩阵**. 对称矩阵的特点是它的元素以主对角线为对称轴对应相等. 如果满足 $\boldsymbol{A}^{\mathrm{T}}=-\boldsymbol{A}$, 即 $a_{ij}=-a_{ji}(i,j=1,2,\cdots,n)$, 则称 $\boldsymbol{A}$ 为**反对称矩阵**. 而反对称矩阵的特点是以主对角线为对称轴的对应元素绝对值相等, 符号相反, 且主对角线上各元素均为 0.

Suppose that $\boldsymbol{A}$ is a square matrix of order n, if it satisfies $\boldsymbol{A}^{\mathrm{T}}=\boldsymbol{A}$, namely, $a_{ij}=a_{ji}(i,j=1,2,\cdots,n)$, then $\boldsymbol{A}$ is said to be a **symmetric matrix**, interestingly, its feature is that the associated elements are equal based on the main diagonal as a symmetric axis. While if a matrix satisfies $\boldsymbol{A}^{\mathrm{T}}=-\boldsymbol{A}$, namely, $a_{ij}=-a_{ji}(i,j=1,2,\cdots,n)$, then $\boldsymbol{A}$ is said to be a **skew-symmetric matrix**, moreover, its feature is that the associated elements are equal with opposite signs based on the main diagonal as a symmetric axis and that the elements on the main diagonal are all zeros.

例如, $\begin{pmatrix}1&2&-1\\2&3&5\\-1&5&0\end{pmatrix}$ 是一个对称矩阵; $\begin{pmatrix}0&-2&1\\2&0&-5\\-1&5&0\end{pmatrix}$ 是一个反对称矩阵.

For example, $\begin{pmatrix}1&2&-1\\2&3&5\\-1&5&0\end{pmatrix}$ is a symmetric matrix; and $\begin{pmatrix}0&-2&1\\2&0&-5\\-1&5&0\end{pmatrix}$ is a skew- symmetric matrix.

例 9 设 $\boldsymbol{A}$ 是 n 阶反对称矩阵, $\boldsymbol{B}$ 是 n 阶对称矩阵. 证明 $\boldsymbol{AB}+\boldsymbol{BA}$ 是 n 阶反对称矩阵.

Example 9 Let $\boldsymbol{A}$ be a skew-symmetric matrix of order n and $\boldsymbol{B}$ be a symmetric matrix of order n. Prove that $\boldsymbol{AB}+\boldsymbol{BA}$ is a skew-symmetric matrix of order n.

证 由 $\boldsymbol{A}^{\mathrm{T}}=-\boldsymbol{A}$, $\boldsymbol{B}^{\mathrm{T}}=\boldsymbol{B}$ 可得

Proof From $\boldsymbol{A}^{\mathrm{T}}=-\boldsymbol{A}$, $\boldsymbol{B}^{\mathrm{T}}=\boldsymbol{B}$, we have

$$\begin{aligned}(\boldsymbol{AB}+\boldsymbol{BA})^{\mathrm{T}}&=(\boldsymbol{AB})^{\mathrm{T}}+(\boldsymbol{BA})^{\mathrm{T}}=\boldsymbol{B}^{\mathrm{T}}\boldsymbol{A}^{\mathrm{T}}+\boldsymbol{A}^{\mathrm{T}}\boldsymbol{B}^{\mathrm{T}}\\&=\boldsymbol{B}(-\boldsymbol{A})+(-\boldsymbol{A})\boldsymbol{B}=-(\boldsymbol{AB}+\boldsymbol{BA})\end{aligned}$$

例 10 设列矩阵

Example 10 Suppose that the column matrix

$$X = \begin{pmatrix} x_1 \\ x_2 \\ \vdots \\ x_n \end{pmatrix}$$

满足 $\boldsymbol{X}^{\mathrm{T}}\boldsymbol{X} = 1, \boldsymbol{E}$ 为 n 阶单位矩阵. 若 $\boldsymbol{H} = \boldsymbol{E} - 2\boldsymbol{X}\boldsymbol{X}^{\mathrm{T}}$, 证明 $\boldsymbol{H}$ 是对称矩阵, 且 $\boldsymbol{H}\boldsymbol{H}^{\mathrm{T}} = \boldsymbol{E}$.

satisfies $\boldsymbol{X}^{\mathrm{T}}\boldsymbol{X} = 1$, and that $\boldsymbol{E}$ is the unit matrix of order n. If $\boldsymbol{H} = \boldsymbol{E} - 2\boldsymbol{X}\boldsymbol{X}^{\mathrm{T}}$, prove that $\boldsymbol{H}$ is a symmetric matrix and that $\boldsymbol{H}\boldsymbol{H}^{\mathrm{T}} = \boldsymbol{E}$.

证 由

Proof Since

$$\boldsymbol{H}^{\mathrm{T}} = (\boldsymbol{E} - 2\boldsymbol{X}\boldsymbol{X}^{\mathrm{T}})^{\mathrm{T}} = \boldsymbol{E}^{\mathrm{T}} - 2(\boldsymbol{X}\boldsymbol{X}^{\mathrm{T}})^{\mathrm{T}} = \boldsymbol{E} - 2\boldsymbol{X}\boldsymbol{X}^{\mathrm{T}} = \boldsymbol{H}$$

可知, $\boldsymbol{H}$ 是对称矩阵.

we know that $\boldsymbol{H}$ is a symmetric matrix.

进一步地,

Further,

$$\begin{aligned}\boldsymbol{H}\boldsymbol{H}^{\mathrm{T}} &= \boldsymbol{H}^2 = (\boldsymbol{E} - 2\boldsymbol{X}\boldsymbol{X}^{\mathrm{T}})^2 = \boldsymbol{E} - 4\boldsymbol{X}\boldsymbol{X}^{\mathrm{T}} + 4(\boldsymbol{X}\boldsymbol{X}^{\mathrm{T}})(\boldsymbol{X}\boldsymbol{X}^{\mathrm{T}}) \\ &= \boldsymbol{E} - 4\boldsymbol{X}\boldsymbol{X}^{\mathrm{T}} + 4\boldsymbol{X}(\boldsymbol{X}^{\mathrm{T}}\boldsymbol{X})\boldsymbol{X}^{\mathrm{T}} = \boldsymbol{E} - 4\boldsymbol{X}\boldsymbol{X}^{\mathrm{T}} + 4\boldsymbol{X}\boldsymbol{X}^{\mathrm{T}} = \boldsymbol{E}\end{aligned}$$ ■

思 考 题

Questions

1. 设下面矩阵都是 n 阶方阵, 下列等式或结论是否成立? 为什么?

1. Suppose that the following matrices are all square matrices of order n, whether the following equations or results are valid, why?

(1) $(\boldsymbol{AB})^k = \boldsymbol{A}^k\boldsymbol{B}^k$;
(2) $(\boldsymbol{A}+\boldsymbol{B})^2 = \boldsymbol{A}^2 + \boldsymbol{B}^2 + 2\boldsymbol{AB}$;
(3) $(\boldsymbol{A}+\boldsymbol{E})^2 = \boldsymbol{A}^2 + 2\boldsymbol{A} + \boldsymbol{E}$;
(4) $(\boldsymbol{A}+\boldsymbol{E})(\boldsymbol{A}-\boldsymbol{E}) = (\boldsymbol{A}-\boldsymbol{E})(\boldsymbol{A}+\boldsymbol{E})$;

(1) $(\boldsymbol{AB})^k = \boldsymbol{A}^k\boldsymbol{B}^k$;
(2) $(\boldsymbol{A}+\boldsymbol{B})^2 = \boldsymbol{A}^2 + \boldsymbol{B}^2 + 2\boldsymbol{AB}$;
(3) $(\boldsymbol{A}+\boldsymbol{E})^2 = \boldsymbol{A}^2 + 2\boldsymbol{A} + \boldsymbol{E}$;
(4) $(\boldsymbol{A}+\boldsymbol{E})(\boldsymbol{A}-\boldsymbol{E}) = (\boldsymbol{A}-\boldsymbol{E})(\boldsymbol{A}+\boldsymbol{E})$;

(5) 若 $\boldsymbol{A}^2 = \boldsymbol{O}$, 则 $\boldsymbol{A} = \boldsymbol{O}$;

(5) If $\boldsymbol{A}^2 = \boldsymbol{O}$, then $\boldsymbol{A} = \boldsymbol{O}$;

(6) 若 $\boldsymbol{A}^2 = \boldsymbol{E}$, 则 $\boldsymbol{A} = \boldsymbol{E}$ 或 $\boldsymbol{A} = -\boldsymbol{E}$;

(6) If $\boldsymbol{A}^2 = \boldsymbol{E}$, then $\boldsymbol{A} = \boldsymbol{E}$ or $\boldsymbol{A} = -\boldsymbol{E}$;

(7) 若 $\boldsymbol{A}^2=\boldsymbol{A}$, 则 $\boldsymbol{A}=\boldsymbol{E}$ 或 $\boldsymbol{A}=\boldsymbol{O}$.

(7) If $\boldsymbol{A}^2=\boldsymbol{A}$, then $\boldsymbol{A}=\boldsymbol{E}$ or $\boldsymbol{A}=\boldsymbol{O}$.

2. 矩阵的运算和数的运算有哪些相同点和不同点?

2. What are the similarity and difference between the operations of matrices and numbers?

2.2 方阵的行列式及其逆矩阵

2.2 Determinants and Inverse Matrices of Square Matrices

本节所讨论的矩阵均为方阵.

The matrices discussed in this section are all square matrices.

2.2.1 方阵的行列式

2.2.1 Determinants of Square Matrices

定义 1 设 $\boldsymbol{A}$ 为 n 阶方阵, 由 $\boldsymbol{A}$ 的元素所构成的行列式 (各元素的位置不变), 称为 $\boldsymbol{A}$ 的行列式, 记为 $|\boldsymbol{A}|$ 或 $\det\boldsymbol{A}$.

Definition 1 Let $\boldsymbol{A}$ be a square matrix of order n. The determinant consisted of the elements of $\boldsymbol{A}$ (the positions of all elements are invariant) is called the determinant of $\boldsymbol{A}$, written as $|\boldsymbol{A}|$ or $\det\boldsymbol{A}$.

例如, 对角方阵

For example, the determinant of the diagonal matrix

$$\boldsymbol{A}=\begin{pmatrix}\lambda_1 & 0 & \cdots & 0\\ 0 & \lambda_2 & \cdots & 0\\ \vdots & \vdots & & \vdots\\ 0 & 0 & \cdots & \lambda_n\end{pmatrix}$$

行列式为 $|\boldsymbol{A}|=\lambda_1\lambda_2\cdots\lambda_n$. 显然, 单位矩阵 $\boldsymbol{E}$ 的行列式 $|\boldsymbol{E}|=1$.

is given by $|\boldsymbol{A}|=\lambda_1\lambda_2\cdots\lambda_n$. Obviously, the determinant of the unit matrix $\boldsymbol{E}$ is $|\boldsymbol{E}|=1$.

当 $|\boldsymbol{A}|\neq 0$ 时, 称 $\boldsymbol{A}$ 是**非奇异矩阵**; 当 $|\boldsymbol{A}|=0$ 时, 称 $\boldsymbol{A}$ 为**奇异矩阵**.

If $|\boldsymbol{A}|\neq 0$, then $\boldsymbol{A}$ is said to be a **nonsingular matrix**; while if $|\boldsymbol{A}|=0$, then $\boldsymbol{A}$ is said to be a **singular matrix**.

设 $\boldsymbol{A},\boldsymbol{B}$ 为 n 阶方阵, λ 为常数. 方阵的行列式具有下面性质:

Suppose that $\boldsymbol{A},\boldsymbol{B}$ are square matrices of order n and that λ is a constant. The determinants of square matrices have the following properties:

(1) 由行列式的性质 1 可以得到

(1) From Property 1 of determinants we obtain

$$\left|\boldsymbol{A}^{\mathrm{T}}\right| = |\boldsymbol{A}|$$

(2) 由矩阵的数乘和行列式的性质 3 可以得到

(2) From scalar multiplication of matrices and Property 3 of determinants we obtain we obtain

$$|\lambda \boldsymbol{A}| = \lambda^n |\boldsymbol{A}|$$

(3) $|\boldsymbol{AB}| = |\boldsymbol{A}||\boldsymbol{B}|$.

(3) $|\boldsymbol{AB}| = |\boldsymbol{A}||\boldsymbol{B}|$.

注 (1) 由性质 (3) 可知, 对于 n 阶方阵 $\boldsymbol{A}, \boldsymbol{B}$, 尽管通常有 $\boldsymbol{AB} \neq \boldsymbol{BA}$, 但 $|\boldsymbol{AB}| = |\boldsymbol{BA}|$. 特别地, $|\boldsymbol{A}^m| = |\boldsymbol{A}|^m$, 其中 m 为正整数.

Note (1) From Property 3 we see that the equation $|\boldsymbol{AB}| = |\boldsymbol{BA}|$ is valid for the square matrices $\boldsymbol{A}, \boldsymbol{B}$ of order n, although $\boldsymbol{AB} \neq \boldsymbol{BA}$ generally. In particular, $|\boldsymbol{A}^m| = |\boldsymbol{A}|^m$, where m is a positive integer.

(2) 性质 3 可以推广到多个 n 阶方阵的情形, 即

(2) Property 3 can be generalized to the case of many square matrices of order n, namely,

$$|\boldsymbol{A}_1\boldsymbol{A}_2\cdots\boldsymbol{A}_m| = |\boldsymbol{A}_1||\boldsymbol{A}_2|\cdots|\boldsymbol{A}_m|$$

(3) 一般地,

(3) In general,

$$|\boldsymbol{A}+\boldsymbol{B}| \neq |\boldsymbol{A}| + |\boldsymbol{B}|$$

(4) 只有方阵才有行列式运算.

(4) Only a square matrix has the determinant operation.

2.2.2 可逆矩阵

2.2.2 Invertible Matrices

在实数运算中, 非零实数 a 的倒数为 a^{-1}, 因此有 $aa^{-1} = a^{-1}a = 1$.

In real arithmetic, the reciprocal of a nonzero real number a is equal to a^{-1}, consequently, $aa^{-1} = a^{-1}a = 1$.

问题 在矩阵运算中, 对于给定的矩阵 $\boldsymbol{A}$, 是否存在矩阵 $\boldsymbol{B}$, 使 $\boldsymbol{AB} = \boldsymbol{BA} = \boldsymbol{E}$? 如果这样的矩阵 $\boldsymbol{B}$ 存在, 如何求得? 下面我们讨论这个问题.

Problem In matrix operations, whether there exists a matrix $\boldsymbol{B}$ such that $\boldsymbol{AB} = \boldsymbol{BA} = \boldsymbol{E}$ for the given matrix $\boldsymbol{A}$? If such a matrix $\boldsymbol{B}$ exists, how to find it? In the following we discuss this problem.

要使 $\boldsymbol{AB}$ 与 $\boldsymbol{BA}$ 都有意义, 由矩阵乘法可知, 矩阵 $\boldsymbol{A}$ 和 $\boldsymbol{B}$ 必须为同阶方阵.

To make $\boldsymbol{AB}$ and $\boldsymbol{BA}$ meaningful, from matrix multiplication we know that the matrices $\boldsymbol{A}$ and $\boldsymbol{B}$ must be square matrices of the same order.

定义 2 设 $\boldsymbol{A}, \boldsymbol{B}$ 均为 n 阶方阵, 若有如下等式成立, 即

Definition 2 Suppose that $\boldsymbol{A}, \boldsymbol{B}$ are all square matrices of order n, if the following equality is valid, namely,

$$\boldsymbol{AB} = \boldsymbol{BA} = \boldsymbol{E} \tag{2.4}$$

则称 $\boldsymbol{A}$**可逆**, $\boldsymbol{B}$ 称为 $\boldsymbol{A}$ 的**逆矩阵**, 记为

then $\boldsymbol{A}$ is said to be **invertible**, and $\boldsymbol{B}$ is called the **inverse matrix** of $\boldsymbol{A}$, written as

$$\boldsymbol{B} = \boldsymbol{A}^{-1}$$

定理 1 如果矩阵 $\boldsymbol{A}$ 可逆, 那么 $\boldsymbol{A}$ 的逆矩阵是唯一的.

Theorem 1 If the matrix $\boldsymbol{A}$ is invertible, then the inverse matrix of $\boldsymbol{A}$ is unique.

证 设矩阵 $\boldsymbol{B}, \boldsymbol{C}$ 都是 $\boldsymbol{A}$ 的逆矩阵, 即有

Proof Suppose that the matrices $\boldsymbol{B}, \boldsymbol{C}$ are all inverse matrices of $\boldsymbol{A}$, namely,

$$\boldsymbol{AB} = \boldsymbol{BA} = \boldsymbol{E}, \quad \boldsymbol{AC} = \boldsymbol{CA} = \boldsymbol{E}$$

则

then

$$\boldsymbol{B} = \boldsymbol{BE} = \boldsymbol{B}(\boldsymbol{AC}) = (\boldsymbol{BA})\boldsymbol{C} = \boldsymbol{EC} = \boldsymbol{C}$$

故 $\boldsymbol{A}$ 的逆矩阵是唯一的. ▲

This means that the inverse matrix of $\boldsymbol{A}$ is unique. ▲

利用逆矩阵的定义容易验证逆矩阵满足下列性质:

Using the definition of inverse matrices, it is easy to illustrate that inverse matrices have the following properties:

(1) 若 $\boldsymbol{A}$ 可逆, 则 $\boldsymbol{A}^{-1}$ 也可逆, 且

(1) If $\boldsymbol{A}$ is invertible, then $\boldsymbol{A}^{-1}$ is also invertible, and

$$(\boldsymbol{A}^{-1})^{-1} = \boldsymbol{A}$$

(2) 若 $\boldsymbol{A}$ 可逆, $\lambda \neq 0$, 则 $\lambda\boldsymbol{A}$ 也可逆, 且

(2) If $\boldsymbol{A}$ is invertible and if $\lambda \neq 0$, then $\lambda\boldsymbol{A}$ is also invertible, and

$$(\lambda\boldsymbol{A})^{-1} = \frac{1}{\lambda}\boldsymbol{A}^{-1}$$

(3) 若 $\boldsymbol{A}, \boldsymbol{B}$ 均可逆, 则 $\boldsymbol{AB}$ 也可逆, 且

(3) If $\boldsymbol{A}, \boldsymbol{B}$ are invertible, then $\boldsymbol{AB}$ is also invertible, and

$$(\boldsymbol{AB})^{-1} = \boldsymbol{B}^{-1}\boldsymbol{A}^{-1}$$

(4) 若 $\boldsymbol{A}$ 可逆, 则 $\boldsymbol{A}^{\mathrm{T}}$ 也可逆, 且

(4) If $\boldsymbol{A}$ is invertible, then $\boldsymbol{A}^{\mathrm{T}}$ is also invertible, and

$$(\boldsymbol{A}^{\mathrm{T}})^{-1} = (\boldsymbol{A}^{-1})^{\mathrm{T}}$$

(5) 若 $\boldsymbol{A}$ 可逆, 则

(5) If $\boldsymbol{A}$ is invertible, then

$$|\boldsymbol{A}^{-1}| = |\boldsymbol{A}|^{-1}$$

证 这里只证明性质 (3), 其余同理. 不难验证下列等式成立:

Proof Here we only prove Property (3), others are in a similar way. It is not difficult to show that the following equalities are valid:

$$(\boldsymbol{AB})(\boldsymbol{B}^{-1}\boldsymbol{A}^{-1}) = \boldsymbol{A}(\boldsymbol{BB}^{-1})\boldsymbol{A}^{-1} = \boldsymbol{AEA}^{-1} = \boldsymbol{AA}^{-1} = \boldsymbol{E}$$
$$(\boldsymbol{B}^{-1}\boldsymbol{A}^{-1})(\boldsymbol{AB}) = \boldsymbol{B}^{-1}(\boldsymbol{AA}^{-1})\boldsymbol{B} = \boldsymbol{B}^{-1}\boldsymbol{EB} = \boldsymbol{B}^{-1}\boldsymbol{B} = \boldsymbol{E}$$

所以

Therefore,

$$(\boldsymbol{AB})^{-1} = \boldsymbol{B}^{-1}\boldsymbol{A}^{-1}$$

注 (1) 性质 (3) 的证明方法很具有代表性, 它相当于在算术中检验除法正确性时, 都是用乘法去验证的.

Note (1) The proof method of Property (3) is quite representative, it is equivalent to that the multiplication is always used to examine the correction of quotient in arithmetic.

(2) 一般地,

(2) In general,

$$(\boldsymbol{A}+\boldsymbol{B})^{-1} \neq \boldsymbol{A}^{-1} + \boldsymbol{B}^{-1}$$

(3) 性质 (3) 可以推广到多个方阵的情形, 即如果 $\boldsymbol{A}_1, \boldsymbol{A}_2, \cdots, \boldsymbol{A}_k$ 均为 n 阶可逆方阵, 则乘积 $\boldsymbol{A}_1\boldsymbol{A}_2\cdots\boldsymbol{A}_k$ 也可逆, 并且

(3) Property (3) can be generalized to the case of many matrices, namely, if $\boldsymbol{A}_1, \boldsymbol{A}_2, \cdots, \boldsymbol{A}_k$ are all invertible square matrices of order n, then the product $\boldsymbol{A}_1\boldsymbol{A}_2\cdots\boldsymbol{A}_k$ is also invertible, and

$$(A_1A_2\cdots A_k)^{-1}=A_k^{-1}A_{k-1}^{-1}\cdots A_2^{-1}A_1^{-1}$$

特别地,

In particular,

$$(A^k)^{-1}=(A^{-1})^k$$

下面我们给出矩阵可逆的充分必要条件和求逆矩阵的方法.

In the following part we give a sufficient and necessary condition of an invertible matrix and a method for finding inverse matrices.

设有 n 阶方阵

Suppose that the square matrix of order n is

$$A=\begin{pmatrix} a_{11} & a_{12} & \cdots & a_{1n} \\ a_{21} & a_{22} & \cdots & a_{2n} \\ \vdots & \vdots & & \vdots \\ a_{n1} & a_{n2} & \cdots & a_{nn} \end{pmatrix}$$

n 阶方阵

the square matrix of order n

$$A^*=\begin{pmatrix} A_{11} & A_{21} & \cdots & A_{n1} \\ A_{12} & A_{22} & \cdots & A_{n2} \\ \vdots & \vdots & & \vdots \\ A_{1n} & A_{2n} & \cdots & A_{nn} \end{pmatrix}$$

称为 A 的**伴随矩阵**, 其中 A_{ij} 为 A 中元素 a_{ij} 的代数余子式, $i,j=1,2,\cdots,n$.

is called the **adjoint matrix** of A, where A_{ij} is the algebraic cofactor associated with the element a_{ij} in A, $i,j=1,2,\cdots,n$.

注

Note

$$A^*=\begin{pmatrix} A_{11} & A_{21} & \cdots & A_{n1} \\ A_{12} & A_{22} & \cdots & A_{n2} \\ \vdots & \vdots & & \vdots \\ A_{1n} & A_{2n} & \cdots & A_{nn} \end{pmatrix}=\begin{pmatrix} A_{11} & A_{12} & \cdots & A_{1n} \\ A_{21} & A_{22} & \cdots & A_{2n} \\ \vdots & \vdots & & \vdots \\ A_{n1} & A_{n2} & \cdots & A_{nn} \end{pmatrix}^{\mathrm{T}}$$

由于

Since

$$a_{i1}A_{j1}+a_{i2}A_{j2}+\cdots+a_{in}A_{jn}=\begin{cases} |A|, & i=j \\ 0, & i\neq j \end{cases}$$

$$a_{1i}A_{1j}+a_{2i}A_{2j}+\cdots+a_{ni}A_{nj}=\begin{cases} |A|, & i=j \\ 0, & i\neq j \end{cases}$$

有

we have

$$AA^* = A^*A = \begin{pmatrix} |A| & 0 & \cdots & 0 \\ 0 & |A| & \cdots & 0 \\ \vdots & \vdots & & \vdots \\ 0 & 0 & \cdots & |A| \end{pmatrix} = |A|\,E \tag{2.5}$$

也就是说, 只要 $|A| \neq 0$, 就有

This is to say, provided $|A| \neq 0$, we have

$$A\left[\frac{1}{|A|}A^*\right] = \left[\frac{1}{|A|}A^*\right]A = E$$

于是得到下面定理.

Consequently, we obtain the following theorem.

定理 2 方阵 A 可逆的充分必要条件是 $|A| \neq 0$, 并且

Theorem 2 A square matrix A is invertible if and only if $|A| \neq 0$, and

$$A^{-1} = \frac{1}{|A|}A^* \tag{2.6}$$

证 **充分性** 若 $|A| \neq 0$, 令 $B = \frac{1}{|A|}A^*$, 则有 $AB = BA = E$, 即 B 是 A 的逆矩阵.

Proof **Sufficiency** If $|A| \neq 0$, let $B = \frac{1}{|A|}A^*$, then we have $AB = BA = E$, that is to say, B is the inverse matrix of A.

必要性 若 A 可逆, 则存在 A^{-1}, 使 $AA^{-1} = E$. 对等式两边取行列式运算得

Necessity If A is invertible, then there exists A^{-1} such that $AA^{-1} = E$. Performing the determinant operation on both sides of the equality yields

$$\left|AA^{-1}\right| = |A|\left|A^{-1}\right| = |E| = 1$$

从而

Therefore,

$$|A| \neq 0$$ ▲

推论 1 设 A, B 均为 n 阶方阵. 若 $AB = E$, 则 A, B 都可逆, 且 $A^{-1} = B, B^{-1} = A$.

Corollary 1 Suppose that A, B are all square matrices of order n. If $AB = E$, then A, B are all invertible, and $A^{-1} = B, B^{-1} = A$.

证 由 $AB = E$, 可得 $|A||B| = 1$, 从而 $|A| \neq 0$, $|B| \neq 0$. 由定理 2 知, A, B 均可逆, 用 A^{-1} 左乘 $AB = E$, 得

Proof Since $AB = E$, we have $|A||B| = 1$, this mean that $|A| \neq 0$ and $|B| \neq 0$. From Theorem 2 we know that A, B are all invertible. Premultiplying $AB = E$ by A^{-1}, we have

$$A^{-1}(AB) = (A^{-1}A)B = EB = B = A^{-1}E = A^{-1}$$

即

namely,

$$B = A^{-1}$$

同理, 用 B^{-1} 右乘 $AB = E$ 得

In a similar way, postmultiplying $AB = E$ by B^{-1} leads to

$$A = B^{-1}$$ ▲

注 (1) 定理 2 不仅给出了判定矩阵可逆的充要条件, 而且提供了一种利用伴随矩阵求逆矩阵的方法.

Note (1) Theorem 2 gives not only a sufficient and necessary condition for deciding whether a matrix is invertible, but also a method for finding the inverse matrix by adjoint matrices.

(2) 由 $AA^* = |A|\,E$, 容易验证 $|A^*| = |A|^{n-1}$(习题 2 17).

(2) Using $AA^* = |A|\,E$, we can easily prove that $|A^*| = |A|^{n-1}$(Exercise 2 17).

对于线性方程组 $Ax = b$, 若对应的系数矩阵 A 为方阵, 有如下定理.

For a linear system $Ax = b$, if the corresponding coefficient matrix A is a square matrix, we have the following theorems.

定理 3 若 $Ax = b$ 的系数矩阵 A 可逆 ($|A| \neq 0$), 则方程组有唯一解, 且

Theorem 3 If the coefficient matrix A associated with $Ax = b$ is invertible ($|A| \neq 0$), then the system has a unique solution, given by

$$x = A^{-1}b$$

定理 3 的逆否定理如下:

The converse-negative theorem of Theorem 3 is as follows:

定理 3′ 若方程组 $Ax = b$ 无解或至少有两个不同的解, 则 A 是奇异的 ($|A| = 0$).

Theorem 3′ If the system $Ax = b$ has no solution or has at least two distinct solutions, then A is singular ($|A| = 0$).

定理 4 若齐次方程组 $Ax = 0$ 的系数矩阵 A 可逆 ($|A| \neq 0$), 则它有唯一零解.

Theorem 4 If the coefficient matrix A associated with the homogeneous system $Ax = 0$ is invertible ($|A| \neq 0$), then it has a unique zero solution.

定理 4 的逆否定理如下:

The converse-negative theorem of Theorem 4 is as follows.

定理 4′ 若方程组 $\boldsymbol{Ax} = \boldsymbol{0}$ 有非零解, 则 $\boldsymbol{A}$ 是奇异的 ($|\boldsymbol{A}| = 0$).

Theorem 4′ If the system $\boldsymbol{Ax} = \boldsymbol{0}$ has nonzero solutions, then $\boldsymbol{A}$ is singular ($|\boldsymbol{A}| = 0$).

例 1 设二阶方阵为

Example 1 Suppose that the square matrix of order 2 is

$$\boldsymbol{A} = \begin{pmatrix} a & b \\ c & d \end{pmatrix}$$

确定 $\boldsymbol{A}$ 可逆的条件, 并求 $\boldsymbol{A}^{-1}$.

Decide the invertible conditions of $\boldsymbol{A}$, and find $\boldsymbol{A}^{-1}$.

解 当 $|\boldsymbol{A}| = ad - bc \neq 0$ 时, $\boldsymbol{A}$ 可逆. $\boldsymbol{A}$ 的伴随矩阵为

Solution As $|\boldsymbol{A}| = ad - bc \neq 0$, $\boldsymbol{A}$ is invertible. The adjoint matrix of $\boldsymbol{A}$ is

$$\boldsymbol{A}^* = \begin{pmatrix} d & -b \\ -c & a \end{pmatrix}$$

所以有

So we have

$$\boldsymbol{A}^{-1} = \frac{1}{|\boldsymbol{A}|}\boldsymbol{A}^* = \frac{1}{ad - bc}\begin{pmatrix} d & -b \\ -c & a \end{pmatrix} \qquad \blacksquare$$

注 例 1 的结果可以作为求二阶方阵逆的公式使用.

Note The result in Example 1 can be used as the formula for finding the inverse matrix of a square matrix of order 2.

例 2 (1) 判断下面矩阵是否可逆? 若可逆, 求出其逆矩阵.

Example 2 (1) Decide whether the following matrix is invertible? If so, find its inverse matrix.

$$\boldsymbol{A} = \begin{pmatrix} 1 & 2 & -1 \\ 3 & 1 & 0 \\ -1 & 0 & -2 \end{pmatrix}$$

(2) 令

(2) Let

$$b=\begin{pmatrix}1\\1\\0\end{pmatrix}$$

判定方程组 $Ax=b$ 是否有唯一解. 若有唯一解, 并求解.

Decide that whether the system $Ax=b$ has a unique solution. If so, find it.

解 (1) 由 $|A|=9\neq 0$ 可知, A 可逆, $|A|$ 中各元素对应的代数余子式分别为

Solution (1) Since $|A|=9\neq 0$, we know that A is invertible and that the algebraic cofactors associated with the elements in $|A|$ are respectively given by

$$\begin{aligned}&A_{11}=-2, \quad A_{21}=4, \quad A_{31}=1\\&A_{12}=6, \quad A_{22}=-3, \quad A_{32}=-3\\&A_{13}=1, \quad A_{23}=-2, \quad A_{33}=-5\end{aligned}$$

于是

Therefore,

$$A^*=\begin{pmatrix}-2&4&1\\6&-3&-3\\1&-2&-5\end{pmatrix}$$

$$A^{-1}=\frac{A^*}{|A|}=\frac{1}{9}\begin{pmatrix}-2&4&1\\6&-3&-3\\1&-2&-5\end{pmatrix}=\begin{pmatrix}-\frac{2}{9}&\frac{4}{9}&\frac{1}{9}\\\frac{2}{3}&-\frac{1}{3}&-\frac{1}{3}\\\frac{1}{9}&-\frac{2}{9}&-\frac{5}{9}\end{pmatrix}$$

(2) 由 (1) 知 A 可逆, 因此方程组 $Ax=b$ 有唯一解, 即

(2) From (1) we know that A is invertible, so the system $Ax=b$ has a unique solution, given by

$$x=A^{-1}b=\frac{1}{9}\begin{pmatrix}-2&4&1\\6&-3&-3\\1&-2&-5\end{pmatrix}\begin{pmatrix}1\\1\\0\end{pmatrix}=\begin{pmatrix}\frac{2}{9}\\\frac{3}{9}\\-\frac{1}{9}\end{pmatrix}$$ ■

例 3 设

Example 3 Suppose that

$$\boldsymbol{A}=\begin{pmatrix} a_1 & 0 & \cdots & 0 \\ 0 & a_2 & \cdots & 0 \\ \vdots & \vdots & & \vdots \\ 0 & 0 & \cdots & a_n \end{pmatrix}$$

其中 $a_i \neq 0\,(i=1,2,\cdots)$. 证明

where $a_i \neq 0\,(i=1,2,\cdots)$. Prove that

$$\boldsymbol{A}^{-1}=\begin{pmatrix} 1/a_1 & 0 & \cdots & 0 \\ 0 & 1/a_2 & \cdots & 0 \\ \vdots & \vdots & & \vdots \\ 0 & 0 & \cdots & 1/a_n \end{pmatrix}$$

证 因为

Proof Since

$$\begin{pmatrix} a_1 & 0 & \cdots & 0 \\ 0 & a_2 & \cdots & 0 \\ \vdots & \vdots & & \vdots \\ 0 & 0 & \cdots & a_n \end{pmatrix}\begin{pmatrix} 1/a_1 & 0 & \cdots & 0 \\ 0 & 1/a_2 & \cdots & 0 \\ \vdots & \vdots & & \vdots \\ 0 & 0 & \cdots & 1/a_n \end{pmatrix}=\begin{pmatrix} 1 & 0 & \cdots & 0 \\ 0 & 1 & \cdots & 0 \\ \vdots & \vdots & & \vdots \\ 0 & 0 & \cdots & 1 \end{pmatrix}$$

得到

we have

$$\boldsymbol{A}^{-1}=\begin{pmatrix} 1/a_1 & 0 & \cdots & 0 \\ 0 & 1/a_2 & \cdots & 0 \\ \vdots & \vdots & & \vdots \\ 0 & 0 & \cdots & 1/a_n \end{pmatrix}$$

同理, 可以证明下式成立.

In a similar way, it can be shown that the following equation is valid.

$$\begin{pmatrix} 0 & \cdots & 0 & a_1 \\ 0 & \cdots & a_2 & 0 \\ \vdots & & \vdots & \vdots \\ a_n & \cdots & 0 & 0 \end{pmatrix}^{-1}=\begin{pmatrix} 0 & \cdots & 0 & 1/a_n \\ 0 & \cdots & 1/a_{n-1} & 0 \\ \vdots & & \vdots & \vdots \\ 1/a_1 & \cdots & 0 & 0 \end{pmatrix}$$

例 4 已知方阵 $\boldsymbol{A}$ 满足 $\boldsymbol{A}^2-2\boldsymbol{A}+\boldsymbol{E}=\boldsymbol{O}$. 证明 $\boldsymbol{A}+\boldsymbol{E}$ 可逆, 并求 $(\boldsymbol{A}+\boldsymbol{E})^{-1}$.

Example 4 Known that the square matrix $\boldsymbol{A}$ satisfies $\boldsymbol{A}^2-2\boldsymbol{A}+\boldsymbol{E}=\boldsymbol{O}$. Prove that $\boldsymbol{A}+\boldsymbol{E}$ is invertible and find $(\boldsymbol{A}+\boldsymbol{E})^{-1}$.

证 由 $A^2-2A+E=O$, 有

Proof Since $A^2-2A+E=O$ we have

$$(A+E)(A-3E)=-4E$$

$$(A+E)\left[-\frac{1}{4}(A-3E)\right]=E$$

由定理 2 的推论 1 知 $A+E$ 可逆, 且

From Corollary 1 of Theorem 2 we know that $A+E$ is invertible, and

$$(A+E)^{-1}=-\frac{1}{4}(A-3E)$$ ■

注 (1) 这一方法不仅可以证明矩阵可逆, 同时也可以求出矩阵的逆.

Note (1) This method not only can prove that an matrix is invertible, but also can find its inverse matrix.

(2) $A^2-2A+E=A(A-2E)+E$ 不能写成 $A^2-2A+E=A(A-2)+E$.

(2) $A^2-2A+E=A(A-2E)+E$ can not be written as $A^2-2A+E=A(A-2)+E$.

例 5 设 A 是 n 阶方阵, 满足 $AA^{\mathrm{T}}=E$, 且 $|A|=-1$. 求 $|A+E|$.

Example 5 Suppose that A is a square matrix of order n satisfying $AA^{\mathrm{T}}=E$, and that $|A|=-1$. Find $|A+E|$.

解 不难验证下面的等式成立, 即

Solution It is not difficult to show that the following equalities are valid, i.e.,

$$|A+E|=\left|A+AA^{\mathrm{T}}\right|=\left|A(E+A^{\mathrm{T}})\right|=|A|\left|E+A^{\mathrm{T}}\right|=-\left|(E+A)^{\mathrm{T}}\right|=-|A+E|$$

因此有

So we have

$$2|A+E|=0$$

即

namely,

$$|A+E|=0$$ ■

思 考 题

Questions

1. 判断下列结论是否成立, 并说明理由.

1. Decide whether or not the following results are valid, and give the reasons.

(1) 若三个方阵 $\boldsymbol{A}$, $\boldsymbol{B}$, $\boldsymbol{C}$ 满足 $\boldsymbol{AB}=\boldsymbol{CA}=\boldsymbol{E}$, 则 $\boldsymbol{B}=\boldsymbol{C}$;

(1) If the three square matrices $\boldsymbol{A}$, $\boldsymbol{B}$, $\boldsymbol{C}$ satisfy $\boldsymbol{AB}=\boldsymbol{CA}=\boldsymbol{E}$, then $\boldsymbol{B}=\boldsymbol{C}$;

(2) 若 $\boldsymbol{AB}=\boldsymbol{O}$, 且 $\boldsymbol{A}$ 可逆, 则 $\boldsymbol{B}=\boldsymbol{O}$;

(2) If $\boldsymbol{AB}=\boldsymbol{O}$ and if $\boldsymbol{A}$ is invertible, then $\boldsymbol{B}=\boldsymbol{O}$;

(3) 若 $\boldsymbol{A}$ 可逆, 由 $\boldsymbol{AX}=\boldsymbol{YA}$, 可得 $\boldsymbol{X}=\boldsymbol{Y}$;

(3) If $\boldsymbol{A}$ is invertible, from$\boldsymbol{AX}=\boldsymbol{YA}$ we obtain $\boldsymbol{X}=\boldsymbol{Y}$;

(4) 若 $\boldsymbol{A}$ 可逆, 且满足 $\boldsymbol{XA}=\boldsymbol{C}$, 则 $\boldsymbol{X}=\boldsymbol{A}^{-1}\boldsymbol{C}$;

(4) If $\boldsymbol{A}$ is invertible and satisfies $\boldsymbol{XA}=\boldsymbol{C}$, then $\boldsymbol{X}=\boldsymbol{A}^{-1}\boldsymbol{C}$;

(5) 若 $\boldsymbol{AB}$ 可逆, 则 $\boldsymbol{A}$ 和 $\boldsymbol{B}$ 都可逆;

(5) If $\boldsymbol{AB}$ is invertible, then $\boldsymbol{A}$ and $\boldsymbol{B}$ are invertible;

(6) 若 $\boldsymbol{AB}$ 可逆, 则 $\boldsymbol{BA}$ 也可逆;

(6) If $\boldsymbol{AB}$ is invertible, then $\boldsymbol{BA}$ is also invertible;

(7) 若 $\boldsymbol{A}^{\mathrm{T}}\boldsymbol{A}$ 可逆, 则 $\boldsymbol{AA}^{\mathrm{T}}$ 也可逆;

(7) If $\boldsymbol{A}^{\mathrm{T}}\boldsymbol{A}$ is invertible, then $\boldsymbol{AA}^{\mathrm{T}}$ is invertible;

(8) 若对称矩阵 $\boldsymbol{A}$ 可逆, 则 $\boldsymbol{A}^{-1}$ 也是对称矩阵;

(8) If the symmetric matrix $\boldsymbol{A}$ is invertible, then $\boldsymbol{A}^{-1}$ is also a symmetric matrix;

(9) 若 $\boldsymbol{A}^*=\boldsymbol{O}$, 则 $\boldsymbol{A}=\boldsymbol{O}$; 若 $|\boldsymbol{A}|=0$, 则 $\boldsymbol{A}=\boldsymbol{O}$;

(9) If $\boldsymbol{A}^*=\boldsymbol{O}$, then $\boldsymbol{A}=\boldsymbol{O}$; while if $|\boldsymbol{A}|=0$, then $\boldsymbol{A}=\boldsymbol{O}$;

(10) $\boldsymbol{A}$ 与 $\boldsymbol{A}^*$, $\boldsymbol{A}^{-1}$ 均可交换;

(10) $\boldsymbol{A}^*$ and $\boldsymbol{A}^{-1}$ can be all commutable with $\boldsymbol{A}$;

(11) $|-\boldsymbol{A}|=-|\boldsymbol{A}|$;

(12) $(\boldsymbol{A}^2+\boldsymbol{A}-2\boldsymbol{E})\dfrac{1}{\boldsymbol{A}+2\boldsymbol{E}}=\boldsymbol{A}-\boldsymbol{E}$.

(11) $|-\boldsymbol{A}|=-|\boldsymbol{A}|$;

(12) $(\boldsymbol{A}^2+\boldsymbol{A}-2\boldsymbol{E})\dfrac{1}{\boldsymbol{A}+2\boldsymbol{E}}=\boldsymbol{A}-\boldsymbol{E}$.

2. 矩阵和行列式有什么区别和联系?

2. What are the difference and relation between a matrix and a determinant?

3. 设 $\boldsymbol{A},\boldsymbol{B}$ 都是 n 阶方阵, 若 $\boldsymbol{AB}=\boldsymbol{BA}=|\boldsymbol{A}|\,\boldsymbol{E}$, 是否必有 $\boldsymbol{B}=\boldsymbol{A}^*$.

3. Suppose that $\boldsymbol{A},\boldsymbol{B}$ are all square matrices of order n, if $\boldsymbol{AB}=\boldsymbol{BA}=|\boldsymbol{A}|\,\boldsymbol{E}$, whether or not there must have $\boldsymbol{B}=\boldsymbol{A}^*$.

2.3 矩阵方程

在实际应用中, 经常需要求解一些简单的、常见的矩阵方程, 如 $\boldsymbol{AX} = \boldsymbol{C}$, $\boldsymbol{XB} = \boldsymbol{C}$ 和 $\boldsymbol{AXB} = \boldsymbol{C}$. 若 $\boldsymbol{A}$ 和 $\boldsymbol{B}$ 都可逆, 则这些矩阵方程的解依次为 $\boldsymbol{X} = \boldsymbol{A}^{-1}\boldsymbol{C}, \boldsymbol{X} = \boldsymbol{CB}^{-1}$ 和 $\boldsymbol{X} = \boldsymbol{A}^{-1}\boldsymbol{CB}^{-1}$. 当然, 在实际问题中还会遇到其他形式的 (或许更复杂的) 矩阵方程.

解矩阵方程的一般步骤是:

第一步 根据矩阵运算规律对矩阵方程化简整理;

第二步 根据矩阵运算定义代入并计算.

例 1 设

2.3 Matrix Equations

In practical applications, we often needs to solve some simple and common matrix equations, such as $\boldsymbol{AX} = \boldsymbol{C}$, $\boldsymbol{XB} = \boldsymbol{C}$ and $\boldsymbol{AXB} = \boldsymbol{C}$. If $\boldsymbol{A}$ and $\boldsymbol{B}$ are all invertible, then the solutions of these matrix equations are in turn given by $\boldsymbol{X} = \boldsymbol{A}^{-1}\boldsymbol{C}$, $\boldsymbol{X} = \boldsymbol{CB}^{-1}$ and $\boldsymbol{X} = \boldsymbol{A}^{-1}\boldsymbol{CB}^{-1}$. Of course, we may also encounter matrix equations of other (maybe, more complicate) forms in practical problems.

General steps for solving matrix equations are as follows:

Step 1 Reduce the matrix equations by the operation rules of matrices;

Step 2 Substitute and calculate them by the definition of matrix operations.

Example 1 Suppose that

$$\boldsymbol{A} = \begin{pmatrix} 1 & 2 & 3 \\ 2 & 2 & 1 \\ 3 & 4 & 3 \end{pmatrix}, \quad \boldsymbol{B} = \begin{pmatrix} 2 & 1 \\ 5 & 3 \end{pmatrix}, \quad \boldsymbol{C} = \begin{pmatrix} 1 & 3 \\ 2 & 0 \\ 3 & 1 \end{pmatrix}$$

求矩阵 $\boldsymbol{X}$ 使得 $\boldsymbol{AXB} = \boldsymbol{C}$.

Find the matrix $\boldsymbol{X}$ such that $\boldsymbol{AXB} = \boldsymbol{C}$.

解 由于 $|\boldsymbol{A}| = 2$, $|\boldsymbol{B}| = 1$, 故 $\boldsymbol{A}^{-1}$, $\boldsymbol{B}^{-1}$ 存在, 且

Solution Since $|\boldsymbol{A}| = 2$ and $|\boldsymbol{B}| = 1$, $\boldsymbol{A}^{-1}$ and $\boldsymbol{B}^{-1}$ exist, moreover,

$$\boldsymbol{A}^{-1} = \begin{pmatrix} 1 & 3 & -2 \\ -\dfrac{3}{2} & -3 & \dfrac{5}{2} \\ 1 & 1 & -1 \end{pmatrix}, \quad \boldsymbol{B}^{-1} = \begin{pmatrix} 3 & -1 \\ -5 & 2 \end{pmatrix}$$

用 $\boldsymbol{A}^{-1}$ 左乘 $\boldsymbol{AXB} = \boldsymbol{C}$, $\boldsymbol{B}^{-1}$ 右乘 $\boldsymbol{AXB} = \boldsymbol{C}$, 有

Premultiplying $\boldsymbol{AXB} = \boldsymbol{C}$ by $\boldsymbol{A}^{-1}$ and postmultiplying $\boldsymbol{AXB} = \boldsymbol{C}$ by $\boldsymbol{B}^{-1}$, we obtain

$$A^{-1}AXBB^{-1} = A^{-1}CB^{-1}$$

即 | namely,

$$X = A^{-1}CB^{-1}$$

于是 | Consequently,

$$X = A^{-1}CB^{-1} = \begin{pmatrix} 1 & 3 & -2 \\ -\frac{3}{2} & -3 & \frac{5}{2} \\ 1 & 1 & -1 \end{pmatrix} \begin{pmatrix} 1 & 3 \\ 2 & 0 \\ 3 & 1 \end{pmatrix} \begin{pmatrix} 3 & -1 \\ -5 & 2 \end{pmatrix}$$

$$= \begin{pmatrix} 1 & 1 \\ 0 & -2 \\ 0 & 2 \end{pmatrix} \begin{pmatrix} 3 & -1 \\ -5 & 2 \end{pmatrix} = \begin{pmatrix} -2 & 1 \\ 10 & -4 \\ -10 & 4 \end{pmatrix}$$ ■

例 2 设 A, B 为 3 阶方阵, 且满足 $AB = A + 2B$, 其中

Example 2 Suppose that A, B are square matrices of order 3 and satisfy $AB = A + 2B$, where

$$A = \begin{pmatrix} 0 & 3 & 3 \\ 1 & 1 & 0 \\ -1 & 2 & 3 \end{pmatrix}$$

求 B. | Find B.

解 由 $AB = A + 2B$ 可得

Solution Using $AB = A + 2B$, we have

$$(A - 2E)B = A$$

故 | Thus,

$$B = (A - 2E)^{-1}A = \begin{pmatrix} -2 & 3 & 3 \\ 1 & -1 & 0 \\ -1 & 2 & 1 \end{pmatrix}^{-1} \begin{pmatrix} 0 & 3 & 3 \\ 1 & 1 & 0 \\ -1 & 2 & 3 \end{pmatrix}$$

$$= \frac{1}{2} \begin{pmatrix} -1 & 3 & 3 \\ -1 & 1 & 3 \\ 1 & 1 & -1 \end{pmatrix} \begin{pmatrix} 0 & 3 & 3 \\ 1 & 1 & 0 \\ -1 & 2 & 3 \end{pmatrix}$$

$$= \begin{pmatrix} 0 & 3 & 3 \\ -1 & 2 & 3 \\ 1 & 1 & 0 \end{pmatrix}$$ ■

注 由 $(\boldsymbol{A}-2\boldsymbol{E})\boldsymbol{B}=\boldsymbol{A}$ 得到 $\boldsymbol{B}=(\boldsymbol{A}-2\boldsymbol{E})^{-1}\boldsymbol{A}$, 而不是 $\boldsymbol{B}=\boldsymbol{A}(\boldsymbol{A}-2\boldsymbol{E})^{-1}$.

Note Since $(\boldsymbol{A}-2\boldsymbol{E})\boldsymbol{B}=\boldsymbol{A}$, we obtain $\boldsymbol{B}=(\boldsymbol{A}-2\boldsymbol{E})^{-1}\boldsymbol{A}$, rather than $\boldsymbol{B}=\boldsymbol{A}(\boldsymbol{A}-2\boldsymbol{E})^{-1}$.

例 3 设 $\boldsymbol{A},\boldsymbol{B}$ 为 3 阶方阵, 且满足 $\boldsymbol{AB}+\boldsymbol{E}=\boldsymbol{A}^2+\boldsymbol{B}$, 其中

Example 3 Suppose that $\boldsymbol{A},\boldsymbol{B}$ are square matrices of order 3 and satisfy $\boldsymbol{AB}+\boldsymbol{E}=\boldsymbol{A}^2+\boldsymbol{B}$, where

$$\boldsymbol{A}=\begin{pmatrix}1&0&1\\0&2&0\\-1&0&1\end{pmatrix}$$

求 $\boldsymbol{B}$.

Find $\boldsymbol{B}$.

解 由 $\boldsymbol{AB}+\boldsymbol{E}=\boldsymbol{A}^2+\boldsymbol{B}$ 可得

Solution Using the known equation $\boldsymbol{AB}+\boldsymbol{E}=\boldsymbol{A}^2+\boldsymbol{B}$, we have

$$(\boldsymbol{A}-\boldsymbol{E})\boldsymbol{B}=\boldsymbol{A}^2-\boldsymbol{E}$$

即

namely,

$$(\boldsymbol{A}-\boldsymbol{E})\boldsymbol{B}=(\boldsymbol{A}-\boldsymbol{E})(\boldsymbol{A}+\boldsymbol{E})$$

$$\boldsymbol{A}-\boldsymbol{E}=\begin{pmatrix}1&0&1\\0&2&0\\-1&0&1\end{pmatrix}-\begin{pmatrix}1&0&0\\0&1&0\\0&0&1\end{pmatrix}=\begin{pmatrix}0&0&1\\0&1&0\\-1&0&0\end{pmatrix}$$

因为 $|\boldsymbol{A}-\boldsymbol{E}|=1\neq 0$, 即 $\boldsymbol{A}-\boldsymbol{E}$ 可逆. 由消去律可得

Since $|\boldsymbol{A}-\boldsymbol{E}|=1\neq 0$, $\boldsymbol{A}-\boldsymbol{E}$ is invertible. Using the cancellation law yields

$$\boldsymbol{B}=\boldsymbol{A}+\boldsymbol{E}=\begin{pmatrix}1&0&1\\0&2&0\\-1&0&1\end{pmatrix}+\begin{pmatrix}1&0&0\\0&1&0\\0&0&1\end{pmatrix}=\begin{pmatrix}2&0&1\\0&3&0\\-1&0&2\end{pmatrix}$$ ■

例 4 设 $\boldsymbol{A},\boldsymbol{B}$ 为 3 阶方阵, 且满足 $\boldsymbol{A}+\boldsymbol{B}=\boldsymbol{AB}$, $\boldsymbol{E}$ 为 3 阶单位矩阵.

Example 4 Suppose that $\boldsymbol{A},\boldsymbol{B}$ are square matrices of order 3 and satisfy $\boldsymbol{A}+\boldsymbol{B}=\boldsymbol{AB}$ and that $\boldsymbol{E}$ is the unit matrix of order 3.

(1) 证明 $\boldsymbol{A}-\boldsymbol{E}$ 可逆;

(1) Prove that $\boldsymbol{A}-\boldsymbol{E}$ is invertible;

(2) 已知

(2) Known that

$$\boldsymbol{B}=\begin{pmatrix}1&-3&0\\2&1&0\\0&0&2\end{pmatrix}$$

求 $\boldsymbol{A}$.

Find $\boldsymbol{A}$.

解 (1) 由 $\boldsymbol{A}+\boldsymbol{B}=\boldsymbol{AB}$ 可得如下等式, 即

Solution (1) Using $\boldsymbol{A}+\boldsymbol{B}=\boldsymbol{AB}$, we have the following equations, namely,

$$\boldsymbol{A}+\boldsymbol{B}-\boldsymbol{AB}=\boldsymbol{O},\quad (\boldsymbol{A}-\boldsymbol{E})-(\boldsymbol{A}-\boldsymbol{E})\boldsymbol{B}=-\boldsymbol{E},\quad (\boldsymbol{A}-\boldsymbol{E})(\boldsymbol{B}-\boldsymbol{E})=\boldsymbol{E}$$

所以 $\boldsymbol{A}-\boldsymbol{E}$ 可逆, 且

So $\boldsymbol{A}-\boldsymbol{E}$ is invertible and

$$(\boldsymbol{A}-\boldsymbol{E})^{-1}=(\boldsymbol{B}-\boldsymbol{E})$$

(2) 由 $(\boldsymbol{A}-\boldsymbol{E})(\boldsymbol{B}-\boldsymbol{E})=\boldsymbol{E}$ 可得 $(\boldsymbol{A}-\boldsymbol{E})=(\boldsymbol{B}-\boldsymbol{E})^{-1}$, 从而

(2) Using$(\boldsymbol{A}-\boldsymbol{E})(\boldsymbol{B}-\boldsymbol{E})=\boldsymbol{E}$, we know that $(\boldsymbol{A}-\boldsymbol{E})=(\boldsymbol{B}-\boldsymbol{E})^{-1}$, consequently,

$$\boldsymbol{A}=\boldsymbol{E}+(\boldsymbol{B}-\boldsymbol{E})^{-1}=\begin{pmatrix}1&0&0\\0&1&0\\0&0&1\end{pmatrix}+\begin{pmatrix}0&-3&0\\2&0&0\\0&0&1\end{pmatrix}^{-1}$$

$$=\begin{pmatrix}1&0&0\\0&1&0\\0&0&1\end{pmatrix}+\begin{pmatrix}0&\frac{1}{2}&0\\-\frac{1}{3}&0&0\\0&0&1\end{pmatrix}=\begin{pmatrix}1&\frac{1}{2}&0\\-\frac{1}{3}&1&0\\0&0&2\end{pmatrix}$$

注 也可直接由 $\boldsymbol{A}+\boldsymbol{B}=\boldsymbol{AB}$ 求得 $\boldsymbol{A}=\boldsymbol{B}(\boldsymbol{B}-\boldsymbol{E})^{-1}$, 然而后面的运算会比较麻烦.

Note We can also obtain $\boldsymbol{A}=\boldsymbol{B}(\boldsymbol{B}-\boldsymbol{E})^{-1}$ from $\boldsymbol{A}+\boldsymbol{B}=\boldsymbol{AB}$ directly, however, the later operations will be more troublesome.

例 5 设 $\boldsymbol{A},\boldsymbol{B}$ 为 3 阶方阵, 且满足 $\boldsymbol{A}^{-1}\boldsymbol{BA}=6\boldsymbol{A}+\boldsymbol{BA}$, 其中

Example 5 Suppose that $\boldsymbol{A},\boldsymbol{B}$ are square matrices of order 3 and satisfy $\boldsymbol{A}^{-1}\boldsymbol{BA}=6\boldsymbol{A}+\boldsymbol{BA}$, where

$$\boldsymbol{A}=\begin{pmatrix}\frac{1}{2}&0&0\\0&\frac{1}{4}&0\\0&0&\frac{1}{7}\end{pmatrix}$$

求 $\boldsymbol{B}$.

Find $\boldsymbol{B}$.

解 由 $A^{-1}BA-BA=6A$ 可得

Solution Using $A^{-1}BA-BA=6A$, we obtain

$$(A^{-1}-E)BA=6A$$

上式两边再右乘 A^{-1} 得

Postmultiplying the above equation by A^{-1} yields

$$(A^{-1}-E)B=6E$$

所以

Thus

$$B=6(A^{-1}-E)^{-1}=6\left(\begin{pmatrix}2&0&0\\0&4&0\\0&0&7\end{pmatrix}-\begin{pmatrix}1&0&0\\0&1&0\\0&0&1\end{pmatrix}\right)^{-1}=6\begin{pmatrix}1&0&0\\0&3&0\\0&0&6\end{pmatrix}^{-1}$$

$$=6\begin{pmatrix}1&0&0\\0&\frac{1}{3}&0\\0&0&\frac{1}{6}\end{pmatrix}=\begin{pmatrix}6&0&0\\0&2&0\\0&0&1\end{pmatrix}$$

例 6* 设 A 可逆, 且 $A^*B=A^{-1}+B$. 证明 B 可逆; 若

Example 6* Suppose that A is invertible and that $A^*B=A^{-1}+B$. Prove that B is invertible; moreover, if

$$A=\begin{pmatrix}2&6&0\\0&2&6\\0&0&2\end{pmatrix}$$

求 B.

find B.

解 由 $A^*B=A^{-1}+B=A^{-1}+EB$ 可得

Solution Using $A^*B=A^{-1}+B=A^{-1}+EB$, we obtain

$$(A^*-E)B=A^{-1}$$

上式两边取行列式运算得

Performing the determinant operation on both sides yields

$$|A^*-E||B|=|A^{-1}|\neq 0$$

所以 $|B|\neq 0$, 即 B 可逆. 此外, 有

it implies that $|B|\neq 0$, namely, B is invertible. Furthermore, we have

$$B=(A^*-E)^{-1}A^{-1}=(A(A^*-E))^{-1}=(|A|\,E-A)^{-1}$$

其中 | where

$$|\boldsymbol{A}|\boldsymbol{E}-\boldsymbol{A}=\begin{pmatrix}8&0&0\\0&8&0\\0&0&8\end{pmatrix}-\begin{pmatrix}2&6&0\\0&2&6\\0&0&2\end{pmatrix}=6\begin{pmatrix}1&-1&0\\0&1&-1\\0&0&1\end{pmatrix}$$

利用逆矩阵的性质 $(\lambda\boldsymbol{A})^{-1}=\frac{1}{\lambda}\boldsymbol{A}^{-1}$ 和求逆公式 $\boldsymbol{A}^{-1}=\frac{\boldsymbol{A}^*}{|\boldsymbol{A}|}$, 易得

Utilizing the formulas $(\lambda\boldsymbol{A})^{-1}=\frac{1}{\lambda}\boldsymbol{A}^{-1}$ and $\boldsymbol{A}^{-1}=\frac{\boldsymbol{A}^*}{|\boldsymbol{A}|}$, we easily get

$$\boldsymbol{B}=\frac{1}{6}\begin{pmatrix}1&1&1\\0&1&1\\0&0&1\end{pmatrix}$$ ■

例 7* 已知 $\boldsymbol{A}$ 的伴随矩阵为

Example 7* Known that the adjoint matrix of $\boldsymbol{A}$ is given by

$$\boldsymbol{A}^*=\begin{pmatrix}1&0&0&0\\0&1&0&0\\1&0&1&0\\0&-3&0&8\end{pmatrix}$$

且有 $\boldsymbol{ABA}^{-1}=\boldsymbol{BA}^{-1}+3\boldsymbol{E}$. 求 $\boldsymbol{B}$.

and that $\boldsymbol{ABA}^{-1}=\boldsymbol{BA}^{-1}+3\boldsymbol{E}$. Find $\boldsymbol{B}$.

解 由 $|\boldsymbol{A}^*|=|\boldsymbol{A}|^3=8$ 得

Solution Since $|\boldsymbol{A}^*|=|\boldsymbol{A}|^3=8$, we know that

$$|\boldsymbol{A}|=2$$

在方程 $\boldsymbol{ABA}^{-1}=\boldsymbol{BA}^{-1}+3\boldsymbol{E}$ 两边右乘 $\boldsymbol{A}$ 得

Postmultiplying both sides of the equation $\boldsymbol{ABA}^{-1}=\boldsymbol{BA}^{-1}+3\boldsymbol{E}$ by $\boldsymbol{A}$ yields

$$\boldsymbol{AB}=\boldsymbol{B}+3\boldsymbol{A}$$

所以有 | so we have

$$\boldsymbol{AB}-\boldsymbol{B}=3\boldsymbol{A},\quad (\boldsymbol{A}-\boldsymbol{E})\boldsymbol{B}=3\boldsymbol{A}$$

于是 | Consequently,

$$\boldsymbol{B}=3(\boldsymbol{A}-\boldsymbol{E})^{-1}\boldsymbol{A}=3[\boldsymbol{A}(\boldsymbol{E}-\boldsymbol{A}^{-1})]^{-1}\boldsymbol{A}=3(\boldsymbol{E}-\boldsymbol{A}^{-1})^{-1}\boldsymbol{A}^{-1}\boldsymbol{A}=3(\boldsymbol{E}-\boldsymbol{A}^{-1})^{-1}$$

$$=3\left(\boldsymbol{E}-\frac{1}{2}\boldsymbol{A}^*\right)^{-1}=6(2\boldsymbol{E}-\boldsymbol{A}^*)^{-1}=6\begin{pmatrix}1&0&0&0\\0&1&0&0\\-1&0&1&0\\0&3&0&-6\end{pmatrix}^{-1}=\begin{pmatrix}6&0&0&0\\0&6&0&0\\6&0&6&0\\0&3&0&-1\end{pmatrix}$$ ■

例 8* 设矩阵

Example 8* Suppose that the matrix

$$A = \begin{pmatrix} 1 & 1 & -1 \\ -1 & 1 & 1 \\ 1 & -1 & 1 \end{pmatrix}$$

满足 $A^*X = A^{-1} + 2X$, 其中 A^* 是 A 的伴随矩阵. 求矩阵 X.

satisfies $A^*X = A^{-1}+2X$, where A^* is the adjoint matrix of A. Find the matrix X.

解 由 $|A| = 4$, 知 A 可逆.

Solution Since $|A| = 4$, we know that A is invertible.

由 $A^*X = A^{-1} + 2X$, 可得

Using $A^*X = A^{-1} + 2X$, we have

$$(A^* - 2E)X = A^{-1}$$

两边左乘 A, 并利用式 $AA^* = |A|\,E$ 得

Premultiplying the equation on both sides by A and using the equation $AA^* = |A|\,E$, we obtain

$$(|A|\,E - 2A)X = E$$

因此

thus,

$$X = (|A|\,E - 2A)^{-1} = (4E - 2A)^{-1} = (2(2E - A))^{-1} = \frac{1}{2}(2E - A)^{-1}$$

$$= \frac{1}{2}\begin{pmatrix} 1 & -1 & 1 \\ 1 & 1 & -1 \\ -1 & 1 & 1 \end{pmatrix}^{-1} = \frac{1}{4}\begin{pmatrix} 1 & 1 & 0 \\ 0 & 1 & 1 \\ 1 & 0 & 1 \end{pmatrix}$$ ■

思 考 题

Question

解矩阵方程时, 求得的未知矩阵是否唯一? 方法是否唯一?

As we solve a matrix equation, whether the obtained unknown matrix is unique, whether the method is unique?

2.4 分块矩阵

2.4 Block Matrices

对于大型矩阵 (即行数和列数较高), 常采用分块方式进行运算, 使大型矩阵的运算化为小型矩阵的运算.

For large scale matrices (namely, the row number and the column number are larger), we always use the partitioned mode such that the operations of large scale matrices reduce to those of small scale matrices.

若将矩阵 $\boldsymbol{A}$ 用若干条纵线和横线分成许多个小矩阵, 每个小矩阵称为 $\boldsymbol{A}$ 的子块. 以子块为元素的矩阵称为 $\boldsymbol{A}$ 的**分块矩阵**.

If the matrix $\boldsymbol{A}$ is partitioned into many small scale matrices by some vertical lines and horizontal lines, each small scale matrix is said to be a subblock. The matrix consisted of subblock matrices is said to be a **block matrix**.

例如, 可将矩阵 $\boldsymbol{A}$ 划分为如下分块矩阵

For example, the matrix $\boldsymbol{A}$ can be partitioned into the following block matrix

$$\boldsymbol{A}=\left(\begin{array}{cc:cc} 1 & 2 & 4 & 1 \\ \hdashline 3 & 0 & 5 & 7 \\ -1 & 2 & 0 & 1 \end{array}\right)=\begin{pmatrix} \boldsymbol{A}_{11} & \boldsymbol{A}_{12} \\ \boldsymbol{A}_{21} & \boldsymbol{A}_{22} \end{pmatrix}$$

其中

where

$$\boldsymbol{A}_{11}=\begin{pmatrix} 1 & 2 \end{pmatrix}, \boldsymbol{A}_{12}=\begin{pmatrix} 4 & 1 \end{pmatrix}, \boldsymbol{A}_{21}=\begin{pmatrix} 3 & 0 \\ -1 & 2 \end{pmatrix}, \boldsymbol{A}_{22}=\begin{pmatrix} 5 & 7 \\ 0 & 1 \end{pmatrix}$$

为分块矩阵 $\boldsymbol{A}$ 的子块.

are subblocks of the block matrix $\boldsymbol{A}$.

根据其特点及不同的需要, 可将一矩阵进行不同形式的分块. 如上面矩阵 $\boldsymbol{A}$ 还可按如下方式分块, 即

A matrix can be partitioned into different forms according to its features and different requirements. The above matrix $\boldsymbol{A}$ can also be partitioned into the following forms, namely,

$$\boldsymbol{A}=\left(\begin{array}{c:ccc} 1 & 2 & 4 & 1 \\ 3 & 0 & 5 & 7 \\ -1 & 2 & 0 & 1 \end{array}\right)=\begin{pmatrix} \boldsymbol{A}_{11} & \boldsymbol{A}_{12} \end{pmatrix}$$

或

or

$$A = \left(\begin{array}{c:c:cc} 1 & 2 & 4 & 1 \\ 3 & 0 & 5 & 7 \\ \hdashline -1 & 2 & 0 & 1 \end{array}\right) = \begin{pmatrix} A_{11} & A_{12} & A_{13} \\ A_{21} & A_{22} & A_{23} \end{pmatrix}$$

在对分块矩阵进行某些运算时, 可先将每个子块当成分块矩阵的元素, 然后按矩阵的运算法则对每个子块进行相应的运算. 但进行分块时应注意以下三点:

As performing certain operations on block matrices, we firstly use each subblock as an element of the block matrix, and then perform the corresponding operations on each subblock by the operation rules of matrices. However, it should note the following points as one perform partitioning:

(1) 在计算 $A \pm B$ 时, A 和 B 的分块方式必须相同, 以保证它们的对应子块同型.

(1) To calculate $A \pm B$, the partitioned forms of A and B must be the same such that their corresponding sublocks have the same forms.

(2) 在计算 AB 时, 对 A 的列的分块方式与 B 的行的分块方式必须一致, 以保证它们对应的子块能够相乘.

(2) To calculate AB, the partitioned forms of the columns of A and those of the rows of B must be the same such that their corresponding sublocks can perform multiplication.

(3) 求 A^{T} 时, 要先将子块作为分块矩阵的元素, 然后分块矩阵转置, 最后再将各子块转置.

(3) To find A^{T}, firstly, each subblock should be viewed as an element of the block matrix; secondly, transpose the block matrix; finally, transpose each subblock.

例如,

For example,

$$\begin{pmatrix} A_{11} & A_{12} & A_{13} \\ A_{21} & A_{22} & A_{23} \end{pmatrix}^{\mathrm{T}} = \begin{pmatrix} A_{11}^{\mathrm{T}} & A_{21}^{\mathrm{T}} \\ A_{12}^{\mathrm{T}} & A_{22}^{\mathrm{T}} \\ A_{13}^{\mathrm{T}} & A_{23}^{\mathrm{T}} \end{pmatrix}$$

例 1 设

Example 1 Suppose that

$$A = \begin{pmatrix} 1 & 0 & 0 & 0 \\ 0 & 1 & 0 & 0 \\ -1 & 2 & 1 & 0 \\ 1 & 1 & 0 & 1 \end{pmatrix}, \quad B = \begin{pmatrix} 1 & 0 & 1 & 0 \\ -1 & 2 & 0 & 1 \\ -1 & 0 & 4 & 1 \\ -1 & -1 & 2 & 0 \end{pmatrix}$$

求 $\boldsymbol{AB}$.

Find $\boldsymbol{AB}$.

解 把 $\boldsymbol{A},\boldsymbol{B}$ 划分为如下的分块矩阵, 并计算 $\boldsymbol{AB}$, 得

Solution Partitioning $\boldsymbol{A},\boldsymbol{B}$ into the following block matrices and calculating $\boldsymbol{AB}$, we obtain

$$\boldsymbol{A}=\left(\begin{array}{cc|cc} 1 & 0 & 0 & 0 \\ 0 & 1 & 0 & 0 \\ \hline -1 & 2 & 1 & 0 \\ 1 & 1 & 0 & 1 \end{array}\right)=\begin{pmatrix} \boldsymbol{E} & \boldsymbol{O} \\ \boldsymbol{A}_1 & \boldsymbol{E} \end{pmatrix}$$

$$\boldsymbol{B}=\left(\begin{array}{cc|cc} 1 & 0 & 1 & 0 \\ -1 & 2 & 0 & 1 \\ \hline -1 & 0 & 4 & 1 \\ -1 & -1 & 2 & 0 \end{array}\right)=\begin{pmatrix} \boldsymbol{B}_{11} & \boldsymbol{E} \\ \boldsymbol{B}_{21} & \boldsymbol{B}_{22} \end{pmatrix}$$

$$\boldsymbol{AB}=\begin{pmatrix} \boldsymbol{E} & \boldsymbol{O} \\ \boldsymbol{A}_1 & \boldsymbol{E} \end{pmatrix}\begin{pmatrix} \boldsymbol{B}_{11} & \boldsymbol{E} \\ \boldsymbol{B}_{21} & \boldsymbol{B}_{22} \end{pmatrix}=\begin{pmatrix} \boldsymbol{B}_{11} & \boldsymbol{E} \\ \boldsymbol{A}_1\boldsymbol{B}_{11}+\boldsymbol{B}_{21} & \boldsymbol{A}_1+\boldsymbol{B}_{22} \end{pmatrix}$$

其中

where

$$\boldsymbol{A}_1\boldsymbol{B}_{11}+\boldsymbol{B}_{21}=\begin{pmatrix} -1 & 2 \\ 1 & 1 \end{pmatrix}\begin{pmatrix} 1 & 0 \\ -1 & 2 \end{pmatrix}+\begin{pmatrix} -1 & 0 \\ -1 & -1 \end{pmatrix}=\begin{pmatrix} -4 & 4 \\ -1 & 1 \end{pmatrix}$$

$$\boldsymbol{A}_1+\boldsymbol{B}_{22}=\begin{pmatrix} -1 & 2 \\ 1 & 1 \end{pmatrix}+\begin{pmatrix} 4 & 1 \\ 2 & 0 \end{pmatrix}=\begin{pmatrix} 3 & 3 \\ 3 & 1 \end{pmatrix}$$

因此, 有

Consequently, we have

$$\boldsymbol{AB}=\begin{pmatrix} 1 & 0 & 1 & 0 \\ -1 & 2 & 0 & 1 \\ -4 & 4 & 3 & 3 \\ -1 & 1 & 3 & 1 \end{pmatrix}$$

■

令 $\boldsymbol{A}$ 为 n 阶方阵. 若 $\boldsymbol{A}$ 可以划分成如下形式

Let $\boldsymbol{A}$ be a square matrix of order n. If $\boldsymbol{A}$ can be partitioned into the following form

$$\boldsymbol{A}=\begin{pmatrix} \boldsymbol{A}_1 & \boldsymbol{O} & \cdots & \boldsymbol{O} \\ \boldsymbol{O} & \boldsymbol{A}_2 & \cdots & \boldsymbol{O} \\ \vdots & \vdots & & \vdots \\ \boldsymbol{O} & \boldsymbol{O} & \cdots & \boldsymbol{A}_s \end{pmatrix}$$

其中 $\boldsymbol{A}_i(i=1,2,\cdots,s)$ 均为方阵 (可以不同型), 则称 $\boldsymbol{A}$ 为**分块对角矩阵**, 且具有如下性质:

where $\boldsymbol{A}_i(i=1,2,\cdots,s)$ are all square matrices (maybe of different sizes), then $\boldsymbol{A}$ is said to be a **block diagonal matrix**, which has the following properties:

(1) $|\boldsymbol{A}|=|\boldsymbol{A}_1||\boldsymbol{A}_2|\cdots|\boldsymbol{A}_s|$.

(2) 若 $|\boldsymbol{A}_i|\neq 0(i=1,2,\cdots,s)$, 则 $\boldsymbol{A}$ 可逆, 且

(2) If $|\boldsymbol{A}_i|\neq 0(i=1,2,\cdots,s)$, then $\boldsymbol{A}$ is invertible, and

$$\boldsymbol{A}^{-1}=\begin{pmatrix}\boldsymbol{A}_1^{-1} & \boldsymbol{O} & \cdots & \boldsymbol{O}\\ \boldsymbol{O} & \boldsymbol{A}_2^{-1} & \cdots & \boldsymbol{O}\\ \vdots & \vdots & & \vdots\\ \boldsymbol{O} & \boldsymbol{O} & \cdots & \boldsymbol{A}_s^{-1}\end{pmatrix}$$

例 2 设

Example 2 Suppose that

$$\boldsymbol{A}=\begin{pmatrix}5&0&0&0&0\\0&3&1&0&0\\0&2&1&0&0\\0&0&0&3&2\\0&0&0&2&1\end{pmatrix}$$

求 $|\boldsymbol{A}^3|$ 与 $\boldsymbol{A}^{-1}$.

Find $|\boldsymbol{A}^3|$ and $\boldsymbol{A}^{-1}$.

解 对 $\boldsymbol{A}$ 进行如下分块

Solution Partitioning $\boldsymbol{A}$ into the following form yields

$$\boldsymbol{A}=\left(\begin{array}{c|cc|cc}5&0&0&0&0\\\hline 0&3&1&0&0\\0&2&1&0&0\\\hline 0&0&0&3&2\\0&0&0&2&1\end{array}\right)=\begin{pmatrix}\boldsymbol{A}_1&\boldsymbol{O}&\boldsymbol{O}\\\boldsymbol{O}&\boldsymbol{A}_2&\boldsymbol{O}\\\boldsymbol{O}&\boldsymbol{O}&\boldsymbol{A}_3\end{pmatrix}$$

其中

where

$$|\boldsymbol{A}_1|=5,\quad |\boldsymbol{A}_2|=1,\quad |\boldsymbol{A}_3|=-1$$

$$\boldsymbol{A}_1^{-1}=\left(\frac{1}{5}\right),\quad \boldsymbol{A}_2^{-1}=\begin{pmatrix}1&-1\\-2&3\end{pmatrix},\quad \boldsymbol{A}_3^{-1}=\begin{pmatrix}-1&2\\2&-3\end{pmatrix}$$

故

So we have

$$|\boldsymbol{A}^3| = |\boldsymbol{A}|^3 = (|\boldsymbol{A}_1|\,|\boldsymbol{A}_2|\,|\boldsymbol{A}_3|)^3 = (5\times 1\times(-1))^3 = -125$$

$$\boldsymbol{A}^{-1} = \begin{pmatrix} \boldsymbol{A}_1^{-1} & \boldsymbol{O} & \boldsymbol{O} \\ \boldsymbol{O} & \boldsymbol{A}_2^{-1} & \boldsymbol{O} \\ \boldsymbol{O} & \boldsymbol{O} & \boldsymbol{A}_3^{-1} \end{pmatrix} = \begin{pmatrix} \frac{1}{5} & 0 & 0 & 0 & 0 \\ 0 & 1 & -1 & 0 & 0 \\ 0 & -2 & 3 & 0 & 0 \\ 0 & 0 & 0 & -1 & 2 \\ 0 & 0 & 0 & 2 & -3 \end{pmatrix}$$ ■

例 3 设

Example 3 Suppose that

$$\boldsymbol{D} = \begin{pmatrix} \boldsymbol{A} & \boldsymbol{O} \\ \boldsymbol{C} & \boldsymbol{B} \end{pmatrix}$$

其中 $\boldsymbol{A}, \boldsymbol{B}$ 分别是 k 阶、r 阶可逆矩阵, $\boldsymbol{C}$ 是 $r\times k$ 矩阵, $\boldsymbol{O}$ 是 $k\times r$ 零矩阵. 求 $\boldsymbol{D}^{-1}$.

where $\boldsymbol{A}, \boldsymbol{B}$ are invertible matrices of order k and order r, respectively, $\boldsymbol{C}$ is an $r\times k$ matrix, $\boldsymbol{O}$ is a $k\times r$ zero matrix. Find $\boldsymbol{D}^{-1}$.

解 设

Solution Suppose that

$$\boldsymbol{D}^{-1} = \begin{pmatrix} \boldsymbol{X}_{11} & \boldsymbol{X}_{12} \\ \boldsymbol{X}_{21} & \boldsymbol{X}_{22} \end{pmatrix}$$

于是

Thus,

$$\begin{pmatrix} \boldsymbol{A} & \boldsymbol{O} \\ \boldsymbol{C} & \boldsymbol{B} \end{pmatrix}\begin{pmatrix} \boldsymbol{X}_{11} & \boldsymbol{X}_{12} \\ \boldsymbol{X}_{21} & \boldsymbol{X}_{22} \end{pmatrix} = \begin{pmatrix} \boldsymbol{E}_k & \boldsymbol{O} \\ \boldsymbol{O} & \boldsymbol{E}_r \end{pmatrix}$$

由上式得

From the above equation we obtain

$$\begin{cases} \boldsymbol{A}\boldsymbol{X}_{11} = \boldsymbol{E}_k \\ \boldsymbol{A}\boldsymbol{X}_{12} = \boldsymbol{O} \\ \boldsymbol{C}\boldsymbol{X}_{11} + \boldsymbol{B}\boldsymbol{X}_{21} = \boldsymbol{O} \\ \boldsymbol{C}\boldsymbol{X}_{12} + \boldsymbol{B}\boldsymbol{X}_{22} = \boldsymbol{E}_r \end{cases}$$

该方程组的解为

The solution of the system is

$$\boldsymbol{X}_{11} = \boldsymbol{A}^{-1},\quad \boldsymbol{X}_{12} = \boldsymbol{A}^{-1}\boldsymbol{O} = \boldsymbol{O},\quad \boldsymbol{X}_{21} = -\boldsymbol{B}^{-1}\boldsymbol{C}\boldsymbol{A}^{-1},\quad \boldsymbol{X}_{22} = \boldsymbol{B}^{-1}$$

因此

Consequently,

$$D^{-1}=\begin{pmatrix} A^{-1} & O \\ -B^{-1}CA^{-1} & B^{-1} \end{pmatrix}$$ ■

例 4 证明: $AB=O$ 的充要条件是 B 的每个列向量都是齐次线性方程组 $Ax=0$ 的解.

Example 4 Prove that $AB=O$ if and only if each column vector of B is a solution of the homogeneous linear system $Ax=0$.

证 不妨设 A 为 $m\times n$ 矩阵, B 为 $n\times s$ 矩阵. 把 B 按列分为 $1\times s$ 分块矩阵 $B=(\beta_1,\beta_2,\cdots,\beta_s)$. 由分块矩阵的乘法有

Proof Not loss of generality, suppose that A is an $m\times n$ matrix and that B is an $n\times s$ matrix. Partition B into a $1\times s$ block matrix by column, $B=(\beta_1,\beta_2,\cdots,\beta_s)$. Using the multiplication of block matrices, we have

$$AB=A(\beta_1,\beta_2,\cdots,\beta_s)=(A\beta_1,A\beta_2,\cdots,A\beta_s)$$

于是

Thus,

$$\begin{aligned} AB=O \Leftrightarrow AB &= A(\beta_1,\beta_2,\cdots,\beta_s)=(A\beta_1,A\beta_2,\cdots,A\beta_s)=(0,0,\cdots,0) \\ &\Leftrightarrow A\beta_1=0, A\beta_2=0,\cdots,A\beta_s=0 \end{aligned}$$

可知 $\beta_1,\beta_2,\cdots,\beta_s$ 都是 $Ax=0$ 的解. ■

we have $\beta_1,\beta_2,\cdots,\beta_s$ are all solutions of $Ax=0$ ■

例 5 如果将方阵 A,E 分块为如下形式

Example 5 If the square matrices A and E are partitioned into the following forms

$$A=\left(\begin{array}{c:c:c:c} a_{11} & a_{12} & \cdots & a_{1n} \\ a_{21} & a_{22} & \cdots & a_{2n} \\ \vdots & \vdots & & \vdots \\ a_{n1} & a_{n2} & \cdots & a_{nn} \end{array}\right)=(A_1,A_2,\cdots,A_n)$$

$$E=\left(\begin{array}{c:c:c:c} 1 & 0 & \cdots & 0 \\ 0 & 1 & \cdots & 0 \\ \vdots & \vdots & & \vdots \\ 0 & 0 & \cdots & 1 \end{array}\right)=(\varepsilon_1,\varepsilon_2,\cdots,\varepsilon_n)$$

则

then

$$AE = A(\varepsilon_1, \varepsilon_2, \cdots, \varepsilon_n) = (A\varepsilon_1, A\varepsilon_2, \cdots, A\varepsilon_n) = (A_1, A_2, \cdots, A_n)$$

所以

consequently,

$$A\varepsilon_j = A_j \quad (j = 1, 2, \cdots, n)$$ ■

思 考 题

Questions

判断下面结论是否正确?

Decide whether or not the following results are valid.

(1) 若 $\boldsymbol{a}$ 和 $\boldsymbol{b}$ 都是 $n \times 1$ 列矩阵, 则 $\boldsymbol{a}\boldsymbol{b}^{\mathrm{T}} = \boldsymbol{b}\boldsymbol{a}^{\mathrm{T}}$;

(1) If $\boldsymbol{a}$ and $\boldsymbol{b}$ are all $n \times 1$ column matrices, then $\boldsymbol{a}\boldsymbol{b}^{\mathrm{T}} = \boldsymbol{b}\boldsymbol{a}^{\mathrm{T}}$;

(2) 若 $\boldsymbol{a}$ 和 $\boldsymbol{b}$ 都是 $n \times 1$ 列矩阵, 则 $\boldsymbol{a}^{\mathrm{T}}\boldsymbol{b} = b^{\mathrm{T}}\boldsymbol{a}$;

(2) If $\boldsymbol{a}$ and $\boldsymbol{b}$ are all $n \times 1$ column matrices, then $\boldsymbol{a}^{\mathrm{T}}\boldsymbol{b} = \boldsymbol{b}^{\mathrm{T}}\boldsymbol{a}$;

习 题 2

Exercise 2

1. 设

1. Suppose that

$$A = \begin{pmatrix} 1 & 1 & 1 \\ 1 & 1 & -1 \\ 1 & -1 & 1 \end{pmatrix}, \quad B = \begin{pmatrix} 1 & 2 & 3 \\ -1 & -2 & 4 \\ 0 & 5 & 1 \end{pmatrix}$$

求

Find

$$(1)\ 3AB - 2A. \ (2)\ AB^{\mathrm{T}}.$$

2. 计算:

2. Calculate

(1) $\begin{pmatrix} 2 & 1 & 4 & 0 \\ 1 & -1 & 3 & 4 \end{pmatrix} \begin{pmatrix} 1 & 3 & 1 \\ 0 & -1 & 3 \\ 1 & -3 & 1 \\ 4 & 0 & -2 \end{pmatrix}$; (2) $(x_1, x_2, x_3) \begin{pmatrix} a_{11} & a_{12} & a_{13} \\ a_{12} & a_{22} & a_{23} \\ a_{13} & a_{23} & a_{33} \end{pmatrix} \begin{pmatrix} x_1 \\ x_2 \\ x_3 \end{pmatrix}$;

(3) $\begin{pmatrix} 1 & \lambda \\ 0 & 1 \end{pmatrix}^n$; (4) $\begin{pmatrix} 4 & 3 & 1 \\ 1 & -2 & 3 \\ 5 & 7 & 0 \end{pmatrix} \begin{pmatrix} 7 \\ 2 \\ 1 \end{pmatrix}$;

(5) $(1, 2, 3) \begin{pmatrix} 3 \\ 2 \\ 1 \end{pmatrix}$; (6) $\begin{pmatrix} 3 \\ 2 \\ 1 \end{pmatrix} \begin{pmatrix} 1, & 2, & 3 \end{pmatrix}$.

3. 设

3. Suppose that

$$A=\begin{pmatrix}-2 & 1\\ \frac{3}{2} & -\frac{1}{2}\end{pmatrix},\quad B=\begin{pmatrix}1 & 1\\ 0 & 0\end{pmatrix},\quad C=\begin{pmatrix}1 & 2\\ 3 & 4\end{pmatrix}$$

且 $D=ABC$. 求

and that $D=ABC$. Find

$$(1)\ B^n;\ (2)\ CA;\ (3)\ D^n.$$

4. 已知

4. Known that

$$f(x)=x^2+2x+1,\quad A=\begin{pmatrix}-1 & 1 & 0\\ 0 & -1 & 0\\ 1 & 0 & -1\end{pmatrix}$$

(1) 求 $f(A)$;

(1) Find $f(A)$;

(2) 若有矩阵 B, 满足 $f(B)=O$, 证明: B 可逆, 并求 B^{-1}.

(2) If there exists a matrix B satisfying $f(B)=O$, prove that B is invertible and find B^{-1}.

5. 设

5. Suppose that

$$A=\begin{pmatrix}1 & 2\\ 1 & 3\end{pmatrix},\quad B=\begin{pmatrix}1 & 0\\ 1 & 2\end{pmatrix}$$

判断下列等式是否成立:

Decide whether or not the following equations are valid:

$$(1)\ AB=BA;\ (2)\ (A+B)^2=A^2+2AB+B^2;\ (3)\ (A+B)(A-B)=A^2-B^2.$$

6. 举反例说明下列命题是错误的:

6. Illustrate the following propositions are not valid by counter examples:

(1) 若 $A^2=O$, 则 $A=O$;
(2) 若 $A^2=A$, 则 $A=O$ 或 $A=E$;
(3) 若 $AX=AY$ 且 $A\neq O$, 则 $X=Y$.

(1) If $A^2=O$, then $A=O$;
(2) If $A^2=A$, then $A=O$ or $A=E$;
(3) If $AX=AY$, and $A\neq O$, then $X=Y$.

7. 用伴随矩阵求下列矩阵的逆矩阵:

7. Find inverse matrices of the following matrices by adjoint matrices:

$$(1)\begin{pmatrix}1 & 2\\ 2 & 5\end{pmatrix};\quad (2)\begin{pmatrix}\cos\theta & -\sin\theta\\ \sin\theta & \cos\theta\end{pmatrix};\quad (3)\begin{pmatrix}1 & 2 & 3\\ 2 & 2 & 1\\ 3 & 4 & 3\end{pmatrix}.$$

8. 解下列矩阵方程:

8. Solve the following matrix equations:

(1) $\begin{pmatrix} 2 & 5 \\ 1 & 3 \end{pmatrix} \boldsymbol{X} = \begin{pmatrix} 4 \\ 2 \end{pmatrix}$; (2) $\boldsymbol{X} \begin{pmatrix} 2 & 1 & -1 \\ 2 & 1 & 0 \\ 1 & -1 & 1 \end{pmatrix} = \begin{pmatrix} 1 & -1 & 3 \\ 4 & 3 & 2 \end{pmatrix}$;

(3) $\begin{pmatrix} 1 & 4 \\ -1 & 2 \end{pmatrix} \boldsymbol{X} \begin{pmatrix} 2 & 0 \\ -1 & 1 \end{pmatrix} = \begin{pmatrix} 3 & 1 \\ 0 & -1 \end{pmatrix}$;

(4) $\boldsymbol{AX} = \boldsymbol{A} + 2\boldsymbol{X}$, 其中

(4) $\boldsymbol{AX} = \boldsymbol{A} + 2\boldsymbol{X}$, where

$$\boldsymbol{A} = \begin{pmatrix} 4 & 2 & 3 \\ 1 & 1 & 0 \\ -1 & 2 & 3 \end{pmatrix}$$

9. 用逆矩阵求解下列线性方程组:

9. Solve the following systems of linear equations by inverse matrices:

(1) $\begin{cases} x_1 + 2x_2 + 3x_3 = 1, \\ 2x_1 + 2x_2 + 5x_3 = 2, \\ 3x_1 + 5x_2 + x_3 = 3; \end{cases}$ (2) $\begin{cases} x_1 - x_2 - x_3 = 2, \\ 2x_1 - x_2 - 3x_3 = 1, \\ 3x_1 + 2x_2 - 5x_3 = 0. \end{cases}$

10. 设 $\boldsymbol{A}^k = \boldsymbol{O}$($k$ 是正整数), 证明:

10. Suppose that $\boldsymbol{A}^k = \boldsymbol{O}$ (k is a positive integer). Prove that

$$(\boldsymbol{E} - \boldsymbol{A})^{-1} = \boldsymbol{E} + \boldsymbol{A} + \boldsymbol{A}^2 + \cdots + \boldsymbol{A}^{k-1}$$

11. 设方阵 $\boldsymbol{A}$ 满足 $\boldsymbol{A}^2 - \boldsymbol{A} - 2\boldsymbol{E} = \boldsymbol{O}$. 证明: $\boldsymbol{A}$ 及 $\boldsymbol{A} + 2\boldsymbol{E}$ 都可逆, 并求 $\boldsymbol{A}^{-1}$, $(\boldsymbol{A} + 2\boldsymbol{E})^{-1}$.

11. Suppose that the square matrix $\boldsymbol{A}$ satisfies $\boldsymbol{A}^2 - \boldsymbol{A} - 2\boldsymbol{E} = \boldsymbol{O}$. Prove that $\boldsymbol{A}$ and $\boldsymbol{A} + 2\boldsymbol{E}$ are all invertible, and find $\boldsymbol{A}^{-1}, (\boldsymbol{A} + 2\boldsymbol{E})^{-1}$.

12. 设 $\boldsymbol{A}$ 和 $\boldsymbol{B}$ 为两个 3 阶方阵, $|\boldsymbol{A}| = 4$, $|\boldsymbol{B}| = -5$. 求

12. Let $\boldsymbol{A}$ and $\boldsymbol{B}$ be two square matrices of order 3, $|\boldsymbol{A}| = 4$, $|\boldsymbol{B}| = -5$. Find

$$|2\boldsymbol{AB}|, \left|\boldsymbol{AB}^{\mathrm{T}}\right|, |\boldsymbol{A}^{-1}\boldsymbol{B}^{-1}|, \ |-\boldsymbol{A}^{-1}\boldsymbol{B}|.$$

13. 设 $\boldsymbol{A}$ 为 n 阶方阵, $|\boldsymbol{A}|=2$, 求

13. Let $\boldsymbol{A}$ be a square matrix of order n, $|\boldsymbol{A}|=2$. Find

$$\left|\left(\frac{1}{2}\boldsymbol{A}\right)^{-1}-3\boldsymbol{A}^*\right|$$

14. 已知

14. Known that

$$\boldsymbol{A}=\begin{pmatrix}1&5&4\\0&2&4\\1&3&1\end{pmatrix}$$

求 $(\boldsymbol{A}^*)^{-1}$.

Find $(\boldsymbol{A}^*)^{-1}$.

15. 设

15. Suppose that

$$\boldsymbol{A}=\begin{pmatrix}3&4&0&0\\4&-3&0&0\\0&0&2&0\\0&0&2&2\end{pmatrix}$$

求 (1) $|\boldsymbol{A}^8|$; (2) $\boldsymbol{A}^4$; (3) $\boldsymbol{A}^{-1}$.

Find (1) $|\boldsymbol{A}^8|$; (2) $\boldsymbol{A}^4$; (3) $\boldsymbol{A}^{-1}$.

16. (1) 设 $\boldsymbol{A}$ 和 $\boldsymbol{B}$ 为两个 n 阶方阵, 且 $\boldsymbol{A}$ 为对称矩阵. 证明: $\boldsymbol{B}^{\mathrm{T}}\boldsymbol{A}\boldsymbol{B}$ 为对称矩阵.

16. (1) Let $\boldsymbol{A}$ and $\boldsymbol{B}$ be two square matrices of order n, and let $\boldsymbol{A}$ be a symmetric matrix. Prove that $\boldsymbol{B}^{\mathrm{T}}\boldsymbol{A}\boldsymbol{B}$ is a symmetric matrix.

(2) 设 $\boldsymbol{A}$ 和 $\boldsymbol{B}$ 都是 n 阶对称矩阵. 证明: $\boldsymbol{AB}$ 是对称的充要条件是 $\boldsymbol{AB}=\boldsymbol{BA}$.

(2) Let $\boldsymbol{A}$ and $\boldsymbol{B}$ be two symmetric matrices of order n. Prove that $\boldsymbol{AB}$ is symmetric if and only if $\boldsymbol{AB}=\boldsymbol{BA}$.

17. 设 $\boldsymbol{A}$ 为 n 阶方阵, 伴随矩阵为 $\boldsymbol{A}^*$, 证明

17. Suppose that $\boldsymbol{A}$ is a square matrix of order n and that $\boldsymbol{A}^*$ is the adjoint matrix of $\boldsymbol{A}$. Prove that

(1) 若 $|\boldsymbol{A}|=0$, 则 $|\boldsymbol{A}^*|=0$;
(2) $|\boldsymbol{A}^*|=|\boldsymbol{A}|^{n-1}$.

(1) If $|\boldsymbol{A}|=0$, then $|\boldsymbol{A}^*|=0$;
(2) $|\boldsymbol{A}^*|=|\boldsymbol{A}|^{n-1}$.

18. 设 $\boldsymbol{A}, \boldsymbol{B}$ 分别为 n 阶和 m 阶可逆方阵. 求

18. Suppose that $\boldsymbol{A}$, $\boldsymbol{B}$ are invertible square matrices of order n and order m, respectively. Find

$$\begin{pmatrix}\boldsymbol{O}&\boldsymbol{A}\\\boldsymbol{B}&\boldsymbol{O}\end{pmatrix}^{-1}$$

19. 设

$$A=\begin{pmatrix}-1&0&0&0\\0&-1&0&0\\-2&1&1&0\\1&-1&0&1\end{pmatrix},\quad B=\begin{pmatrix}-1&0&1&0\\1&3&0&1\\1&0&3&1\\-1&-1&2&1\end{pmatrix}$$

采用分块矩阵方法求 AB.

19. Suppose that (matrices A, B above). Find AB by the method of block matrices.

补充题 2

Supplement Exercise 2

1. 证明任一方阵 A 可表示为一对称矩阵与一反对称矩阵之和.

1. Prove that an arbitrary square matrix A can be expressed by the sum of a symmetric matrix and a skew-symmetric matrix.

2. 设 α 为 3×1 列矩阵, 且

2. Let α be a 3×1 column matrix, and

$$\alpha\alpha^{\mathrm{T}}=\begin{pmatrix}1&-1&1\\-1&1&-1\\1&-1&1\end{pmatrix}$$

求 $\alpha^{\mathrm{T}}\alpha$.

Find $\alpha^{\mathrm{T}}\alpha$.

3. 设

3. Suppose that

$$A=\begin{pmatrix}2&4&6\\1&2&3\\1&2&3\end{pmatrix}$$

求 A^n.

Find A^n.

4. 设

4. Suppose that

$$A=\begin{pmatrix}0&-1&0\\1&0&0\\0&0&-1\end{pmatrix}$$

$B=P^{-1}AP$, 其中 P 为 3 阶可逆矩阵. 求 $B^{2004}-2A^2$.

and that $B=P^{-1}AP$, where P is an invertible matrix of order 3. Find $B^{2004}-2A^2$.

5. 设

5. Suppose that

$$A=\begin{pmatrix}1&4&6\\0&2&5\\0&0&3\end{pmatrix}$$

求 $\left|A^{-1}(A^*)^{-1}\right|$.

Find $\left|A^{-1}(A^*)^{-1}\right|$.

6. 设两个 3 阶方阵 A, B 满足 $A^2B - A - B = E$, 其中

6. Suppose that A, B are two square matrices of order 3 and satisfy $A^2B - A - B = E$, where

$$A = \begin{pmatrix} 1 & 0 & 1 \\ 0 & 2 & 0 \\ -1 & 0 & 1 \end{pmatrix}$$

求 $|B|$.

Find $|B|$.

7. 设

7. Suppose that

$$A = \begin{pmatrix} 1 & 2 & 1 \\ 1 & 0 & 1 \end{pmatrix}, \quad B = \begin{pmatrix} 1 & 2 \\ 1 & 1 \\ 1 & 1 \end{pmatrix}$$

求 $|AB|$ 和 $|BA|$.

Find $|AB|$ and $|BA|$.

8. 设 A, B 为 4 阶方阵, $A = (\alpha, \gamma_2, \gamma_3, \gamma_4)$, $B = (\beta, \gamma_2, \gamma_3, \gamma_4)$, 其中 $\alpha, \beta, \gamma_2, \gamma_3, \gamma_4$ 均为列矩阵, 且 $|A| = 4$, $|B| = 1$. 求 $|A + B|$.

8. Suppose that A, B are matrices of order 4, given by $A = (\alpha, \gamma_2, \gamma_3, \gamma_4)$ and $B = (\beta, \gamma_2, \gamma_3, \gamma_4)$, where $\alpha, \beta, \gamma_2, \gamma_3, \gamma_4$ are all column matrices, and $|A| = 4$, $|B| = 1$. Find $|A + B|$.

9. 设

9. Suppose that

$$A = \frac{1}{2}\begin{pmatrix} 1 & 3 & 0 \\ 2 & 5 & 0 \\ 1 & -1 & 2 \end{pmatrix}$$

求 $(A^{-1})^*$.

Find $(A^{-1})^*$.

10. 证明

10. Prove that

$$(A^*)^{\mathrm{T}} = (A^{\mathrm{T}})^*, (A^*)^{-1} = (A^{-1})^*$$

11. 设矩阵 A, B 满足 $A^*BA = 2BA - 8E$, 其中

11. Suppose that the matrices A, B satisfy $A^*BA = 2BA - 8E$, where

$$A = \begin{pmatrix} 1 & 0 & 0 \\ 0 & -2 & 0 \\ 0 & 0 & 1 \end{pmatrix}$$

求 $\boldsymbol{B}$.

Find $\boldsymbol{B}$.

12. 已知

12. Known that

$$\boldsymbol{A}^2=\boldsymbol{A},\ 2\boldsymbol{A}-\boldsymbol{B}-\boldsymbol{AB}=\boldsymbol{E}$$

(1) 证明: $\boldsymbol{A}-\boldsymbol{B}$ 可逆;

(1) Prove that $\boldsymbol{A}-\boldsymbol{B}$ is invertible;

(2) 若

(2) If

$$\boldsymbol{A}=\begin{pmatrix}1&0&0\\0&3&-1\\0&6&-2\end{pmatrix}$$

求 $\boldsymbol{B}$.

Find $\boldsymbol{B}$.

13. 设 $\boldsymbol{A}$ 是 n 阶可逆矩阵. 证明: $(\boldsymbol{A}^*)^*=|\boldsymbol{A}|^{n-2}\boldsymbol{A}$, 并求 $|(\boldsymbol{A}^*)^*|$.

13. Let $\boldsymbol{A}$ be an invertible matrix of order n. Prove that $(\boldsymbol{A}^*)^*=|\boldsymbol{A}|^{n-2}\boldsymbol{A}$ and find $|(\boldsymbol{A}^*)^*|$.

第 3 章
Chapter 3

矩阵的初等变换
Elementary Operations on Matrices

在前面两章中，我们利用行列式和矩阵的相关知识研究了未知数的个数和方程的个数相等的特殊线性方程组 $\boldsymbol{A}_{n\times n}\boldsymbol{X} = \boldsymbol{b}$. 为了讨论更一般的线性方程组 $\boldsymbol{A}_{m\times n}\boldsymbol{X} = \boldsymbol{b}$($m$ 和 n 不一定相等), 我们引入矩阵的初等变换.

In the first two chapters, we use the related knowledge of determinants and matrices to study the special system of linear equations with the same numbers of unknowns and equations, namely, $\boldsymbol{A}_{n\times n}\boldsymbol{X} = \boldsymbol{b}$. To discuss more general systems of linear equations, i.e., $\boldsymbol{A}_{m\times n}\boldsymbol{X} = \boldsymbol{b}$, maybe, $m \neq n$, we introduce elementary operations performing on matrices.

初等变换理论在化简矩阵、求矩阵的逆和秩、解线性方程组等问题中起着非常重要的作用，它在线性代数中占有很重要的地位.

The theory of elementary operations plays a very important role in the problems of simplifying matrices, finding inverse and rank of matrices, solving linear systems, and so on. Moreover, it possesses an indispensible status in Linear Algebra.

➢本章内容提要

Headline of this chapter

1. 初等变换 (矩阵) 的定义、作用、性质;

1. Definitions, functions and properties of elementary operations (matrices);

2. 初等变换的应用: 求矩阵的逆和秩、求解线性方程组.

2. Applications of elementary operations: finding inverse and rank of matrices, finding solutions of linear systems.

3.1 初等变换与初等矩阵

3.1 Elementary Operations and Elementary Matrices

3.1.1 矩阵的初等变换

3.1.1 Elementary Operations on Matrices

在初等代数里, 用消元法求解二元、三元线性方程组时, 常需要对方程组进行下列同解变形:

In elementary algebra, using elimination to solve linear system in two or three unknowns, we always need to perform the following equivalent operations:

(1) 交换两个方程的位置;

(1) Interchange the positions of two equations;

(2) 用一个非零数乘以某一个方程;

(2) Multiply a certain equation by a nonzero number;

(3) 将某个方程乘以一个常数后加到另外一个方程上去.

(3) Add a constant times an equation to another equation.

例如, 考虑如下的线性方程组

For example, consider the following linear system

$$\begin{cases} 2x_1 - x_2 + 2x_3 = 4 \\ x_1 + x_2 + 2x_3 = 1 \\ 4x_1 + x_2 + 4x_3 = 2 \end{cases} \tag{3.1}$$

将方程组 (3.1) 中第一、二个方程的位置互换, 得

Interchanging the positions of the first equation and the second equation in System (3.1) yields

$$\begin{cases} x_1 + x_2 + 2x_3 = 1 \\ 2x_1 - x_2 + 2x_3 = 4 \\ 4x_1 + x_2 + 4x_3 = 2 \end{cases} \tag{3.2}$$

将方程组 (3.2) 中第一个方程的 -2 倍加到第二个方程上, -4 倍加到第三个方程上, 得

Adding (-2) times the first equation to the second equation and adding (-4) times the first equation to the third equation in System (3.2), we obtain

$$\begin{cases} x_1 + x_2 + 2x_3 = 1 \\ \quad -3x_2 - 2x_3 = 2 \\ \quad -3x_2 - 4x_3 = -2 \end{cases} \tag{3.3}$$

将方程组 (3.3) 中第二个方程的 -1 倍加到第三个方程上, 得

Adding (-1) times the second equation to the third equation in System (3.3) leads to

$$\begin{cases} x_1 + x_2 + 2x_3 = 1 \\ \quad -3x_2 - 2x_3 = 2 \\ \qquad\quad -2x_3 = -4 \end{cases} \tag{3.4}$$

再将方程组 (3.4) 中第三个方程的 -1 倍加到第二个方程上, 1 倍加到第一个方程上, 得

Adding (-1) times the third equation to the second equation and adding the third equation to the first equation in System (3.4), we get

$$\begin{cases} x_1 + x_2 = -3 \\ -3x_2 = 6 \\ -2x_3 = -4 \end{cases} \tag{3.5}$$

最后, 将方程组 (3.5) 中第二个方程的 $\frac{1}{3}$ 倍加到第一个方程上, 将第二、第三个方程分别乘以数 $-\frac{1}{3}$ 和 $-\frac{1}{2}$, 得

Finally, adding $\frac{1}{3}$ times the second equation to the first equation in System (3.5), then performing $\left(-\frac{1}{3}\right)$ times the second equation and $\left(-\frac{1}{2}\right)$ times the third equation, respectively, we have

$$\begin{cases} x_1 = -1 \\ x_2 = -2 \\ x_3 = \ \ 2 \end{cases} \tag{3.6}$$

由初等代数可知, 以上各方程组同解, 故方程组 (3.1) 的解为

From elementary algebra we know that the above systems are equivalent, so the solution of System (3.1) is given by

$$x_1 = -1, \quad x_2 = -2, \quad x_3 = 2$$

若用矩阵术语来讨论方程组的求解过程, 则上述运算实质上是对方程组的增广矩阵实施行的运算. 这种行的运算就是矩阵的初等行变换.

If we discuss the solving process of a system by matrix terminology, then the above operations are actually equivalent to performing row operations on the augmented matrix of the system. The row operations are just elementary row operations on matrices.

定义 1 如下的三种变换称为对矩阵实施的**初等行变换**:

(1) 互换矩阵的第 i, j 两行 (记为 $r_i \leftrightarrow r_j$), 简称为**换行**;

(2) 将矩阵的第 i 行乘以非零常数 k(记为 kr_i), 简称为**数乘**;

(3) 将矩阵的第 j 行乘以数 k 后加到第 i 行上 (记为 $r_i + kr_j$), 简称为**倍加**.

若将定义 1 中的 "行" 换成 "列", 即得初等列变换的定义 (所用记号是将 "r" 换成 "c"). 对矩阵进行的初等行变换和初等列变换, 统称为**初等变换**.

如果 $\boldsymbol{A}$ 经过有限次初等变换变成 $\boldsymbol{B}$, 则称 $\boldsymbol{A}$ 与 $\boldsymbol{B}$ **等价**, 记为 $\boldsymbol{A} \to \boldsymbol{B}$ 或 $\boldsymbol{A} \sim \boldsymbol{B}$. 容易证明, 矩阵的等价关系具有下列性质:

(1) **反身性** $\boldsymbol{A}$ 与 $\boldsymbol{A}$ 等价;

(2) **对称性** 如果 $\boldsymbol{A}$ 与 $\boldsymbol{B}$ 等价, 那么 $\boldsymbol{B}$ 与 $\boldsymbol{A}$ 等价;

(3) **传递性** 如果 $\boldsymbol{A}$ 与 $\boldsymbol{B}$ 等价, $\boldsymbol{B}$ 与 $\boldsymbol{C}$ 等价, 那么 $\boldsymbol{A}$ 与 $\boldsymbol{C}$ 等价.

Definition 1 The following three operations are called **elementary row operations** performing on matrices:

(1) Interchange the i-th row and the j-th row of a matrix (written as $r_i \leftrightarrow r_j$), for short, **row-interchanging**;

(2) Multiply the i-th row of a matrix by a nonzero constant k (written as kr_i), for short, **scalar-multiplication**;

(3) Add a nonzero constant k times the j-th row of a matrix to the i-th row (written as $r_i + kr_j$), for short, **multiple-adding**.

If the "row" in Definition 1 is replaced by the "column", then we have the definition of elementary column operations (the notations related to "r" are replaced by "c"). Elementary row operations and elementary column operations performing on matrices are called **elementary operations** collectively.

We say that $\boldsymbol{B}$ is equivalent to $\boldsymbol{A}$ if $\boldsymbol{B}$ is obtained from $\boldsymbol{A}$ by a finite sequence of elementary operations, written as $\boldsymbol{A} \to \boldsymbol{B}$ or $\boldsymbol{A} \sim \boldsymbol{B}$. It is easy to show that the equivalence relation of matrices has the following properties:

(1) **Reflexivity** $\boldsymbol{A}$ is equivalent to itself;

(2) **Symmetry** If $\boldsymbol{A}$ is equivalent to $\boldsymbol{B}$, then $\boldsymbol{B}$ is equivalent to $\boldsymbol{A}$;

(3) **Transitivity** If $\boldsymbol{A}$ is equivalent to $\boldsymbol{B}$ and if $\boldsymbol{B}$ is equivalent to $\boldsymbol{C}$, then $\boldsymbol{A}$ is equivalent to $\boldsymbol{C}$.

由于对矩阵的初等行变换对应于线性方程组的同解变换, 所以线性方程组 (3.1) 的求解过程用矩阵的初等行变换描述如下:

Since elementary row operations performing on matrices correspond to the equivalent operations on a system of linear equations, the solution process of the linear system (3.1) can be described by elementary row operations performing on matrices, as follows:

$$\overline{\boldsymbol{A}}=\begin{pmatrix}2&-1&2&4\\1&1&2&1\\4&1&4&2\end{pmatrix}\xrightarrow{r_1\leftrightarrow r_2}\begin{pmatrix}1&1&2&1\\2&-1&2&4\\4&1&4&2\end{pmatrix}$$

$$\xrightarrow[r_3-4r_1]{r_2-2r_1}\begin{pmatrix}1&1&2&1\\0&-3&-2&2\\0&-3&-4&-2\end{pmatrix}\xrightarrow{r_3-r_2}\begin{pmatrix}1&1&2&1\\0&-3&-2&2\\0&0&-2&-4\end{pmatrix}$$

$$\xrightarrow[r_1+r_3]{r_2-r_3}\begin{pmatrix}1&1&0&-3\\0&-3&0&6\\0&0&-2&-4\end{pmatrix}\xrightarrow[\frac{-1}{3}r_2,\frac{-1}{2}r_3]{r_1+\frac{1}{3}r_2}\begin{pmatrix}1&0&0&-1\\0&1&0&-2\\0&0&1&2\end{pmatrix}$$

以上矩阵依次对应线性方程组 (3.1)~(3.6).

The above matrices in turn correspond to the linear systems (3.1)~(3.6).

因此, 初等行变换是**同解变换**, 即原矩阵所对应的方程组和经过有限次初等行变换得到的新矩阵所对应的方程组是同解的. 于是有如下定理.

Consequently, elementary row operations are **equivalent operations**, namely, the systems corresponding to the original matrices are equivalent to the systems corresponding to the new matrices obtained by a finite sequence of elementary row operations. We have the following theorem.

定理 1 若线性方程组 $\boldsymbol{Ax}=\boldsymbol{b}$ 的增广矩阵 $\overline{\boldsymbol{A}}=(\boldsymbol{A},\boldsymbol{b})$, 经过有限次初等行变换后所得矩阵为 $\overline{\boldsymbol{B}}=(\boldsymbol{B},\boldsymbol{d})$, 则 $\overline{\boldsymbol{B}}$ 所对应的线性方程组 $\boldsymbol{Bx}=\boldsymbol{d}$ 与 $\boldsymbol{Ax}=\boldsymbol{b}$ 同解.

Theorem 1 If the matrix $\overline{\boldsymbol{B}}=(\boldsymbol{B},\boldsymbol{d})$ is obtained from the augmented matrix $\overline{\boldsymbol{A}}=(\boldsymbol{A},\boldsymbol{b})$ associated with the linear system $\boldsymbol{Ax}=\boldsymbol{b}$ by a finite sequence of elementary row operations, then the solutions of the linear system $\boldsymbol{Bx}=\boldsymbol{d}$ corresponding to $\overline{\boldsymbol{B}}$ is equivalent to those of $\boldsymbol{Ax}=\boldsymbol{b}$.

由于初等列变换会改变对应方程组未知数的位置, 从而导致解的位置发生变化, 所以, 我们通常只用初等行变换, 不用初等列变换求解线性方程组.

Since elementary column operations may change the positions of unknowns in the corresponding system, which may change the positions of solutions, this is the reason that we always use elementary row operations and do not use elementary column operations to solve a linear system.

注 用初等变换研究矩阵的基本思路如下:

Note The basic ideas for studying matrices by elementary operations are as follows:

(1) 证明初等变换前后的矩阵保持某些性质不变, 例如, 矩阵的初等变换不改变对应的方程组的解;

(1) Prove that certain properties of the matrices before and after elementary operations are invariant, such as, elementary operations on matrices do not change the solution of the corresponding system;

(2) 用初等变换对矩阵进行约化, 然后通过新矩阵来研究原矩阵所对应的问题.

(2) Reduce a matrix by elementary operations, and then study the corresponding problems of the original matrix by the new matrix.

一般地, 矩阵 $\boldsymbol{A}_{m\times n}$ 可以经过有限次的初等行变换化为形如

In general, a matrix $\boldsymbol{A}_{m\times n}$ can be reduced to the following form by a finite sequence of elementary row operations, namely,

$$\begin{pmatrix} c_{11} & c_{12} & \cdots & c_{1r} & c_{1,r+1} & \cdots & c_{1n} \\ 0 & c_{22} & \cdots & c_{2r} & c_{2,r+1} & \cdots & c_{2n} \\ \vdots & \vdots & & \vdots & \vdots & & \vdots \\ 0 & 0 & \cdots & c_{rr} & c_{r,r+1} & \cdots & c_{rn} \\ 0 & 0 & \cdots & 0 & 0 & \cdots & 0 \\ 0 & 0 & \cdots & 0 & 0 & \cdots & 0 \end{pmatrix}$$

的矩阵, 称为**行阶梯形矩阵**. 它具有如下特点:

it is called a **row echelon matrix**. It has the following features:

(1) 每个阶梯只占一行;

(1) Each echelon only occupies a row;

(2) 非零行 (即元素不全为零的行) 的第一个非零元素的列标随着行标的增大而严格增大, 即列标一定不小于行标;

(2) The column index of the first nonzero element of the nonzero row (namely, the row containing some nonzero elements) increases along with the increasing row index strictly, namely, the column index must be no less than the row index;

(3) 元素全为零的行 (如果存在) 必位于矩阵的最下面几行.

(3) The zero rows (if there exist), namely, the elements are all zeros, must locate at the bottom rows of the matrix.

例如, 下列矩阵

For example, the following matrices

$$\boldsymbol{A}=\begin{pmatrix}1&0&-1\\0&2&1\\0&0&3\end{pmatrix},\quad \boldsymbol{B}=\begin{pmatrix}1&2&1&2&-1\\0&0&1&0&2\\0&0&0&0&3\end{pmatrix}$$

$$\boldsymbol{C}=\begin{pmatrix}1&1&2&-1\\0&0&0&1\\0&0&0&0\\0&0&0&0\end{pmatrix},\quad \boldsymbol{D}=\begin{pmatrix}1&0&-1&0&4\\0&0&-1&0&3\\0&0&0&1&-3\\0&0&0&0&0\end{pmatrix}$$

均为行阶梯形矩阵.

are all row echelon matrices.

进一步地, 经过有限次的初等行变换可以化为如下的矩阵

Further, after a finite sequence of elementary row operations, the matrix may be reduced to the following form

$$\begin{pmatrix}1&0&\cdots&0&b_{1r+1}&\cdots&b_{1n}\\0&1&\cdots&0&b_{2r+1}&\cdots&b_{2n}\\\vdots&\vdots&&\vdots&\vdots&&\vdots\\0&0&\cdots&1&b_{rr+1}&\cdots&b_{rn}\\0&0&\cdots&0&0&\cdots&0\\\vdots&\vdots&&\vdots&\vdots&&\vdots\\0&0&&0&0&\cdots&0\end{pmatrix}$$

称为**行最简形**. 它的特点是: 在行最简形矩阵中, 每一非零行的第一个非零元素全为 1; 且它所在的列中其余元素全为零.

which is called the **row-reduced form**. The features are that the first nonzero element on each nonzero row in a row-reduced matrix is equal to 1 and that the other elements located at its column are all zeros.

例如, 下列矩阵

For example, the following matrices

$$A=\begin{pmatrix}1&0&0\\0&1&0\\0&0&1\end{pmatrix},\quad B=\begin{pmatrix}1&2&0&2&0\\0&0&1&0&0\\0&0&0&0&1\end{pmatrix}$$

$$C=\begin{pmatrix}1&1&2&0\\0&0&0&1\\0&0&0&0\\0&0&0&0\end{pmatrix},\quad D=\begin{pmatrix}1&0&0&0&4\\0&0&1&0&3\\0&0&0&1&-3\\0&0&0&0&0\end{pmatrix}$$

均为行最简形矩阵.

are all row-reduced matrices.

矩阵 $A_{m\times n}$ 经过初等行变换可化为行阶梯形以及行最简形. 若经过有限次初等列变换可进一步化为下面的最简单形式

The matrix $A_{m\times n}$ can be reduced to a row echelon form and a row-reduced form by elementary row operations. Moreover, it can be reduced to the following simplest form after a finite sequence of elementary column operations

$$\begin{pmatrix}1&0&\cdots&0&0&\cdots&0\\0&1&\cdots&0&0&\cdots&0\\\vdots&\vdots&&\vdots&\vdots&&\vdots\\0&0&\cdots&1&0&\cdots&0\\0&0&\cdots&0&0&\cdots&0\\\vdots&\vdots&&\vdots&\vdots&&\vdots\\0&0&\cdots&0&0&\cdots&0\end{pmatrix}$$

称为 $A_{m\times n}$ 的**标准形**.

which is called the **canonical form** of $A_{m\times n}$.

定理 2 任一矩阵可经有限次初等行变换化为行阶梯形矩阵.

Theorem 2 Any matrix can be reduced to a row echelon matrix by a finite sequence of elementary row operations.

证 令 $A=(a_{ij})_{m\times n}$.

Proof Let $A=(a_{ij})_{m\times n}$.

若 A 中的元素 a_{ij} 都等于零, 那么 A 是行阶梯形矩阵.

If the elements a_{ij} in A are all zeros, A then is a row echelon matrix.

若 $\boldsymbol{A}$ 中至少有一元素 a_{ij} 不为零, 且不为零的元素在第 1 列时, 不妨设 $a_{11}\neq 0$, 否则对 $\boldsymbol{A}$ 施以第一种初等行变换总可将不为零的元素换到 a_{11} 的位置上; 然后用 $-\dfrac{a_{i1}}{a_{11}}$ 乘以第 1 行各元素加到第 i 行 $(i=2,3,\cdots,m)$ 的对应元素上, 得矩阵 $\boldsymbol{A}_1$, 即

If there exists at least an element a_{ij} in $\boldsymbol{A}$ is nonzero, and if the nonzero element is on the first column, not loss of generality, suppose that $a_{11}\neq 0$; otherwise, performing the elementary row operation of the first kind, one can interchange the nonzero element to the position of a_{11}. Then adding $\left(-\dfrac{a_{i1}}{a_{11}}\right)$ times the elements on the first row to the corresponding elements on the i-th row $(i=2,3,\cdots,m)$, we obtain a new matrix $\boldsymbol{A}_1$, namely,

$$\boldsymbol{A}\longrightarrow\boldsymbol{A}_1=\begin{pmatrix} a_{11} & a_{12} & a_{13} & \cdots & a_{1n}\\ 0 & a'_{22} & a'_{23} & \cdots & a'_{2n}\\ \vdots & \vdots & \vdots & & \vdots\\ 0 & a'_{m2} & a'_{m3} & \cdots & a'_{mn}\end{pmatrix}$$

类似地, 若 $\boldsymbol{A}_1$ 中除第 1 行外其余各行元素全为零, 那么 $\boldsymbol{A}_1$ 即为行阶梯形矩阵. 如若不然, 不妨设 $a'_{22}\neq 0$, 可仿照上面的方法将 $\boldsymbol{A}_1$ 的第 3 行至第 m 行的第 2 列元素化为零, 即

Similarly, if the elements on the other rows in $\boldsymbol{A}_1$ except for the first row are all zero, then $\boldsymbol{A}_1$ is a row echelon matrix; if not, suppose that $a'_{22}\neq 0$, we can reduce the elements on the second column from the third row to the m-th row to zeros by copying the above processes, i.e.,

$$\boldsymbol{A}_1\longrightarrow\begin{pmatrix} a_{11} & a_{12} & a_{13} & \cdots & a_{1n}\\ 0 & a'_{22} & a'_{23} & \cdots & a'_{2n}\\ 0 & 0 & a''_{33} & \cdots & a''_{3n}\\ \vdots & \vdots & \vdots & & \vdots\\ 0 & 0 & a''_{m3} & \cdots & a''_{mn}\end{pmatrix}$$

按上述规律及方法继续下去, 最后可将 $\boldsymbol{A}$ 化为行阶梯形矩阵. 如果 $\boldsymbol{A}$ 的第一列元素全为零, 那么依次考虑它的第 2 列等. ▲

Performing the above rules and methods continuously, we can finally reduce $\boldsymbol{A}$ to a row echelon matrix. If the elements on the first column of $\boldsymbol{A}$ are all zeros, we can in turn consider the second column, and so on. ▲

同理可以证明如下结论.

In a similar way, we can prove the following results.

推论 1 任一矩阵可经过有限次初等行变换化成行最简形.

Corollary 1 Any matrix can be reduced to a row-reduced form by a finite sequence of elementary row operations.

推论 2 任一可逆方阵可经过有限次初等行变换化成单位矩阵.

Corollary 2 Any invertible square matrix can be reduced to the unit matrix by a finite sequence of elementary row operations.

证 设 $\boldsymbol{A}$ 为可逆方阵, 则 $|\boldsymbol{A}| \neq 0$. 由于方阵 $\boldsymbol{B}$ 是 $\boldsymbol{A}$ 经过一次初等行变换得到, 根据行列式的性质可知 $|\boldsymbol{B}| \neq 0$. 所以, $\boldsymbol{A}$ 经过有限次初等行变换的标准形必为单位矩阵 $\boldsymbol{E}$. ▲

Proof Suppose that $\boldsymbol{A}$ is an invertible matrix, then $|\boldsymbol{A}| \neq 0$. Since $\boldsymbol{B}$ is obtained from $\boldsymbol{A}$ by an elementary row operation, from the properties of determinants we know that $|\boldsymbol{B}| \neq 0$. That is to say, the canonical form of $\boldsymbol{A}$ after a finite sequence of elementary row operations must be the unit matrix $\boldsymbol{E}$. ▲

例 1 用初等变换将下面的矩阵化为行阶梯形、行最简形矩阵和标准形.

Example 1 Reduce the following matrix to a row echelon form, a row-reduced form and a canonical form by elementary operations.

$$\boldsymbol{A} = \begin{pmatrix} 1 & 1 & 2 & 2 & 1 \\ 0 & 2 & 1 & 5 & -1 \\ 2 & 0 & 3 & -1 & 3 \\ 1 & 1 & 2 & 4 & -1 \end{pmatrix}$$

解 对 $\boldsymbol{A}$ 实施初等行变换, 得

Solution Performing elementary row operations on $\boldsymbol{A}$ yields

$$\boldsymbol{A} \xrightarrow[r_4 - r_1]{r_3 - 2r_1} \begin{pmatrix} 1 & 1 & 2 & 2 & 1 \\ 0 & 2 & 1 & 5 & -1 \\ 0 & -2 & -1 & -5 & 1 \\ 0 & 0 & 0 & 2 & -2 \end{pmatrix}$$

$$\xrightarrow{r_3+r_2}\begin{pmatrix}1&1&2&2&1\\0&2&1&5&-1\\0&0&0&0&0\\0&0&0&2&-2\end{pmatrix}\xrightarrow{r_3\leftrightarrow r_4}\begin{pmatrix}1&1&2&2&1\\0&2&1&5&-1\\0&0&0&2&-2\\0&0&0&0&0\end{pmatrix}=\boldsymbol{B}$$

$\boldsymbol{B}$ 为行阶梯形矩阵. 继续对 $\boldsymbol{B}$ 进行初行变换, 得

$\boldsymbol{B}$ is a row echelon matrix. Performing elementary row operations on $\boldsymbol{B}$ continuously, we have

$$\boldsymbol{B}=\begin{pmatrix}1&1&2&2&1\\0&2&1&5&-1\\0&0&0&2&-2\\0&0&0&0&0\end{pmatrix}\xrightarrow{\frac{1}{2}r_3}\begin{pmatrix}1&1&2&2&1\\0&2&1&5&-1\\0&0&0&1&-1\\0&0&0&0&0\end{pmatrix}$$

$$\xrightarrow[r_2-5r_3]{r_1-2r_3}\begin{pmatrix}1&1&2&0&3\\0&2&1&0&4\\0&0&0&1&-1\\0&0&0&0&0\end{pmatrix}\xrightarrow[\frac{1}{2}r_2]{r_1-\frac{1}{2}r_2}\begin{pmatrix}1&0&\frac{3}{2}&0&1\\0&1&\frac{1}{2}&0&2\\0&0&0&1&-1\\0&0&0&0&0\end{pmatrix}=\boldsymbol{C}$$

$\boldsymbol{C}$ 为行最简形矩阵. 继续对 $\boldsymbol{C}$ 进行初列变换, 得

$\boldsymbol{C}$ is a row-reduced matrix. Performing the elementary column operations on $\boldsymbol{C}$ continuously, we have

$$\boldsymbol{C}=\begin{pmatrix}1&0&\frac{3}{2}&0&1\\0&1&\frac{1}{2}&0&2\\0&0&0&1&-1\\0&0&0&0&0\end{pmatrix}\xrightarrow{c_3\leftrightarrow c_4}\begin{pmatrix}1&0&0&\frac{3}{2}&1\\0&1&0&\frac{1}{2}&2\\0&0&1&0&-1\\0&0&0&0&0\end{pmatrix}$$

$$\xrightarrow[c_4-\frac{1}{2}c_2,c_5-2c_2,c_5+c_3]{c_4-\frac{3}{2}c_1,c_5-c_1}\begin{pmatrix}1&0&0&0&0\\0&1&0&0&0\\0&0&1&0&0\\0&0&0&0&0\end{pmatrix}$$

显然, $\boldsymbol{D}$ 为标准形矩阵. ■

Obviously, $\boldsymbol{D}$ is a canonical matrix. ■

3.1.2 初等矩阵

定义 2 由单位矩阵 $\boldsymbol{E}$ 经过一次初等变换得到的矩阵称为**初等矩阵**.

事实上, 矩阵的三种初等变换对应于三种初等矩阵.

(1) 互换单位矩阵 $\boldsymbol{E}$ 的第 i 行与第 j 行 (或第 i 列与第 j 列) 可以得到**第一类初等矩阵**, 即

3.1.2 Elementary Matrices

Definition 2 The matrix is called an **elementary matrix** if it is obtained by performing an elementary operation on the unit matrix $\boldsymbol{E}$.

In fact, the three kinds of elementary operations on matrices correspond to three kinds of elementary matrices.

(1) The **elementary matrix of the first kind** is obtained by interchanging the i-th row and the j-th row (or the i-th column and the j-th column) of the unit matrix $\boldsymbol{E}$, namely,

$$\boldsymbol{E}(i,j)=\begin{pmatrix} 1 & & & & & & & & \\ & \ddots & & & & & & & \\ & & 1 & & & & & & \\ & & & 0 & \cdots & 1 & & & \\ & & & & \ddots & & & & \\ & & & \vdots & 1 & \vdots & & & \\ & & & & & \ddots & & & \\ & & & 1 & \cdots & 0 & & & \\ & & & & & & 1 & & \\ & & & & & & & \ddots & \\ & & & & & & & & 1 \end{pmatrix}\begin{matrix} \\ \\ \\ 第i行 \\ \\ \\ \\ 第j行 \\ \\ \\ \\ \end{matrix}$$

(2) 将单位矩阵 $\boldsymbol{E}$ 的第 i 行 (或列) 乘以非零常数 k 可以得到**第二类初等矩阵**, 即

(2) The **elementary matrix of the second kind** is obtained by multiplying the i-th row (or column) of the unit matrix $\boldsymbol{E}$ by a nonzero constant k, namely,

$$E(i(k)) = \begin{pmatrix} 1 & & & & & & \\ & \ddots & & & & & \\ & & 1 & & & & \\ & & & k & & & \\ & & & & 1 & & \\ & & & & & \ddots & \\ & & & & & & 1 \end{pmatrix} \text{第}i\text{行}$$

(3) 将单位矩阵 $\boldsymbol{E}$ 的第 j 行 (或第 i 列) 乘以常数 k 加到第 i 行 (或第 j 列) 的对应元素上, 可以得到**第三类初等矩阵**, 即

(3) The **elementary matrix of the third kind** is obtained by adding a nonzero constant k times the j-th row (or the i-th column) to the i-th row (or the j-th column) of the unit matrix $\boldsymbol{E}$, namely,

$$E(ij(k)) = \begin{pmatrix} 1 & & & & & & \\ & \ddots & & & & & \\ & & 1 & \cdots & k & & \\ & & & \ddots & \vdots & & \\ & & & & 1 & & \\ & & & & & \ddots & \\ & & & & & & 1 \end{pmatrix} \begin{matrix} \text{第}i\text{行} \\ \\ \text{第}j\text{行} \end{matrix}$$

初等变换与初等矩阵建立起对应关系后, 可以得到如下关于初等矩阵的结论.

Establishing the relation between an elementary operation and an elementary matrix, we can obtain the following results on elementary matrices.

定理 3 设 $\boldsymbol{A}$ 是一个 $m \times n$ 矩阵. 对 $\boldsymbol{A}$ 施以一次初等行变换, 相当于在 $\boldsymbol{A}$ 的左边乘以相应的 m 阶初等矩阵; 对 $\boldsymbol{A}$ 施以一次初等列变换, 相当于在 $\boldsymbol{A}$ 的右边乘以相应的 n 阶初等矩阵. 简称为**左乘变行, 右乘变列**.

Theorem 3 Let $\boldsymbol{A}$ be an $m \times n$ matrix. Performing an elementary row operation on $\boldsymbol{A}$ is equivalent to multiplying $\boldsymbol{A}$ on the left by the corresponding elementary matrix of order m; interestingly, performing an elementary column operation on $\boldsymbol{A}$ is equivalent to multiplying $\boldsymbol{A}$ on the right by the corresponding elementary matrix of order n. For short, **premultiplying once changes a row** and **postmultiplying once changes a column**.

注 对 $\boldsymbol{A}$ 施以一次初等行变换得到的矩阵 $\boldsymbol{B}$ 和 $\boldsymbol{A}$ 是等价的; 而在 $\boldsymbol{A}$ 的左边乘以相应的 m 阶初等矩阵等于矩阵 $\boldsymbol{B}$.

Note The matrix $\boldsymbol{B}$ obtained by performing an elementary row operation on $\boldsymbol{A}$ is equivalent to $\boldsymbol{A}$; moreover, the matrix $\boldsymbol{B}$ is equal to multiplying $\boldsymbol{A}$ on the left by the corresponding elementary matrix of order m.

例如, 对矩阵 $\boldsymbol{A}$ 施以一次初等行变换

For example, performing an elementary row operation on the matrix $\boldsymbol{A}$ yields

$$\boldsymbol{A}=\begin{pmatrix} a_{11} & a_{12} & a_{13} & a_{14} \\ a_{21} & a_{22} & a_{23} & a_{24} \\ a_{31} & a_{32} & a_{33} & a_{34} \end{pmatrix} \xrightarrow{r_1\leftrightarrow r_2} \begin{pmatrix} a_{21} & a_{22} & a_{23} & a_{24} \\ a_{11} & a_{12} & a_{13} & a_{14} \\ a_{31} & a_{32} & a_{33} & a_{34} \end{pmatrix}$$

相应地

correspondingly,

$$\boldsymbol{E}(1,2)\boldsymbol{A}=\begin{pmatrix} 0 & 1 & 0 \\ 1 & 0 & 0 \\ 0 & 0 & 1 \end{pmatrix}\begin{pmatrix} a_{11} & a_{12} & a_{13} & a_{14} \\ a_{21} & a_{22} & a_{23} & a_{24} \\ a_{31} & a_{32} & a_{33} & a_{34} \end{pmatrix}=\begin{pmatrix} a_{21} & a_{22} & a_{23} & a_{24} \\ a_{11} & a_{12} & a_{13} & a_{14} \\ a_{31} & a_{32} & a_{33} & a_{34} \end{pmatrix}$$

对 $\boldsymbol{A}$ 施以一次初等列变换

Performing an elementary column operation on $\boldsymbol{A}$ yields

$$\boldsymbol{A}=\begin{pmatrix} a_{11} & a_{12} & a_{13} & a_{14} \\ a_{21} & a_{22} & a_{23} & a_{24} \\ a_{31} & a_{32} & a_{33} & a_{34} \end{pmatrix} \xrightarrow{c_2+kc_3} \begin{pmatrix} a_{11} & a_{12}+ka_{13} & a_{13} & a_{14} \\ a_{21} & a_{22}+ka_{23} & a_{23} & a_{24} \\ a_{31} & a_{32}+ka_{33} & a_{33} & a_{34} \end{pmatrix}$$

相应地

correspondingly,

$$\boldsymbol{A}\boldsymbol{E}(32(k))=\begin{pmatrix} a_{11} & a_{12} & a_{13} & a_{14} \\ a_{21} & a_{22} & a_{23} & a_{24} \\ a_{31} & a_{32} & a_{33} & a_{34} \end{pmatrix}\begin{pmatrix} 1 & 0 & 0 & 0 \\ 0 & 1 & 0 & 0 \\ 0 & k & 1 & 0 \\ 0 & 0 & 0 & 1 \end{pmatrix}$$

$$=\begin{pmatrix} a_{11} & a_{12}+ka_{13} & a_{13} & a_{14} \\ a_{21} & a_{22}+ka_{23} & a_{23} & a_{24} \\ a_{31} & a_{32}+ka_{33} & a_{33} & a_{34} \end{pmatrix}$$

因此根据定理 2, 可以将矩阵 $\boldsymbol{A}$ 的等价关系用矩阵的乘法表示出来.

Therefore, from Theorem 2 we know that the equivalence relation of a matrix $\boldsymbol{A}$ can be represented by matrix multiplication.

推论 1 两个 $m \times n$ 矩阵 $\boldsymbol{A}$ 与 $\boldsymbol{B}$ 等价的充要条件是存在 m 阶初等矩阵 $\boldsymbol{P}_1, \boldsymbol{P}_2, \cdots, \boldsymbol{P}_l$ 及 n 阶初等矩阵 $\boldsymbol{Q}_1, \boldsymbol{Q}_2, \cdots, \boldsymbol{Q}_t$, 使得

Corollary 1 Two $m \times n$ matrices $\boldsymbol{A}$ and $\boldsymbol{B}$ are equivalent if and only if there exist elementary matrices of order m, given by $\boldsymbol{P}_1, \boldsymbol{P}_2, \cdots, \boldsymbol{P}_l$, and elementary matrices of order n, given by $\boldsymbol{Q}_1, \boldsymbol{Q}_2, \cdots, \boldsymbol{Q}_t$, such that

$$\boldsymbol{P}_l \boldsymbol{P}_{l-1} \cdots \boldsymbol{P}_1 \boldsymbol{A} \boldsymbol{Q}_1 \boldsymbol{Q}_2 \cdots \boldsymbol{Q}_t = \boldsymbol{B}$$

对于初等矩阵 $\boldsymbol{E}(i,j), \boldsymbol{E}(i(k)), \boldsymbol{E}(ij(k))$, 由于其行列式

For the elementary matrices $\boldsymbol{E}(i,j)$, $\boldsymbol{E}(i(k))$, $\boldsymbol{E}(ij(k))$, since their determinants, given by

$$|\boldsymbol{E}(i,j)| = -1, \quad |\boldsymbol{E}(i(k))| = k, \quad |\boldsymbol{E}(ij(k))| = 1$$

均不为零, 故可逆. 此外, 容易验证

are all nonzero, this means that they are invertible. Moreover, it is easy to show that

$$(1)\ (\boldsymbol{E}(i,j))^{-1} = \boldsymbol{E}(i,j), (\boldsymbol{E}(i(k)))^{-1} = \boldsymbol{E}\left(i\left(\frac{1}{k}\right)\right), (\boldsymbol{E}(ij(k)))^{-1} = \boldsymbol{E}(ij(-k)).$$

(2) 若 $|\boldsymbol{A}| = a$, 则

(2) If $|\boldsymbol{A}| = a$, then

$$|\boldsymbol{E}(i,j)\boldsymbol{A}| = -a, |\boldsymbol{E}(i(k))\boldsymbol{A}| = ka, |\boldsymbol{E}(ij(k))\boldsymbol{A}| = a.$$

定理 4 初等矩阵均可逆, 而且初等矩阵的逆矩阵仍为同类型的初等矩阵.

Theorem 4 Elementary matrices are all invertible and their inverse matrices are all elementary matrices of the same types.

定理 5 若 $|\boldsymbol{A}| \neq 0$, 则与 $\boldsymbol{A}$ 等价的 $\boldsymbol{B}$ 的行列式不为零, 即 $|\boldsymbol{B}| \neq 0$.

Theorem 5 If $|\boldsymbol{A}| \neq 0$, then the determinant of $\boldsymbol{B}$ equivalent to $\boldsymbol{A}$ is nonzero, i.e., $|\boldsymbol{B}| \neq 0$.

例 2 对于给定的矩阵

Example 2 For the given matrix

$$\boldsymbol{A} = \begin{pmatrix} 3 & 0 & 1 \\ 1 & -1 & 2 \\ 0 & 1 & 1 \end{pmatrix}$$

令

let

$$\boldsymbol{E}(1,2) = \begin{pmatrix} 0 & 1 & 0 \\ 1 & 0 & 0 \\ 0 & 0 & 1 \end{pmatrix}, \quad \boldsymbol{E}(31(2)) = \begin{pmatrix} 1 & 0 & 0 \\ 0 & 1 & 0 \\ 2 & 0 & 1 \end{pmatrix}$$

则

then

$$\boldsymbol{E}(1,2)\boldsymbol{A} = \begin{pmatrix} 0 & 1 & 0 \\ 1 & 0 & 0 \\ 0 & 0 & 1 \end{pmatrix} \cdot \begin{pmatrix} 3 & 0 & 1 \\ 1 & -1 & 2 \\ 0 & 1 & 1 \end{pmatrix} = \begin{pmatrix} 1 & -1 & 2 \\ 3 & 0 & 1 \\ 0 & 1 & 1 \end{pmatrix}$$

上式表明, 用 $\boldsymbol{E}(1,2)$ 左乘 $\boldsymbol{A}$ 相当于交换矩阵 $\boldsymbol{A}$ 的第 1 行与第 2 行. 同理

the above equation shows that multiplying $\boldsymbol{A}$ on the left by $\boldsymbol{E}(1,2)$ is equivalent to interchanging the first and the second rows of $\boldsymbol{A}$. Similarly,

$$\boldsymbol{A}\boldsymbol{E}(31(2)) = \begin{pmatrix} 3 & 0 & 1 \\ 1 & -1 & 2 \\ 0 & 1 & 1 \end{pmatrix} \begin{pmatrix} 1 & 0 & 0 \\ 0 & 1 & 0 \\ 2 & 0 & 1 \end{pmatrix} = \begin{pmatrix} 5 & 0 & 1 \\ 5 & -1 & 2 \\ 2 & 1 & 1 \end{pmatrix}$$

上式表明, 用 $\boldsymbol{E}(31(2))$ 右乘 $\boldsymbol{A}$ 相当于将矩阵 $\boldsymbol{A}$ 的第 3 列乘 2 加到第 1 列. ■

the above equation shows that multiplying $\boldsymbol{A}$ on the right by $\boldsymbol{E}(31(2))$ is equivalent to adding 2 times the third column to the first column of $\boldsymbol{A}$. ■

3.1.3 用初等行变换求逆矩阵

3.1.3 Finding Inverse Matrices by Elementary Operations

虽然我们可以通过伴随矩阵求逆矩阵, 但对高阶矩阵来说, 计算量非常大. 下面介绍利用初等变换求逆矩阵的方法.

Although we can find inverse matrices by adjoint matrices, the computational cost will be very large for higher order matrices. In the following part we introduce a method for finding inverse matrices by elementary operations.

定理 6 方阵 $\boldsymbol{A}$ 可逆的充分必要条件是它能表示成有限个初等矩阵的乘积, 即

Theorem 6 A square matrix $\boldsymbol{A}$ is invertible if and only if it can be represented by a product of finite elementary matrices, namely,

$$\boldsymbol{A} = \boldsymbol{P}_1\boldsymbol{P}_2\cdots\boldsymbol{P}_s$$

其中 $\boldsymbol{P}_1, \boldsymbol{P}_2, \cdots, \boldsymbol{P}_s$ 均为初等矩阵.

where $\boldsymbol{P}_1, \boldsymbol{P}_2, \cdots, \boldsymbol{P}_s$ are all elementary matrices.

证 充分性 若

Proof Sufficiency If

$$\boldsymbol{A} = \boldsymbol{P}_1\boldsymbol{P}_2\cdots\boldsymbol{P}_s,$$

则

then

$$|\boldsymbol{A}| = |\boldsymbol{P}_1|\cdots|\boldsymbol{P}_s| \neq 0,$$

即 $\boldsymbol{A}$ 可逆.

it shows that $\boldsymbol{A}$ is invertible.

必要性 若 $\boldsymbol{A}$ 可逆, 则由定理 1 的推论 2 可知 $\boldsymbol{A}$ 一定可以经过有限次初等行变换变为单位矩阵 $\boldsymbol{E}$, 即存在初等矩阵 $\boldsymbol{Q}_1, \boldsymbol{Q}_2, \cdots, \boldsymbol{Q}_s$, 使

Necessity If $\boldsymbol{A}$ is invertible, then from Corollary 2 associated with Theorem 1 we know that $\boldsymbol{A}$ can be reduced to the unit matrix $\boldsymbol{E}$ by a finite sequence of elementary row operations, namely, there exist the elementary matrices $\boldsymbol{Q}_1, \boldsymbol{Q}_2, \cdots, \boldsymbol{Q}_s$ such that

$$\boldsymbol{Q}_s \cdots \boldsymbol{Q}_2 \boldsymbol{Q}_1 \boldsymbol{A} = \boldsymbol{E} \tag{3.7}$$

或

or

$$\boldsymbol{A} = \boldsymbol{Q}_1^{-1}\boldsymbol{Q}_2^{-1}\cdots\boldsymbol{Q}_s^{-1}\boldsymbol{E} = \boldsymbol{Q}_1^{-1}\boldsymbol{Q}_2^{-1}\cdots\boldsymbol{Q}_s^{-1}$$

令

Let

$$\boldsymbol{P}_1 = \boldsymbol{Q}_1^{-1}, \boldsymbol{P}_2 = \boldsymbol{Q}_2^{-1}, \cdots, \boldsymbol{P}_s = \boldsymbol{Q}_s^{-1},$$

则由定理 3 可知, $\boldsymbol{P}_1, \boldsymbol{P}_2, \cdots, \boldsymbol{P}_s$ 仍为初等矩阵, 故

From Theorem 3 we know that $\boldsymbol{P}_1, \boldsymbol{P}_2, \cdots, \boldsymbol{P}_s$ are still elementary matrices, so

$$\boldsymbol{A} = \boldsymbol{P}_1 \boldsymbol{P}_2 \cdots \boldsymbol{P}_s$$ ▲

由式 (3.7) 可知

From Equation (3.7) we have

$$\boldsymbol{Q}_s \cdots \boldsymbol{Q}_2 \boldsymbol{Q}_1 = \boldsymbol{A}^{-1}$$

又因为

Also from

$$\boldsymbol{Q}_s \cdots \boldsymbol{Q}_2 \boldsymbol{Q}_1 \boldsymbol{E} = \boldsymbol{Q}_s \cdots \boldsymbol{Q}_2 \boldsymbol{Q}_1$$

可知, 若矩阵 $\boldsymbol{A}$ 经过一系列初等行变换变为单位矩阵 $\boldsymbol{E}$, 则单位矩阵 $\boldsymbol{E}$ 经过同样的初等行变换变为 $\boldsymbol{A}^{-1}$, 其过程可表示为

we know that if $\boldsymbol{A}$ is transformed into the unit matrix $\boldsymbol{E}$ by a finite sequence of elementary row operations, then the unit matrix $\boldsymbol{E}$ can be reduced to $\boldsymbol{A}^{-1}$ by the same elementary row operations, process is represented by

$$(\boldsymbol{A} \,\vdots\, \boldsymbol{E}) \xrightarrow[\text{elementary row operations}]{\text{初等行变换}} (\boldsymbol{E} \,\vdots\, \boldsymbol{A}^{-1}) \tag{3.8}$$

其中 $(\boldsymbol{A} \mid \boldsymbol{E})$ 为 $n \times 2n$ 矩阵.

where $(\boldsymbol{A} \mid \boldsymbol{E})$ is an $n \times 2n$ matrix.

类似地, 若对 $2n \times n$ 矩阵 $\begin{pmatrix} \boldsymbol{A} \\ \boldsymbol{E} \end{pmatrix}$ 施行初等列变换, 有

Similarly, performing elementary column operations on the $2n \times n$ matrix $\begin{pmatrix} \boldsymbol{A} \\ \boldsymbol{E} \end{pmatrix}$, we have

$$\begin{pmatrix} \boldsymbol{A} \\ \boldsymbol{E} \end{pmatrix} \xrightarrow[\text{elementary column operations}]{\text{初等列变换}} \begin{pmatrix} \boldsymbol{E} \\ \boldsymbol{A}^{-1} \end{pmatrix} \tag{3.9}$$

例 3 求 $\boldsymbol{A}$ 的逆矩阵, 其中

Example 3 Find the inverse matrix of $\boldsymbol{A}$, where

$$\boldsymbol{A} = \begin{pmatrix} 2 & 2 & 3 \\ 1 & -1 & 0 \\ -1 & 2 & 1 \end{pmatrix}$$

解 对 $(\boldsymbol{A} \mid \boldsymbol{E})$ 实施初等行变换, 得

Solution Performing elementary row operations on $(\boldsymbol{A} \mid \boldsymbol{E})$ yields

$$(\boldsymbol{A} \mid \boldsymbol{E}) = \left(\begin{array}{ccc|ccc} 2 & 2 & 3 & 1 & 0 & 0 \\ 1 & -1 & 0 & 0 & 1 & 0 \\ -1 & 2 & 1 & 0 & 0 & 1 \end{array}\right) \xrightarrow[r_2 \leftrightarrow r_3]{r_1 \leftrightarrow r_2} \left(\begin{array}{ccc|ccc} 1 & -1 & 0 & 0 & 1 & 0 \\ -1 & 2 & 1 & 0 & 0 & 1 \\ 2 & 2 & 3 & 1 & 0 & 0 \end{array}\right)$$

$$\xrightarrow[r_3 - 2r_1]{r_2 + r_1} \left(\begin{array}{ccc|ccc} 1 & -1 & 0 & 0 & 1 & 0 \\ 0 & 1 & 1 & 0 & 1 & 1 \\ 0 & 4 & 3 & 1 & -2 & 0 \end{array}\right) \xrightarrow{r_3 - 4r_2} \left(\begin{array}{ccc|ccc} 1 & -1 & 0 & 0 & 1 & 0 \\ 0 & 1 & 1 & 0 & 1 & 1 \\ 0 & 0 & -1 & 1 & -6 & -4 \end{array}\right)$$

$$\xrightarrow{r_2 + r_3} \left(\begin{array}{ccc|ccc} 1 & -1 & 0 & 0 & 1 & 0 \\ 0 & 1 & 0 & 1 & -5 & -3 \\ 0 & 0 & -1 & 1 & -6 & -4 \end{array}\right) \xrightarrow[(-1)r_3]{r_1 + r_2} \left(\begin{array}{ccc|ccc} 1 & 0 & 0 & 1 & -4 & -3 \\ 0 & 1 & 0 & 1 & -5 & -3 \\ 0 & 0 & 1 & -1 & 6 & 4 \end{array}\right)$$

所以有

So we have

$$\boldsymbol{A}^{-1} = \begin{pmatrix} 1 & -4 & -3 \\ 1 & -5 & -3 \\ -1 & 6 & 4 \end{pmatrix}$$ ■

注 对矩阵 $(\boldsymbol{A} \mid \boldsymbol{E})$ 只能实施初等行变换, 且在进行初等行变换时, 必须将右边单位矩阵 $\boldsymbol{E}$ 所在的块同时进行.

Note For the matrix $(\boldsymbol{A} \mid \boldsymbol{E})$, we can only perform elementary row operations, and must perform the same operations on the unit matrix $\boldsymbol{E}$ located at the right block simultaneously.

以上用初等变换求逆矩阵的方法也可用于解某些特殊的矩阵方程.

The above method for finding inverse matrices by elementary operations can also be used to solve some special matrix equations.

设矩阵方程 $\boldsymbol{AX}=\boldsymbol{B}$. 若 $\boldsymbol{A}$ 可逆, 则 $\boldsymbol{X}=\boldsymbol{A}^{-1}\boldsymbol{B}$, 即

Suppose that the matrix equation is given by $\boldsymbol{AX}=\boldsymbol{B}$. If $\boldsymbol{A}$ is invertible, then $\boldsymbol{X}=\boldsymbol{A}^{-1}\boldsymbol{B}$, namely,

$$(\boldsymbol{A}\,\vdots\,\boldsymbol{B})\xrightarrow[\text{elementary row operations}]{\text{初等行变换}}(\boldsymbol{E}\,\vdots\,\boldsymbol{X})$$

例 4 解矩阵方程 $\boldsymbol{AX}=\boldsymbol{B}$, 其中

Example 4 Solve the matrix equation $\boldsymbol{AX}=\boldsymbol{B}$, where

$$\boldsymbol{A}=\begin{pmatrix}1&0&1\\-1&1&1\\2&-1&1\end{pmatrix},\quad \boldsymbol{B}=\begin{pmatrix}1&1\\0&1\\-1&0\end{pmatrix}$$

解 因为 $\boldsymbol{A}$ 可逆, 所以 $\boldsymbol{X}=\boldsymbol{A}^{-1}\boldsymbol{B}$. 对 $(\boldsymbol{A}\,\vdots\,\boldsymbol{B})$ 实施初等行变换, 可得

Solution Since $\boldsymbol{A}$ is invertible, we have $\boldsymbol{X}=\boldsymbol{A}^{-1}\boldsymbol{B}$. Performing elementary row operations on $(\boldsymbol{A}\,\vdots\,\boldsymbol{B})$ yields

$$(\boldsymbol{A}\,\vdots\,\boldsymbol{B})=\begin{pmatrix}1&0&1&1&1\\-1&1&1&0&1\\2&-1&1&-1&0\end{pmatrix}\xrightarrow[r_3-2r_1]{r_2+r_1}\begin{pmatrix}1&0&1&1&1\\0&1&2&1&2\\0&-1&-1&-3&-2\end{pmatrix}$$

$$\xrightarrow{r_3+r_2}\begin{pmatrix}1&0&1&1&1\\0&1&2&1&2\\0&0&1&-2&0\end{pmatrix}\xrightarrow[r_1-r_3]{r_2-2r_3}\begin{pmatrix}1&0&0&3&1\\0&1&0&5&2\\0&0&1&-2&0\end{pmatrix}=(\boldsymbol{E}\,\vdots\,\boldsymbol{X})$$

所以有

So we obtain

$$\boldsymbol{X}=\begin{pmatrix}3&1\\5&2\\-2&0\end{pmatrix}$$ ■

例 5* 将可逆矩阵

Example 5* Decompose the invertible matrix

$$\boldsymbol{A}=\begin{pmatrix}1&2&0\\-1&1&1\\3&-2&0\end{pmatrix}$$

分解为初等矩阵的乘积.

into a product of elementary matrices.

解 对 $\boldsymbol{A}$ 进行如下初等变换

Solution Performing the following elementary operations on $\boldsymbol{A}$ leads to

$$\boldsymbol{A}=\begin{pmatrix}1&2&0\\-1&1&1\\3&-2&0\end{pmatrix}\xrightarrow{c_2-2c_1}\begin{pmatrix}1&0&0\\-1&3&1\\3&-8&0\end{pmatrix}\xrightarrow{r_2+r_1}\begin{pmatrix}1&0&0\\0&3&1\\3&-8&0\end{pmatrix}$$

$$\xrightarrow{r_3-3r_1}\begin{pmatrix}1&0&0\\0&3&1\\0&-8&0\end{pmatrix}\xrightarrow{c_2\leftrightarrow c_3}\begin{pmatrix}1&0&0\\0&1&3\\0&0&-8\end{pmatrix}\xrightarrow{c_3-3c_2}\begin{pmatrix}1&0&0\\0&1&0\\0&0&-8\end{pmatrix}$$

$$\xrightarrow{\frac{-1}{8}r_3}\begin{pmatrix}1&0&0\\0&1&0\\0&0&1\end{pmatrix}$$

与初等变换对应的初等矩阵分别为

The elementary matrices corresponding to the elementary operations are respectively given by

$$\boldsymbol{Q}_1=\begin{pmatrix}1&-2&0\\0&1&0\\0&0&1\end{pmatrix},\quad \boldsymbol{P}_1=\begin{pmatrix}1&0&0\\1&1&0\\0&0&1\end{pmatrix},\quad \boldsymbol{P}_2=\begin{pmatrix}1&0&0\\0&1&0\\-3&0&1\end{pmatrix}$$

$$\boldsymbol{Q}_2=\begin{pmatrix}1&0&0\\0&0&1\\0&1&0\end{pmatrix},\quad \boldsymbol{Q}_3=\begin{pmatrix}1&0&0\\0&1&-3\\0&0&1\end{pmatrix},\quad \boldsymbol{P}_3=\begin{pmatrix}1&0&0\\0&1&0\\0&0&-\frac{1}{8}\end{pmatrix}$$

其中初等矩阵 $\boldsymbol{P}_i$ 和 $\boldsymbol{Q}_i$ 分别对应于初等行和列变换. 它们的逆矩阵分别为

where the elementary matrices $\boldsymbol{P}_i$ and $\boldsymbol{Q}_i$ correspond to the elementary row and column operations, respectively. Their inverse matrices are respectively given by

$$\boldsymbol{P}_1^{-1}=\begin{pmatrix}1&0&0\\-1&1&0\\0&0&1\end{pmatrix},\quad \boldsymbol{P}_2^{-1}=\begin{pmatrix}1&0&0\\0&1&0\\3&0&1\end{pmatrix},\quad \boldsymbol{P}_3^{-1}=\begin{pmatrix}1&0&0\\0&1&0\\0&0&-8\end{pmatrix},$$

$$\boldsymbol{Q}_1^{-1}=\begin{pmatrix}1&2&0\\0&1&0\\0&0&1\end{pmatrix},\quad \boldsymbol{Q}_2^{-1}=\begin{pmatrix}1&0&0\\0&0&1\\0&1&0\end{pmatrix},\quad \boldsymbol{Q}_3^{-1}=\begin{pmatrix}1&0&0\\0&1&3\\0&0&1\end{pmatrix}$$

由 $P_3P_2P_1AQ_1Q_2Q_3 = E$, 得到

Using $P_3P_2P_1AQ_1Q_2Q_3 = E$, we obtain

$$A = P_1^{-1}P_2^{-1}P_3^{-1}Q_3^{-1}Q_2^{-1}Q_1^{-1}$$
$$= \begin{pmatrix} 1 & 0 & 0 \\ -1 & 1 & 0 \\ 0 & 0 & 1 \end{pmatrix}\begin{pmatrix} 1 & 0 & 0 \\ 0 & 1 & 0 \\ 3 & 0 & 1 \end{pmatrix}\begin{pmatrix} 1 & 0 & 0 \\ 0 & 1 & 0 \\ 0 & 0 & -8 \end{pmatrix}\cdot$$
$$\begin{pmatrix} 1 & 0 & 0 \\ 0 & 1 & 3 \\ 0 & 0 & 1 \end{pmatrix}\begin{pmatrix} 1 & 0 & 0 \\ 0 & 0 & 1 \\ 0 & 1 & 0 \end{pmatrix}\begin{pmatrix} 1 & 2 & 0 \\ 0 & 1 & 0 \\ 0 & 0 & 1 \end{pmatrix}$$

■

例 6 设 A 为 n 阶可逆矩阵, 且 A 的第 i 行乘一个常数 $k\,(\neq 0)$ 后得到 B.

Example 6 Suppose that A is an invertible matrix of order n and that B is obtained by multiplying the i-th row of A by a constant $k\,(\neq 0)$.

(1) 证明 B 可逆, 并且 B^{-1} 是由 A^{-1} 的第 i 列乘一个常数 $\dfrac{1}{k}$ 后得到的矩阵;

(1) Prove that B is invertible and that B^{-1} is obtained by multiplying the i-th column of A^{-1} by a constant $\dfrac{1}{k}$;

(2) 求 AB^{-1} 及 BA^{-1}.

(2) Find AB^{-1} and BA^{-1}.

解 由已知条件可得 $E\,(i(k))\,A = B$.

Solution From the known condition we have $E\,(i(k))\,A = B$.

(1) 由于 $E\,(i(k))\,, A$ 均可逆, 故 $B = E(i(k))A$ 也可逆, 且有

(1) Since $E\,(i(k))\,, A$ are all invertible, we know that $B = E(i(k))A$, it is also invertible, and that

$$B^{-1} = A^{-1}E^{-1}\,(i(k)) = A^{-1}E\left(i\left(\frac{1}{k}\right)\right)$$

可见 B^{-1} 即为 A^{-1} 的第 i 列乘常数 $\dfrac{1}{k}$ 后得到的矩阵.

This shows that B^{-1} is the matrix obtained by multiplying the i-th column of A^{-1} by a constant $\dfrac{1}{k}$.

(2)
$$AB^{-1} = AA^{-1}E\left(i\left(\frac{1}{k}\right)\right) = E\left(i\left(\frac{1}{k}\right)\right)$$
$$BA^{-1} = \left(AB^{-1}\right)^{-1} = E^{-1}\left(i\left(\frac{1}{k}\right)\right) = E\,(i(k))$$

■

例 7 设

Example 7 Suppose that

$$\begin{pmatrix} 0 & 1 & 0 \\ 1 & 0 & 0 \\ 0 & 0 & 1 \end{pmatrix} \boldsymbol{A} \begin{pmatrix} 1 & 0 & 1 \\ 0 & 1 & 0 \\ 0 & 0 & 1 \end{pmatrix} = \begin{pmatrix} 1 & 2 & 3 \\ 4 & 5 & 6 \\ 7 & 8 & 9 \end{pmatrix}$$

求 $\boldsymbol{A}$.

Find $\boldsymbol{A}$.

解 注意到

Solution Note that

$$\begin{pmatrix} 0 & 1 & 0 \\ 1 & 0 & 0 \\ 0 & 0 & 1 \end{pmatrix}$$

是初等矩阵 $\boldsymbol{E}(1,2)$, 其逆矩阵就是其本身.

is the elementary matrix $\boldsymbol{E}(1,2)$, and the inverse matrix is itself. The matrix

$$\begin{pmatrix} 1 & 0 & 1 \\ 0 & 1 & 0 \\ 0 & 0 & 1 \end{pmatrix}$$

是初等矩阵 $\boldsymbol{E}(13(1))$, 其逆矩阵是

is the elementary matrix $\boldsymbol{E}(13(1))$ and its inverse matrix is

$$\boldsymbol{E}(13(-1)) = \begin{pmatrix} 1 & 0 & -1 \\ 0 & 1 & 0 \\ 0 & 0 & 1 \end{pmatrix}$$

所以

So

$$\begin{aligned} \boldsymbol{A} &= \begin{pmatrix} 0 & 1 & 0 \\ 1 & 0 & 0 \\ 0 & 0 & 1 \end{pmatrix}^{-1} \begin{pmatrix} 1 & 2 & 3 \\ 4 & 5 & 6 \\ 7 & 8 & 9 \end{pmatrix} \begin{pmatrix} 1 & 0 & 1 \\ 0 & 1 & 0 \\ 0 & 0 & 1 \end{pmatrix}^{-1} \\ &= \begin{pmatrix} 0 & 1 & 0 \\ 1 & 0 & 0 \\ 0 & 0 & 1 \end{pmatrix} \begin{pmatrix} 1 & 2 & 3 \\ 4 & 5 & 6 \\ 7 & 8 & 9 \end{pmatrix} \begin{pmatrix} 1 & 0 & -1 \\ 0 & 1 & 0 \\ 0 & 0 & 1 \end{pmatrix} \\ &= \begin{pmatrix} 4 & 5 & 6 \\ 1 & 2 & 3 \\ 7 & 8 & 9 \end{pmatrix} \begin{pmatrix} 1 & 0 & -1 \\ 0 & 1 & 0 \\ 0 & 0 & 1 \end{pmatrix} = \begin{pmatrix} 4 & 5 & 2 \\ 1 & 2 & 2 \\ 7 & 8 & 2 \end{pmatrix} \end{aligned}$$

■

思 考 题

设 $\begin{pmatrix} 1 & 2 \\ -1 & 3 \end{pmatrix} \xrightarrow{r_2+r_1, r_1+r_2} \boldsymbol{B}$, 试给出 $\boldsymbol{B}$. 注意初等变换与它们的变换顺序有关, 并且后面的变换是在前面变换完成的基础上进行的.

Question

Suppose that $\begin{pmatrix} 1 & 2 \\ -1 & 3 \end{pmatrix} \xrightarrow{r_2+r_1, r_1+r_2} \boldsymbol{B}$, try to give $\boldsymbol{B}$. It should note that elementary operations are connected with their operation sequences and that the later operation is based on the former operation.

3.2 矩阵的秩

3.2 Ranks of Matrices

3.2.1 矩阵秩的概念

3.2.1 Concept of Rank of a Matrix

矩阵的秩是线性代数中的重要概念之一, 在分析线性方程组解的结构和向量组的线性关系中起着重要的作用. 下面先介绍 k 阶子式的概念.

The rank of a matix is one of the important concepts in Linear Algebra, and it plays an important role during the course of analyzing the solution structures of a system of linear equations and the linear relation of a vector set. In the following part we first introduce the concept of k-th order subdeterminants.

定义 1 设 $\boldsymbol{A}$ 为 $m \times n$ 矩阵, 在 $\boldsymbol{A}$ 中任取 k 行与 k 列 ($k \leqslant \min\{m, n\}$), 取位于这些行、列交叉处的 k^2 个元素, 不改变它们在 $\boldsymbol{A}$ 中所处的位置, 构成的 k 阶行列式, 称为 $\boldsymbol{A}$ 的一个 **k 阶子式.**

Definition 1 Let $\boldsymbol{A}$ be an $m \times n$ matrix. Extracting k rows and k columns ($k \leqslant \min\{m, n\}$) in $\boldsymbol{A}$ arbitrarily, we obtain k^2 elements locating at the intersection points of the extracted rows and columns. The k-th order determinant composed of the k^2 elements in the original order of arrangement is called a **k-th order subdeterminant** of $\boldsymbol{A}$.

由排列组合性质知, $m \times n$ 矩阵的 k 阶子式共有 $\mathrm{C}_n^k \mathrm{C}_m^k$ 个.

From the properties of permutation and combination, we know that an $m \times n$ matrix has totally $\mathrm{C}_n^k \mathrm{C}_m^k$ subdeterminants of order k.

例如, 由

For example, from

$$\mathrm{C}_2^1\mathrm{C}_3^1 = 6, \mathrm{C}_2^2\mathrm{C}_3^2 = 3$$

可知矩阵

we know that the matrix

$$\boldsymbol{A} = \begin{pmatrix} 1 & 2 & 3 \\ 0 & 1 & 2 \end{pmatrix}$$

有 6 个一阶子式, 有 3 个二阶子式.

has 6 subdeterminants of order one and has 3 subdeterminants of order two.

定义 2 矩阵 $\boldsymbol{A}$ 中不为零的子式的最高阶数称为 **$\boldsymbol{A}$ 的秩**, 记为 $R(\boldsymbol{A})$.

Definition 2 The highest order of the nonzero subdeterminant of a matrix $\boldsymbol{A}$ is called the **rank** of $\boldsymbol{A}$, written as $R(\boldsymbol{A})$.

设 $\boldsymbol{A}$ 为 n 阶方阵, 若 $|\boldsymbol{A}| \neq 0(\Leftrightarrow \boldsymbol{A}$ 可逆), 则 $R(\boldsymbol{A}) = n$, 称 $\boldsymbol{A}$ 为**满秩矩阵**; 若 $|\boldsymbol{A}| = 0$ ($\Leftrightarrow \boldsymbol{A}$ 不可逆), 则 $R(\boldsymbol{A}) < n$, 称 $\boldsymbol{A}$ 为**降秩矩阵**.

Let $\boldsymbol{A}$ be a square matrix of order n, if $|\boldsymbol{A}| \neq 0(\Leftrightarrow \boldsymbol{A}$ is invertible), then $R(\boldsymbol{A}) = n$, and $\boldsymbol{A}$ is called a **nonsingular matrix**; while if $|\boldsymbol{A}| = 0(\Leftrightarrow \boldsymbol{A}$ is not invertible), then $R(\boldsymbol{A}) < n$, and $\boldsymbol{A}$ is called a **singular matrix**.

注 任意阶零矩阵的秩均为零.

Note The ranks of zero matrices of any order are all zero.

由矩阵的秩的定义可以得到如下结论:

From the definition of ranks of matrices we have the following results:

(1) 若 $\boldsymbol{A}$ 为 $m \times n$ 矩阵, 则 $R(\boldsymbol{A}) \leqslant \min\{m, n\}$, 即 $\boldsymbol{A}$ 的秩既不超过其行数, 又不超过其列数.

(1) If $\boldsymbol{A}$ is an $m \times n$ matrix, then $R(\boldsymbol{A}) \leqslant \min\{m, n\}$, namely, the rank of $\boldsymbol{A}$ does not exceed either its row number or its column number.

(2) 若 $\boldsymbol{A}$ 有一个 r 阶子式不等于零, 则 $R(\boldsymbol{A}) \geqslant r$; 若 $\boldsymbol{A}$ 的所有 $r+1$ 阶子式都为零, 则 $R(\boldsymbol{A}) \leqslant r$. 因此若这两个条件同时满足, 则 $R(\boldsymbol{A}) = r$.

(2) If $\boldsymbol{A}$ has a nonzero r-th order subdeterminant, then $R(\boldsymbol{A}) \geqslant r$; while if the $(r+1)$-th order subdeterminants of $\boldsymbol{A}$ are all zeros, then $R(\boldsymbol{A}) \leqslant r$. Therefore, if the two conditions are satisfied simultaneously, then $R(\boldsymbol{A}) = r$.

(3) $R(\boldsymbol{A}) = R(\boldsymbol{A}^{\mathrm{T}})$.

例 1 求下列矩阵的秩:

Example 1 Find the ranks of the following matrices:

$$(1)\ \boldsymbol{A}=\begin{pmatrix}1&3&-9&3\\1&4&-12&7\\-1&0&0&9\end{pmatrix};\quad (2)\ \boldsymbol{B}=\begin{pmatrix}1&2&1&3&4\\0&2&5&1&2\\0&0&0&1&1\\0&0&0&0&0\end{pmatrix}.$$

解 (1) $\boldsymbol{A}$ 的最高阶子式为三阶, 且有 4 个 3 阶子式,

Solution (1) The highest order of the subdeterminants of $\boldsymbol{A}$ is three, and $\boldsymbol{A}$ has 4 subdeterminants of order three, as follows,

$$\begin{vmatrix}1&3&-9\\1&4&-12\\-1&0&0\end{vmatrix}=0,\quad \begin{vmatrix}1&3&3\\1&4&7\\-1&0&9\end{vmatrix}=0,\quad \begin{vmatrix}1&-9&3\\1&-12&7\\-1&0&9\end{vmatrix}=0,\quad \begin{vmatrix}3&-9&3\\4&-12&7\\0&0&9\end{vmatrix}=0$$

易见, $\boldsymbol{A}$ 的所有三阶子式均为零, 而 $\boldsymbol{A}$ 有一个 2 阶子式 $\begin{vmatrix}1&3\\1&4\end{vmatrix}\neq 0$, 故 $R(\boldsymbol{A})=2$.

Obviously, the subdeterminants of order three of $\boldsymbol{A}$ are all equal to zero, but $\boldsymbol{A}$ has a nonzero second order subdeterminant, i.e., $\begin{vmatrix}1&3\\1&4\end{vmatrix}\neq 0$, it means that $R(\boldsymbol{A})=2$.

(2) 容易看出 $\boldsymbol{B}$ 的所有四阶子式均为零, 但存在一个不为零的三阶子式, 即

(2) It is easy to see that the forth order subdeterminants of $\boldsymbol{B}$ are all zeros, but $\boldsymbol{B}$ has a nonzero third order subdeterminant, namely,

$$\begin{vmatrix}1&2&3\\0&2&1\\0&0&1\end{vmatrix}=2\neq 0$$

故 $R(\boldsymbol{B})=3$. ■

So $R(\boldsymbol{B})=3$. ■

由例 1 看出, 用定义 2 求阶数较高的矩阵的秩显得比较困难, 但阶梯形矩阵 $\boldsymbol{B}$ 的秩较容易求.

From Example 1 we see that using Definition 2 to find the rank of a matrix of higher order should be very difficult, by contrast, the rank of an echelon matrix should be easier to find.

3.2.2 用初等变换求矩阵的秩

由阶梯形矩阵的特点可知, 它的秩等于阶梯上非零行的个数, 即阶梯数.

问题 矩阵 $\boldsymbol{A}$ 的秩与 $\boldsymbol{A}$ 的阶行梯形矩阵 $\boldsymbol{B}$ 的秩是否相等?

定理 1 初等变换不改变矩阵的秩, **即初等变换是保秩变换**.

证 只需证明 $R(\boldsymbol{A}) = R(\boldsymbol{B})$, 其中矩阵 $\boldsymbol{B}$ 由 $\boldsymbol{A}$ 经过一次初等行变换得到.

设 $R(\boldsymbol{A}) = r$, 即矩阵 $\boldsymbol{A}$ 中有一个不为零的 r 阶子式, 记为 A_r. 根据行列式的性质和初等变换的定义, 在矩阵 $\boldsymbol{B}$ 中也能找到不为零的 r 阶子式 B_r(B_r 和 A_r 的关系只能满足三种情形之一: 即相等、反号、倍数), 从而 $r = R(\boldsymbol{A}) \leqslant R(\boldsymbol{B})$. 由于初等变换都是可逆的, 有 $R(\boldsymbol{A}) \geqslant R(\boldsymbol{B})$. 因此 $R(\boldsymbol{A}) = R(\boldsymbol{B})$. ▲

对例 1 中的矩阵 $\boldsymbol{A}$ 进行初等行变换, 得

3.2.2 Finding Ranks of Matrices by Elementary Operations

From the features of an echelon matrix we know that the rank is equal to the number of nonzero rows at the echelons, namely, the echelon number.

Problem Whether or not the rank of $\boldsymbol{A}$ is equal to the rank of the echelon matrix $\boldsymbol{B}$ associated with $\boldsymbol{A}$?

Theorem 1 Elementary operations do not change the rank of a matrix, namely, **elementary operations are rank-preserving operations**.

Proof It only needs to prove that $R(\boldsymbol{A}) = R(\boldsymbol{B})$, where the matrix $\boldsymbol{B}$ is obtained by performing an elementary row operation on $\boldsymbol{A}$.

Suppose that $R(\boldsymbol{A}) = r$, namely, there exists a nonzero r-th order subdeterminant of $\boldsymbol{A}$, written as A_r. From the properties of determinants and the definition of elementary operations, we know that there also exists a nonzero r-th order subdeterminant of $\boldsymbol{B}$, written as B_r (the relation between B_r and A_r only satisfies one of the three cases, namely, equal, equal with opposite sign, multiple), therefore, $r = R(\boldsymbol{A}) \leqslant R(\boldsymbol{B})$. Since elementary operations are all invertible, it implies that $R(\boldsymbol{A}) \geqslant R(\boldsymbol{B})$. Consequently, $R(\boldsymbol{A}) = R(\boldsymbol{B})$. ▲

Performing elementary row operations on the matrix $\boldsymbol{A}$ in Example 1 yields

$$\boldsymbol{A} = \begin{pmatrix} 1 & 3 & -9 & 3 \\ 1 & 4 & -12 & 7 \\ -1 & 0 & 0 & 9 \end{pmatrix} \xrightarrow{r_3 \leftrightarrow r_1} \begin{pmatrix} -1 & 0 & 0 & 9 \\ 1 & 4 & -12 & 7 \\ 1 & 3 & -9 & 3 \end{pmatrix}$$

$$\xrightarrow[r_3+r_1]{r_2+r_1}\begin{pmatrix}-1&0&0&9\\0&4&-12&16\\0&3&-9&12\end{pmatrix}\xrightarrow[\frac{1}{3}r_3]{\frac{1}{4}r_2}\begin{pmatrix}-1&0&0&9\\0&1&-3&4\\0&1&-3&4\end{pmatrix}$$

$$\xrightarrow{r_3-r_2}\begin{pmatrix}-1&0&0&9\\0&1&-3&4\\0&0&0&0\end{pmatrix}$$

因此, $R(\boldsymbol{A})=2$. 它的标准形为

Thus, $R(\boldsymbol{A})=2$. Its canonical form is

$$\begin{pmatrix}1&0&0&0\\0&1&0&0\\0&0&0&0\end{pmatrix}$$

例 2 设

Example 2 Suppose that

$$\boldsymbol{A}=\begin{pmatrix}1&-2&2&-1\\2&-4&8&0\\-2&4&-2&3\\3&-6&0&-6\end{pmatrix},\quad \boldsymbol{b}=\begin{pmatrix}1\\2\\3\\4\end{pmatrix}$$

求 $\boldsymbol{A}$ 及 $\overline{\boldsymbol{A}}=(\boldsymbol{A},\boldsymbol{b})$ 的秩.

Find the ranks of $\boldsymbol{A}$ and $\overline{\boldsymbol{A}}=(\boldsymbol{A},\boldsymbol{b})$.

解 对 $\overline{\boldsymbol{A}}$ 实施初等行变换, 得到如下行阶梯形矩阵

Solution Performing elementary row operations on $\overline{\boldsymbol{A}}$ yields the following row echelon matrix

$$\overline{\boldsymbol{A}}=\begin{pmatrix}1&-2&2&-1&1\\2&-4&8&0&2\\-2&4&-2&3&3\\3&-6&0&-6&4\end{pmatrix}\xrightarrow[r_4-3r_1]{\substack{r_2-2r_1\\r_3+2r_1}}\begin{pmatrix}1&-2&2&-1&1\\0&0&4&2&0\\0&0&2&1&5\\0&0&-6&-3&1\end{pmatrix}$$

$$\xrightarrow[r_4+3r_2]{\substack{(1/2)r_2\\r_3-r_2}}\begin{pmatrix}1&-2&2&-1&1\\0&0&2&1&0\\0&0&0&0&5\\0&0&0&0&1\end{pmatrix}\xrightarrow[r_4-r_3]{(1/5)r_3}\begin{pmatrix}1&-2&2&-1&1\\0&0&2&1&0\\0&0&0&0&1\\0&0&0&0&0\end{pmatrix}$$

所以 $R(\boldsymbol{A})=2, R(\overline{\boldsymbol{A}})=3$. ■

So $R(\boldsymbol{A})=2, R(\overline{\boldsymbol{A}})=3$. ■

例 3 设

Example 3 Suppose that

$$\boldsymbol{A}=\begin{pmatrix}3&2&0&5&0\\3&-2&3&6&-1\\2&0&1&5&-3\\1&6&-4&-1&4\end{pmatrix}$$

求 $\boldsymbol{A}$ 的秩.

Find the rank of $\boldsymbol{A}$.

解 对 $\boldsymbol{A}$ 作初等行变换, 得到如下行阶梯形矩阵

Solution Performing elementary row operations on $\boldsymbol{A}$ yields the following row echelon matrix

$$\boldsymbol{A} \xrightarrow{r_1 \leftrightarrow r_4} \begin{pmatrix} 1 & 6 & -4 & -1 & 4 \\ 3 & -2 & 3 & 6 & -1 \\ 2 & 0 & 1 & 5 & -3 \\ 3 & 2 & 0 & 5 & 0 \end{pmatrix} \xrightarrow{r_2 - r_4} \begin{pmatrix} 1 & 6 & -4 & -1 & 4 \\ 0 & -4 & 3 & 1 & -1 \\ 2 & 0 & 1 & 5 & -3 \\ 3 & 2 & 0 & 5 & 0 \end{pmatrix}$$

$$\xrightarrow[r_4 - 3r_1]{r_3 - 2r_1} \begin{pmatrix} 1 & 6 & -4 & -1 & 4 \\ 0 & -4 & 3 & 1 & -1 \\ 0 & -12 & 9 & 7 & -11 \\ 0 & -16 & 12 & 8 & -12 \end{pmatrix} \xrightarrow[r_4 - 4r_2]{r_3 - 3r_2} \begin{pmatrix} 1 & 6 & -4 & -1 & 4 \\ 0 & -4 & 3 & 1 & -1 \\ 0 & 0 & 0 & 4 & -8 \\ 0 & 0 & 0 & 4 & -8 \end{pmatrix}$$

$$\xrightarrow{r_4 - r_3} \begin{pmatrix} 1 & 6 & -4 & -1 & 4 \\ 0 & -4 & 3 & 1 & -1 \\ 0 & 0 & 0 & 4 & -8 \\ 0 & 0 & 0 & 0 & 0 \end{pmatrix}$$

易见, 行阶梯形矩阵有三个非零行. 因此, $R(\boldsymbol{A}) = 3$. ■

Obviously, the row echelon matrix has three nonzero rows. This means that $R(\boldsymbol{A}) = 3$.■

例 4 设 $\boldsymbol{A}$ 为 n 阶可逆矩阵, $\boldsymbol{B}$ 为 $n \times m$ 矩阵. 证明 $R(\boldsymbol{AB}) = R(\boldsymbol{B})$.

Example 4 Let $\boldsymbol{A}$ be an invertible matrix of order n and $\boldsymbol{B}$ be an $n \times m$ matrix. Prove that $R(\boldsymbol{AB}) = R(\boldsymbol{B})$.

证 因为 $\boldsymbol{A}$ 可逆, 它可以表示成若干初等矩阵之积, 即 $\boldsymbol{A} = \boldsymbol{P}_1\boldsymbol{P}_2\cdots\boldsymbol{P}_s$, 其中 $\boldsymbol{P}_i(i = 1, 2, \cdots, s)$ 皆为初等矩阵, $\boldsymbol{AB} = \boldsymbol{P}_1\boldsymbol{P}_2\cdots\boldsymbol{P}_s\boldsymbol{B}$, 即 $\boldsymbol{AB}$ 是 $\boldsymbol{B}$ 经 s 次初等行变换后得出的, 由定理 1 知

Proof Since $\boldsymbol{A}$ is invertible, it can be represented by some elementary matrices, namely, $\boldsymbol{A} = \boldsymbol{P}_1\boldsymbol{P}_2\cdots\boldsymbol{P}_s$, where $\boldsymbol{P}_i(i = 1, 2, \cdots, s)$ are all elementary matrices, this shows that $\boldsymbol{AB} = \boldsymbol{P}_1\boldsymbol{P}_2\cdots\boldsymbol{P}_s\boldsymbol{B}$, in which $\boldsymbol{AB}$ is obtained by performing elementary row operations s times on $\boldsymbol{B}$. From Theorem 1 we know that

$$R(\boldsymbol{AB}) = R(\boldsymbol{B})$$ ■

例 5 设

Example 5 Suppose that

$$\boldsymbol{A} = \begin{pmatrix} 1 & -1 & 1 & 2 \\ 3 & \lambda & -1 & 2 \\ 5 & 3 & \mu & 6 \end{pmatrix}$$

若 $R(\boldsymbol{A})=2$, 求 λ 与 μ 的值.

If $R(\boldsymbol{A})=2$, find the values of λ and μ.

解 对 $\boldsymbol{A}$ 作初等行变换, 得到如下行阶梯形矩阵

Solution Performing elementary row operations on $\boldsymbol{A}$ yields the following row echelon matrix

$$\boldsymbol{A}\xrightarrow[r_3-5r_1]{r_2-3r_1}\begin{pmatrix}1&-1&1&2\\0&\lambda+3&-4&-4\\0&8&\mu-5&-4\end{pmatrix}\xrightarrow{r_3-r_2}\begin{pmatrix}1&-1&1&2\\0&\lambda+3&-4&-4\\0&5-\lambda&\mu-1&0\end{pmatrix}$$
$$\xrightarrow[c_2\leftrightarrow c_4]{}\begin{pmatrix}1&2&1&-1\\0&-4&-4&\lambda+3\\0&0&\mu-1&5-\lambda\end{pmatrix}$$

因为 $R(\boldsymbol{A})=2$, 所以有

Since $R(\boldsymbol{A})=2$, we have

$$\begin{cases}5-\lambda=0\\ \mu-1=0\end{cases}$$

即

namely,

$$\begin{cases}\lambda=5\\ \mu=1\end{cases}$$ ■

例 6 设

Example 6 Suppose that

$$\boldsymbol{A}=\begin{pmatrix}1&-2&3k\\-1&2k&-3\\k&-2&3\end{pmatrix}$$

问 k 为何值, 可使

What values should k take on such that

(1) $R(\boldsymbol{A})=1$; (2) $R(\boldsymbol{A})=2$; (3) $R(\boldsymbol{A})=3$?

解 容易求得 $\boldsymbol{A}$ 的行阶梯形矩阵如下:

Solution It is easy to find that the echelon matrix of $\boldsymbol{A}$ is as follows:

$$\boldsymbol{A}=\begin{pmatrix}1&-2&3k\\-1&2k&-3\\k&-2&3\end{pmatrix}\to\begin{pmatrix}1&-1&k\\0&k-1&k-1\\0&0&-(k-1)(k+2)\end{pmatrix}$$

有

We have

(1) 当 $k=1$ 时, $R(\boldsymbol{A})=1$;

(2) 当 $k=-2$ 时, $R(\boldsymbol{A})=2$;

(3) 当 $k\neq 1$ 且 $k\neq -2$ 时, $R(\boldsymbol{A})=3$. ■

(1) If $k=1$, $R(\boldsymbol{A})=1$;

(2) if $k=-2$, $R(\boldsymbol{A})=2$;

(3) If $k\neq 1$ and $k\neq -2$, $R(\boldsymbol{A})=3$. ■

思 考 题

Questions

1. 判断下列结论是否正确:

(1) 若 $R(\boldsymbol{A})=r$, 则 $\boldsymbol{A}$ 中所有的 r 阶子式不为零;

(2) 若 $\boldsymbol{A}$ 中有 r 阶子式不等于 0, 则 $R(\boldsymbol{A})=r$;

(3) 两个同型矩阵等价的充分必要条件是它们有相同的秩;

(4) 方阵 $\boldsymbol{A}$ 可逆的充分必要条件是 $\boldsymbol{A}$ 与单位矩阵等价.

2. 一个非零矩阵的行最简形和行阶梯形有什么区别和联系?

1. Decide whether or not the following results are valid:

(1) If $R(\boldsymbol{A})=r$, then the r-th order subdeterminants of $\boldsymbol{A}$ are all nonzero;

(2) If there exists an r-th order subdeterminant of $\boldsymbol{A}$ is nonzero, then $R(\boldsymbol{A})=r$;

(3) Two matrices of the same size are equivalent if and only if they have the same rank;

(4) A square matrix $\boldsymbol{A}$ is invertible if and only if $\boldsymbol{A}$ is equivalent to the unit matrix.

2. What are the difference and relation between a row-reduced form and a row echelon form of a nonzero matrix?

3.3 线性方程组的解

3.3 Solutions of Linear Systems

问题 对于给定的线性方程组 $\boldsymbol{Ax}=\boldsymbol{b}$, 如何利用矩阵的秩判定 $\boldsymbol{Ax}=\boldsymbol{b}$ 的解?

Problem For the given linear system $\boldsymbol{Ax}=\boldsymbol{b}$, how to decide the solutions of $\boldsymbol{Ax}=\boldsymbol{b}$ by the ranks of matrices?

设含有 n 个未知量, m 个方程的线性方程组

Suppose that a linear system of m equations in n unknowns is given by

$$\begin{cases} a_{11}x_1 + a_{12}x_2 + \cdots + a_{1n}x_n = b_1 \\ a_{21}x_1 + a_{22}x_2 + \cdots + a_{2n}x_n = b_2 \\ \cdots\cdots \\ a_{m1}x_1 + a_{m2}x_2 + \cdots + a_{mn}x_n = b_m \end{cases} \tag{3.10}$$

令

Let

$$\boldsymbol{A} = \begin{pmatrix} a_{11} & a_{12} & \cdots & a_{1n} \\ a_{21} & a_{22} & \cdots & a_{2n} \\ \vdots & \vdots & & \vdots \\ a_{m1} & a_{m2} & \cdots & a_{mn} \end{pmatrix}, \quad \boldsymbol{x} = \begin{pmatrix} x_1 \\ x_2 \\ \vdots \\ x_n \end{pmatrix}, \quad \boldsymbol{b} = \begin{pmatrix} b_1 \\ b_2 \\ \vdots \\ b_m \end{pmatrix}$$

$$\overline{\boldsymbol{A}} = \begin{pmatrix} a_{11} & a_{12} & \cdots & a_{1n} & b_1 \\ a_{21} & a_{22} & \cdots & a_{2n} & b_2 \\ \vdots & \vdots & & \vdots & \vdots \\ a_{m1} & a_{m2} & \cdots & a_{mn} & b_m \end{pmatrix}$$

$\boldsymbol{A}$ 和 $\overline{\boldsymbol{A}}$ 分别称为方程组 (3.10) 的系数矩阵和增广矩阵, 则 (3.10) 写成如下的矩阵形式

$\boldsymbol{A}$ and $\overline{\boldsymbol{A}}$ are called the coefficient matrix and the augmented matrix associated with System (3.10), then (3.10) can be written as the following matrix form

$$\boldsymbol{A}_{m\times n}\boldsymbol{x} = \boldsymbol{b} \tag{3.11}$$

若令

If let

$$\boldsymbol{\alpha}_j = \begin{pmatrix} a_{1j} \\ a_{2j} \\ \vdots \\ a_{mj} \end{pmatrix}, \quad j = 1, 2, \cdots, n$$

则

then

$$\boldsymbol{A} = (\boldsymbol{\alpha}_1, \boldsymbol{\alpha}_2, \cdots, \boldsymbol{\alpha}_n), \quad \bar{\boldsymbol{A}} = (\boldsymbol{\alpha}_1, \boldsymbol{\alpha}_2, \cdots, \boldsymbol{\alpha}_n, \boldsymbol{b})$$

若 $b_1, b_2, \cdots, b_m$ 不全为零, 即 $\boldsymbol{b} \neq \boldsymbol{0}$, 则称方程组 $\boldsymbol{Ax} = \boldsymbol{b}$ 为**非齐次线性方程组**; 而称 $\boldsymbol{Ax} = \boldsymbol{0}$ 为**齐次线性方程组**.

If $b_1, b_2, \cdots, b_m$ are not all zeros, namely, $\boldsymbol{b} \neq 0$, then the system $\boldsymbol{Ax} = \boldsymbol{b}$ is called a **non-homogeneous linear system**; correspondingly, the system $\boldsymbol{Ax} = \boldsymbol{0}$ is called a **homogeneous linear system**.

对于线性方程组 (3.10), 核心问题是判定方程组是否有解? 如果有解, 有多少组解, 并且如何求解? 我们有如下定理.

For the linear system (3.10), the essential problem is to decide whether or not the system has solutions? If so, how many solutions does it have and how to find them? We have the following theorem.

定理 1 对于由 (3.10) 给定的线性方程组 $\boldsymbol{A}_{m\times n}\boldsymbol{x}=\boldsymbol{b}$,

Theorem 1 For the linear system $\boldsymbol{A}_{m\times n}\boldsymbol{x}=\boldsymbol{b}$ given by (3.10),

(1) 若 $R(\boldsymbol{A})\neq R(\overline{\boldsymbol{A}})$, (3.10) 无解;

(1) (3.10) has no solution if $R(\boldsymbol{A})\neq R(\overline{\boldsymbol{A}})$;

(2) 若 $R(\boldsymbol{A})=R(\overline{\boldsymbol{A}})=n$, (3.10) 有解, 且有唯一解;

(2) (3.10) has a unique solution if $R(\boldsymbol{A})=R(\overline{\boldsymbol{A}})=n$;

(3) 若 $R(\boldsymbol{A})=R(\overline{\boldsymbol{A}})=r<n$, (3.10) 有解, 且有无穷多个解.

(3) (3.10) has infinitely many solutions if $R(\boldsymbol{A})=R(\overline{\boldsymbol{A}})=r<n$.

证 为了叙述方便, 不妨假设 $\overline{\boldsymbol{A}}=(\boldsymbol{\alpha}_1,\boldsymbol{\alpha}_2,\cdots,\boldsymbol{\alpha}_n,\boldsymbol{b})$, 其行最简形为

Proof For stating conveniently, suppose that $\overline{\boldsymbol{A}}=(\boldsymbol{\alpha}_1,\boldsymbol{\alpha}_2,\cdots,\boldsymbol{\alpha}_n,\boldsymbol{b})$, the row-reduced form is given by

$$\begin{pmatrix} 1 & 0 & \cdots & 0 & b_{1,r+1} & \cdots & b_{1n} & d_1 \\ 0 & 1 & \cdots & 0 & b_{2,r+1} & \cdots & b_{2n} & d_2 \\ \vdots & \vdots & & \vdots & \vdots & & \vdots & \vdots \\ 0 & 0 & \cdots & 1 & b_{r,r+1} & \cdots & b_{rn} & d_r \\ 0 & 0 & \cdots & 0 & 0 & \cdots & 0 & d_{r+1} \\ 0 & 0 & \cdots & 0 & 0 & \cdots & 0 & 0 \\ \vdots & \vdots & & \vdots & \vdots & & \vdots & \vdots \\ 0 & 0 & \cdots & 0 & 0 & \cdots & 0 & 0 \end{pmatrix}$$

(1) 若 $R(\boldsymbol{A})\neq R(\overline{\boldsymbol{A}})$, 即 $d_{r+1}\neq 0$. 此时第 $r+1$ 行对应于矛盾方程 $0=d_{r+1}$, 故 (3.10) 无解.

(1) If $R(\boldsymbol{A})\neq R(\overline{\boldsymbol{A}})$, namely, $d_{r+1}\neq 0$, the equation $0=d_{r+1}$ on the $(r+1)$-th row is contradictory, so (3.10) has no solution.

(2) 当 $R(\boldsymbol{A})=R(\overline{\boldsymbol{A}})=n$ 时, $\overline{\boldsymbol{A}}$ 的行最简形为

(2) If $R(\boldsymbol{A})=R(\overline{\boldsymbol{A}})=n$, the row-reduced form of $\overline{\boldsymbol{A}}$ is

$$\begin{pmatrix} 1 & 0 & \cdots & 0 & d_1 \\ 0 & 1 & \cdots & 0 & d_2 \\ \vdots & \vdots & & \vdots & \vdots \\ 0 & 0 & \cdots & 1 & d_n \end{pmatrix}$$

此时, 方程组 (3.10) 有解, 且有唯一解

In this case, System (3.10) has a unique solution, namely,

$$\begin{cases} x_1 = d_1 \\ x_2 = d_2 \\ \cdots\cdots \\ x_n = d_n \end{cases}$$

或写为

or written as

$$\begin{pmatrix} x_1 \\ x_2 \\ \vdots \\ x_n \end{pmatrix} = \begin{pmatrix} d_1 \\ d_2 \\ \vdots \\ d_n \end{pmatrix}$$

(3) 当 $R(\boldsymbol{A}) = R(\overline{\boldsymbol{A}}) = r < n$, $\overline{\boldsymbol{A}}$ 的行最简形为

(3) If $R(\boldsymbol{A}) = R(\overline{\boldsymbol{A}}) = r < n$, the row-reduced form of $\overline{\boldsymbol{A}}$ is

$$\begin{pmatrix} 1 & 0 & \cdots & 0 & b_{1,r+1} & \cdots & b_{1n} & d_1 \\ 0 & 1 & \cdots & 0 & b_{2,r+1} & \cdots & b_{2n} & d_2 \\ \vdots & \vdots & & \vdots & \vdots & & \vdots & \vdots \\ 0 & 0 & \cdots & 1 & b_{r,r+1} & \cdots & b_{rn} & d_r \\ 0 & 0 & \cdots & 0 & 0 & \cdots & 0 & 0 \\ 0 & 0 & \cdots & 0 & 0 & \cdots & 0 & 0 \\ \vdots & \vdots & & \vdots & \vdots & & \vdots & \vdots \\ 0 & 0 & \cdots & 0 & 0 & \cdots & 0 & 0 \end{pmatrix}$$

与 $\boldsymbol{Ax} = \boldsymbol{b}$ 同解的线性方程组为

The linear system equivalent to $\boldsymbol{Ax} = \boldsymbol{b}$ is as follows

$$\begin{cases} x_1 \qquad\qquad + b_{1,r+1}x_{r+1} + \cdots + b_{1n}x_n = d_1 \\ \qquad x_2 \qquad + b_{2,r+1}x_{r+1} + \cdots + b_{2n}x_n = d_2 \\ \qquad\qquad \cdots\cdots \\ \qquad\qquad x_r + b_{r,r+1}x_{r+1} + \cdots + b_{rn}x_n = d_r \end{cases}$$

即

namely,

$$\begin{cases} x_1 = -b_{1,r+1}x_{r+1} - \cdots - b_{1n}x_n + d_1 \\ x_2 = -b_{2,r+1}x_{r+1} - \cdots - b_{2n}x_n + d_2 \\ \cdots\cdots \\ x_r = -b_{r,r+1}x_{r+1} - \cdots - b_{rn}x_n + d_r \end{cases} \tag{3.12}$$

显然, 方程组 (3.10) 与 (3.12) 同解.

Obviously, System (3.10) has the same solutions as (3.12).

在方程组 (3.12) 中, 任意给定 $x_{r+1},\cdots,x_n$ 一组值, 可唯一确定 $x_1,x_2,\cdots,x_r$ 的值, 从而得到 (3.10) 的一组解. $x_1,x_2,\cdots,x_r$ 称为**主变量**, $x_{r+1},\cdots,x_n$ 称为**自由未知量**, 可取任意实数.

In System (3.12), for any given values of $x_{r+1},\cdots,x_n$, we can uniquely decide the values of $x_1,x_2,\cdots,x_r$, therefore, we obtain a group of solutions of (3.10). $x_1,x_2,\cdots,x_r$ are called **principal variables** and $x_{r+1},\cdots,x_n$ are called **free unknowns** which can be arbitrary real numbers.

令 $x_{r+1}=k_1,\cdots,x_n=k_{n-r}$, 得到方程组 (3.12) 的一组解. 因此, 方程组 (3.10) 不仅有解, 且有无穷多个解

Let $x_{r+1}=k_1,\cdots,x_n=k_{n-r}$, then we obtain a group of solutions of System (3.12). Consequently, System (3.10) not only has solutions, but also has infinitely many solutions, as follows,

$$\begin{cases} x_1=-b_{1,r+1}k_1-\cdots-b_{1n}k_{n-r}+d_1 \\ \qquad\qquad\cdots\cdots \\ x_r=-b_{r,r+1}k_1-\cdots-b_{rn}k_{n-r}+d_r \\ x_{r+1}=\qquad k_1 \\ \qquad\qquad\cdots\cdots \\ x_n=\qquad\qquad\qquad k_{n-r} \end{cases}$$

或写为

or written as

$$\begin{pmatrix} x_1 \\ \vdots \\ x_r \\ x_{r+1} \\ \vdots \\ x_n \end{pmatrix}=\begin{pmatrix} -b_{1,r+1}k_1-\cdots-b_{1n}k_{n-r}+d_1 \\ \vdots \\ -b_{r,r+1}k_1-\cdots-b_{rn}k_{n-r}+d_r \\ k_1 \\ \vdots \\ k_{n-r} \end{pmatrix}$$

$$=k_1\begin{pmatrix} -b_{1,r+1} \\ \vdots \\ -b_{r,r+1} \\ 1 \\ \vdots \\ 0 \end{pmatrix}+\cdots+k_{n-r}\begin{pmatrix} -b_{1n} \\ \vdots \\ -b_{rn} \\ 0 \\ \vdots \\ 1 \end{pmatrix}+\begin{pmatrix} d_1 \\ \vdots \\ d_r \\ 0 \\ \vdots \\ 0 \end{pmatrix}$$

其中 $k_1, k_2, \cdots, k_{n-r}$ 为任意实数. ▲

where $k_1, k_2, \cdots, k_{n-r}$ are arbitrary real numbers. ▲

特别地, 有如下定理.

In particular, we have the following theorem.

定理 2 对于齐次线性方程组 $\boldsymbol{A}_{m\times n}\boldsymbol{x}=\boldsymbol{0}$,

(1) 若 $R(\boldsymbol{A})=\boldsymbol{n}$, 齐次线性方程组 $\boldsymbol{A}_{m\times n}\boldsymbol{x}=\boldsymbol{0}$ 有唯一零解;

(2) 若 $R(\boldsymbol{A})<n$, 齐次线性方程组 $\boldsymbol{A}_{m\times n}\boldsymbol{x}=\boldsymbol{0}$ 有 (无穷多个) 非零解.

Theorem 2 For the homogeneous linear system $\boldsymbol{A}_{m\times n}\boldsymbol{x}=\boldsymbol{0}$,

(1) $\boldsymbol{A}_{m\times n}\boldsymbol{x}=\boldsymbol{0}$ has a unique zero solution if $R(\boldsymbol{A})=\boldsymbol{n}$;

(2) $\boldsymbol{A}_{m\times n}\boldsymbol{x}=\boldsymbol{0}$ has (infinitely many) nonzero solutions if $R(\boldsymbol{A})<n$.

例 1 求解线性方程组

Example 1 Solve the linear system

$$\begin{cases} x_1-2x_2+3x_3-\ x_4=1\\ 3x_1-\ x_2+5x_3-3x_4=2\\ 2x_1+\ x_2+2x_3-2x_4=3 \end{cases}$$

解 对方程组的增广矩阵 $\overline{\boldsymbol{A}}$ 施以初等行变换

Solution Performing elementary row operations on the augmented matrix $\overline{\boldsymbol{A}}$ associated with the system, we have

$$\overline{\boldsymbol{A}}=\begin{pmatrix} 1 & -2 & 3 & -1 & 1\\ 3 & -1 & 5 & -3 & 2\\ 2 & 1 & 2 & -2 & 3 \end{pmatrix} \xrightarrow[r_3-2r_1]{r_2-3r_1} \begin{pmatrix} 1 & -2 & 3 & -1 & 1\\ 0 & 5 & -4 & 0 & -1\\ 0 & 5 & -4 & 0 & 1 \end{pmatrix}$$
$$\xrightarrow{r_3-r_2} \begin{pmatrix} 1 & -2 & 3 & -1 & 1\\ 0 & 5 & -4 & 0 & -1\\ 0 & 0 & 0 & 0 & 2 \end{pmatrix}$$

显然, $R(\boldsymbol{A})=2, R(\overline{\boldsymbol{A}})=3, R(\boldsymbol{A})\neq R(\overline{\boldsymbol{A}})$. 因此方程组无解. ■

Obviously, $R(\boldsymbol{A})=2, R(\overline{\boldsymbol{A}})=3, R(\boldsymbol{A})\neq R(\overline{\boldsymbol{A}})$. This means that the system has no solution. ■

例 2 求解线性方程组

Example 2 Solve the linear system

$$\begin{cases} x_1+\ x_2+2x_3=1\\ 2x_1-\ x_2+2x_3=4\\ x_1-2x_2\qquad\ =3\\ 4x_1+\ x_2+4x_3=2 \end{cases}$$

解 对方程组的增广矩阵 $\overline{\boldsymbol{A}}$ 施以初等行变换

Solution Performing elementary row operations on the augmented matrix $\overline{\boldsymbol{A}}$ associated with the system, we get

$$\overline{\boldsymbol{A}}=\begin{pmatrix}1&1&2&1\\2&-1&2&4\\1&-2&0&3\\4&1&4&2\end{pmatrix}\xrightarrow[\substack{r_3-r_1\\r_4-4r_1}]{r_2-2r_1}\begin{pmatrix}1&1&2&1\\0&-3&-2&2\\0&-3&-2&2\\0&-3&-4&-2\end{pmatrix}$$

$$\xrightarrow[r_4-r_2]{r_3-r_2}\begin{pmatrix}1&1&2&1\\0&-3&-2&2\\0&0&0&0\\0&0&-2&-4\end{pmatrix}\xrightarrow[-\frac{1}{2}r_3]{r_3\leftrightarrow r_4}\begin{pmatrix}1&1&2&1\\0&-3&-2&2\\0&0&1&2\\0&0&0&0\end{pmatrix}=\boldsymbol{B}$$

显然, $R(\boldsymbol{A})=R(\overline{\boldsymbol{A}})=3$, 因此方程组有唯一解. 为了更方便求方程组的解, 对 $\overline{\boldsymbol{A}}$ 进一步施以初等行变换, 并将其化为行最简形

Obviously, $R(\boldsymbol{A})=R(\overline{\boldsymbol{A}})=3$, it implies that the system has a unique solution. To solve the system conveniently, we perform elementary row operations on $\overline{\boldsymbol{A}}$, and then reduce it to a row-reduced form

$$\boldsymbol{B}\xrightarrow[r_1-2r_3]{r_2+2r_3}\begin{pmatrix}1&1&0&-3\\0&-3&0&6\\0&0&1&2\\0&0&0&0\end{pmatrix}\xrightarrow[-\frac{1}{3}r_2]{r_1+\frac{1}{3}r_2}\begin{pmatrix}1&0&0&-1\\0&1&0&-2\\0&0&1&2\\0&0&0&0\end{pmatrix}$$

与原方程组同解的方程组为

The system equivalent to the original system is

$$\begin{cases}x_1=-1\\x_2=-2\\x_3=2\end{cases}$$

这也是原方程组的唯一解

This is also the unique solution of the original system, given by

$$\begin{pmatrix}x_1\\x_2\\x_3\end{pmatrix}=\begin{pmatrix}-1\\-2\\2\end{pmatrix}$$ ■

例 3 求解线性方程组

Example 3 Solve the linear system

$$\begin{cases} x_1 - 2x_2 + 3x_3 - 4x_4 = 4 \\ \quad\quad x_2 - x_3 + x_4 = -3 \\ x_1 + 3x_2 \quad\quad - 3x_4 = 1 \\ \quad - 7x_2 + 3x_3 + x_4 = -3 \end{cases}$$

解 对方程组的增广矩阵 $\overline{A}$ 施以初等行变换

Solution Performing elementary row operations on the augmented matrix $\overline{A}$ associated with the system, we obtain

$$\overline{\boldsymbol{A}} = \begin{pmatrix} 1 & -2 & 3 & -4 & 4 \\ 0 & 1 & -1 & 1 & -3 \\ 1 & 3 & 0 & -3 & 1 \\ 0 & -7 & 3 & 1 & -3 \end{pmatrix} \xrightarrow{r_3 - r_1} \begin{pmatrix} 1 & -2 & 3 & -4 & 4 \\ 0 & 1 & -1 & 1 & -3 \\ 0 & 5 & -3 & 1 & -3 \\ 0 & -7 & 3 & 1 & -3 \end{pmatrix}$$

$$\xrightarrow[r_4 + 7r_2]{r_3 - 5r_2} \begin{pmatrix} 1 & -2 & 3 & -4 & 4 \\ 0 & 1 & -1 & 1 & -3 \\ 0 & 0 & 2 & -4 & 12 \\ 0 & 0 & -4 & 8 & -24 \end{pmatrix} \xrightarrow[\frac{1}{2}r_3]{r_4 + 2r_3} \begin{pmatrix} 1 & -2 & 3 & -4 & 4 \\ 0 & 1 & -1 & 1 & -3 \\ 0 & 0 & 1 & -2 & 6 \\ 0 & 0 & 0 & 0 & 0 \end{pmatrix}$$

$$\xrightarrow[r_1 - 3r_3]{r_2 + r_3} \begin{pmatrix} 1 & -2 & 0 & 2 & -14 \\ 0 & 1 & 0 & -1 & 3 \\ 0 & 0 & 1 & -2 & 6 \\ 0 & 0 & 0 & 0 & 0 \end{pmatrix} \xrightarrow{r_1 + 2r_2} \begin{pmatrix} 1 & 0 & 0 & 0 & -8 \\ 0 & 1 & 0 & -1 & 3 \\ 0 & 0 & 1 & -2 & 6 \\ 0 & 0 & 0 & 0 & 0 \end{pmatrix}$$

显然, $R(\boldsymbol{A}) = R(\overline{\boldsymbol{A}}) = 3 < 4$, 因此, 方程组有无穷多解. 与原方程组同解的方程组

Obviously, $R(\boldsymbol{A}) = R(\overline{\boldsymbol{A}}) = 3 < 4$, so the system has infinitely many solutions. The equivalent system to the original system is as follows

$$\begin{cases} x_1 = -8 \\ x_2 - x_4 = 3 \\ x_3 - 2x_4 = 6 \end{cases}$$

即

namely,

$$\begin{cases} x_1 = -8 \\ x_2 = 3 + x_4 \\ x_3 = 6 + 2x_4 \end{cases}$$

令 $x_4 = k$, 则

Let $x_4 = k$, then

$$\begin{cases} x_1 = -8 \\ x_2 = 3 + k \\ x_3 = 6 + 2k \\ x_4 = k \end{cases}$$

即

namely,

$$\begin{pmatrix} x_1 \\ x_2 \\ x_3 \\ x_4 \end{pmatrix} = k\begin{pmatrix} 0 \\ 1 \\ 2 \\ 1 \end{pmatrix} + \begin{pmatrix} -8 \\ 3 \\ 6 \\ 0 \end{pmatrix}$$

其中 k 为任意实数. ■

where k is an arbitrary real number. ■

例 4 求解线性方程组

Example 4 Solve the linear system

$$\begin{cases} x_1 + x_2 + x_3 + 4x_4 - 3x_5 = 0 \\ x_1 - x_2 + 3x_3 - 2x_4 - x_5 = 0 \\ 2x_1 + x_2 + 3x_3 + 5x_4 - 5x_5 = 0 \\ 3x_1 + x_2 + 5x_3 + 6x_4 - 7x_5 = 0 \end{cases}$$

解 因为 $m = 4, n = 5, m < n$, $R(\boldsymbol{A}) < n$, 所以所给方程组有无穷多个非零解. 对系数矩阵 $\boldsymbol{A}$ 施以初等行变换

Solution Since $m = 4, n = 5, m < n, R(\boldsymbol{A}) < n$, this implies that the system have infinitely many nonzero solutions. Performing elementary row operations on the coefficient matrix $\boldsymbol{A}$ yields

$$\boldsymbol{A} = \begin{pmatrix} 1 & 1 & 1 & 4 & -3 \\ 1 & -1 & 3 & -2 & -1 \\ 2 & 1 & 3 & 5 & -5 \\ 3 & 1 & 5 & 6 & -7 \end{pmatrix} \xrightarrow[r_3-2r_1, r_4-3r_1]{r_2-r_1} \begin{pmatrix} 1 & 1 & 1 & 4 & -3 \\ 0 & -2 & 2 & -6 & 2 \\ 0 & -1 & 1 & -3 & 1 \\ 0 & -2 & 2 & -6 & 2 \end{pmatrix}$$

$$\xrightarrow[r_2+2r_3, r_4+2r_3, r_1-r_3]{-r_3} \begin{pmatrix} 1 & 0 & 2 & 1 & -2 \\ 0 & 0 & 0 & 0 & 0 \\ 0 & 1 & -1 & 3 & -1 \\ 0 & 0 & 0 & 0 & 0 \end{pmatrix}$$

$R(\boldsymbol{A}) = 2 < 5$. 原方程组与下面方程组同解

$R(\boldsymbol{A}) = 2 < 5$. The original system and the following system are equivalent

$$\begin{cases} x_1 = -2x_3 - x_4 + 2x_5 \\ x_2 = x_3 - 3x_4 + x_5 \end{cases}$$

令 $x_3 = k_1, x_4 = k_2, x_5 = k_3$, 则

Let $x_3 = k_1, x_4 = k_2, x_5 = k_3$, then

$$\begin{cases} x_1 = -2k_1 - k_2 + 2k_3 \\ x_2 = k_1 - 3k_2 + k_3 \end{cases}$$

即

namely,

$$\begin{pmatrix} x_1 \\ x_2 \\ x_3 \\ x_4 \\ x_5 \end{pmatrix} = k_1 \begin{pmatrix} -2 \\ 1 \\ 1 \\ 0 \\ 0 \end{pmatrix} + k_2 \begin{pmatrix} -1 \\ -3 \\ 0 \\ 1 \\ 0 \end{pmatrix} + k_3 \begin{pmatrix} 2 \\ 1 \\ 0 \\ 0 \\ 1 \end{pmatrix}$$

其中 k_1, k_2, k_3 为任意实数. ■

where k_1, k_2, k_3 are arbitrary real numbers. ■

例 5* 当 c, d 取何值时, 线性方程组

Example 5* What values should c, d take on such that the linear system

$$\begin{cases} x_1 + x_2 + x_3 + x_4 + x_5 = 1 \\ 3x_1 + 2x_2 + x_3 + x_4 - 3x_5 = c \\ x_2 + 2x_3 + 2x_4 + 6x_5 = 3 \\ 5x_1 + 4x_2 + 3x_3 + 3x_4 - x_5 = d \end{cases}$$

有解? 若有解, 求它的解.

has solutions? If so, find them.

解 对方程组的增广矩阵 $\overline{\boldsymbol{A}}$ 施以初等行变换得

Solution Performing elementary row operations on the augmented matrix $\overline{\boldsymbol{A}}$ associated with the system we obtain

$$\overline{\boldsymbol{A}} = \begin{pmatrix} 1 & 1 & 1 & 1 & 1 & 1 \\ 3 & 2 & 1 & 1 & -3 & c \\ 0 & 1 & 2 & 2 & 6 & 3 \\ 5 & 4 & 3 & 3 & -1 & d \end{pmatrix} \to \begin{pmatrix} 1 & 1 & 1 & 1 & 1 & 1 \\ 0 & -1 & -2 & -2 & -6 & c-3 \\ 0 & 1 & 2 & 2 & 6 & 3 \\ 0 & -1 & -2 & -2 & -6 & d-5 \end{pmatrix}$$

$$\to \begin{pmatrix} 1 & 1 & 1 & 1 & 1 & 1 \\ 0 & 0 & 0 & 0 & 0 & c \\ 0 & 1 & 2 & 2 & 6 & 3 \\ 0 & 0 & 0 & 0 & 0 & d-2 \end{pmatrix} \to \begin{pmatrix} 1 & 0 & -1 & -1 & -5 & -2 \\ 0 & 1 & 2 & 2 & 6 & 3 \\ 0 & 0 & 0 & 0 & 0 & c \\ 0 & 0 & 0 & 0 & 0 & d-2 \end{pmatrix}$$

由此可得, 当 $c=0, d=2$ 时, $R(\boldsymbol{A})=R(\overline{\boldsymbol{A}})=2<5$ 方程组有无穷多组解, 其解为

This means that if $c=0, d=2$, $R(\boldsymbol{A})=R(\overline{\boldsymbol{A}})=2<5$, the system has infinitely many solutions and the solution is given by

$$\begin{cases} x_1=-2+x_3+x_4+5x_5 \\ x_2=3-2x_3-2x_4-6x_5 \end{cases}$$

令 $x_3=k_1$, $x_4=k_2$, $x_5=k_3$, 则

Let $x_3=k_1$, $x_4=k_2$, $x_5=k_3$, then

$$\begin{cases} x_1=-2+k_1+k_2+5k_3 \\ x_2=3-2k_1-2k_2-6k_3 \end{cases}$$

即

namely,

$$\begin{pmatrix} x_1 \\ x_2 \\ x_3 \\ x_4 \\ x_5 \end{pmatrix}=k_1\begin{pmatrix} 1 \\ -2 \\ 1 \\ 0 \\ 0 \end{pmatrix}+k_2\begin{pmatrix} 1 \\ -2 \\ 0 \\ 1 \\ 0 \end{pmatrix}+k_3\begin{pmatrix} 5 \\ -6 \\ 0 \\ 0 \\ 1 \end{pmatrix}+\begin{pmatrix} -2 \\ 3 \\ 0 \\ 0 \\ 0 \end{pmatrix}$$

其中 k_1, k_2, k_3 为任意实数. ■

where k_1, k_2, k_3 are arbitrary real numbers. ■

例 6 当 k 取何值时, 线性方程组

Example 6 What values should k take on such that the linear system

$$\begin{cases} x_1+\ \ x_2+kx_3=4 \\ -x_1+kx_2+\ \ x_3=k^2 \\ x_1-\ \ x_2+2x_3=-4 \end{cases}$$

(1) 有唯一解; (2) 无解; (3) 有无穷多个解? 在方程组有解的情形下, 求出解.

(1) has a unique solution; (2) has no solution; (3) has infinitely many solutions? In the case that the system has solutions, find the solutions.

解 方程组的系数行列式为

Solution The coefficient determinant associated with the system is

$$D=\begin{vmatrix} 1 & 1 & k \\ -1 & k & 1 \\ 1 & -1 & 2 \end{vmatrix}=-(k^2-3k-4)=-(k+1)(k-4)$$

(1) 当 $-(k+1)(k-4) \neq 0$, 即 $k \neq -1$ 且 $k \neq 4$ 时, 由克拉默法则方程组有唯一解.

(1) As $-(k+1)(k-4) \neq 0$, namely, $k \neq -1$ and $k \neq 4$, from Cramer's rule we know that this system has a unique solution.

(2) 当 $k=-1$ 时, 对方程组的增广矩阵作初等行变换得

(2) As $k=-1$, performing elementary row operations on the augmented matrix associated with the system yields

$$\overline{\boldsymbol{A}} = \begin{pmatrix} 1 & 1 & -1 & 4 \\ -1 & -1 & 1 & 1 \\ 1 & -1 & 2 & -4 \end{pmatrix} \to \begin{pmatrix} 1 & 1 & -1 & 4 \\ 0 & -2 & 3 & -8 \\ 0 & 0 & 0 & 5 \end{pmatrix}$$

显然, $R(\overline{\boldsymbol{A}})=3, R(\boldsymbol{A})=2$, 因此方程组无解.

Obviously, $R(\boldsymbol{A})=3, R(\boldsymbol{A})=2$, so the system has no solution.

(3) 当 $k=4$ 时, 对方程组的增广矩阵作初等行变换得到

(3) As $k=4$, performing elementary row operations on the augmented matrix associated with the system yields

$$\overline{\boldsymbol{A}} = \begin{pmatrix} 1 & 1 & 4 & 4 \\ -1 & 4 & 1 & 16 \\ 1 & -1 & 2 & -4 \end{pmatrix} \to \begin{pmatrix} 1 & 1 & 4 & 4 \\ 0 & 1 & 1 & 4 \\ 0 & 0 & 0 & 0 \end{pmatrix} \to \begin{pmatrix} 1 & 0 & 3 & 0 \\ 0 & 1 & 1 & 4 \\ 0 & 0 & 0 & 0 \end{pmatrix}$$

显然, $R(\overline{\boldsymbol{A}})=R(\boldsymbol{A})=2<3$, 因此方程组有无穷多个解. 其解为

Obviously, $R(\overline{\boldsymbol{A}})=R(\boldsymbol{A})=2<3$, so this system has infinitely many solutions, given by

$$\begin{cases} x_1 = -3x_3 \\ x_2 = -x_3 + 4 \end{cases}$$

令 $x_3=k$ 则

Let $x_3=k$, then

$$\begin{cases} x_1 = -3k \\ x_2 = -k + 4 \end{cases}$$

即

namely,

$$\begin{pmatrix} x_1 \\ x_2 \\ x_3 \end{pmatrix} = k \begin{pmatrix} -3 \\ -1 \\ 1 \end{pmatrix} + \begin{pmatrix} 0 \\ 4 \\ 0 \end{pmatrix}$$

其中 k 为任意实数. ■

where k is an arbitrary real number. ■

思 考 题

Questions

判断下列结论是否正确? 为什么?

Decide whether or not the following results are valid, why?

(1) 若 $R(\boldsymbol{A}_{m\times n})=m$, 则 $\boldsymbol{Ax}=\boldsymbol{0}$ 只有零解;

(1) If $R(\boldsymbol{A}_{m\times n})=m$, then $\boldsymbol{Ax}=\boldsymbol{0}$ has only a zero solution;

(2) 若 $\boldsymbol{Ax}=\boldsymbol{b}$ $(\boldsymbol{b}\neq\boldsymbol{0})$ 有无穷多个解, 则 $\boldsymbol{Ax}=\boldsymbol{0}$ 也有无穷多个解;

(2) If $\boldsymbol{Ax}=\boldsymbol{b}$ $(\boldsymbol{b}\neq\boldsymbol{0})$ has infinitely many solutions, then $\boldsymbol{Ax}=\boldsymbol{0}$ also has infinitely many solutions;

(3) 若 $\boldsymbol{A}_{m\times n}\boldsymbol{x}=\boldsymbol{0}$, 且 $R(\boldsymbol{A}_{m\times n})=m$, 则 $\boldsymbol{Ax}=\boldsymbol{0}$ 一定有解;

(3) If $\boldsymbol{A}_{m\times n}\boldsymbol{x}=\boldsymbol{0}$ and if $R(\boldsymbol{A}_{m\times n})=m$, then $\boldsymbol{Ax}=\boldsymbol{0}$ must have solutions;

(4) 若 $\boldsymbol{A}_{m\times n}\boldsymbol{x}=\boldsymbol{b}$, 且 $R(\boldsymbol{A}_{m\times n})=m$, 则 $\boldsymbol{Ax}=\boldsymbol{b}$ 一定有解;

(4) If $\boldsymbol{A}_{m\times n}\boldsymbol{x}=\boldsymbol{b}$ and if $R(\boldsymbol{A}_{m\times n})=m$, then $\boldsymbol{Ax}=\boldsymbol{b}$ must have solutions;

(5) 若 $\boldsymbol{A}_{m\times n}\boldsymbol{x}=\boldsymbol{0}$, 且 $m<n$, 则 $\boldsymbol{Ax}=\boldsymbol{0}$ 有非零解;

(5) If $\boldsymbol{A}_{m\times n}\boldsymbol{x}=\boldsymbol{0}$ and if $m<n$, then $\boldsymbol{Ax}=\boldsymbol{0}$ has nonzero solutions.

习 题 3

Exercise 3

1. 用初等行变换, 将下列矩阵化为行阶梯形及行最简形, 并求出矩阵的秩.

1. Reduce the following matrices to row echelon forms and row-reduced forms by elementary row operations, and find the ranks.

$$(1)\ \boldsymbol{A}=\begin{pmatrix}1&2&3\\2&2&1\\3&4&3\end{pmatrix};\quad (2)\ \boldsymbol{B}=\begin{pmatrix}1&-1&3&0\\-2&1&-2&1\\-1&-1&5&2\end{pmatrix};$$

$$(3)\ \mathbf{C}=\begin{pmatrix}0&0&1&2&-1\\1&3&-2&2&-1\\2&6&-4&5&7\\-1&-3&4&0&5\end{pmatrix}.$$

2. 对于给定的矩阵

2. For the given matrix

$$\boldsymbol{A}=\begin{pmatrix}1&1&2&a&3\\2&2&3&1&4\\1&0&1&1&5\\2&3&5&5&4\end{pmatrix}$$

若 $R(\boldsymbol{A})=3$, 求 a 的值.

if $R(\boldsymbol{A})=3$, find the value of a.

3. 对于 λ 的不同取值, 计算矩阵的秩

3. For different values of λ, calculate the rank of the matrix

$$\begin{pmatrix} 1 & \lambda & -1 & 2 \\ 2 & -1 & \lambda & 5 \\ 1 & 10 & -6 & 1 \end{pmatrix}$$

4. 如果 $R(\boldsymbol{A})=r$, 矩阵 $\boldsymbol{A}$ 中能否有等于零的 $r-1$ 阶子式? 能否有等于零的 r 阶子式? 能否有不为零的 $r+1$ 阶子式?

4. If $R(\boldsymbol{A})=r$, whether there exists a $(r-1)$-th order subdeterminant of $\boldsymbol{A}$ equal to zero, there exists a r-th order subdeterminant equal to zero, there exists a $(r+1)$-th order subdeterminant not equal to zero?

5. 利用初等行变换求下列矩阵的逆矩阵:

5. Find the inverse matrices of the following matrices by elementary row operations:

(1) $\begin{pmatrix} 1 & 2 & -1 \\ 3 & 4 & -2 \\ 5 & 4 & 1 \end{pmatrix}$; (2) $\begin{pmatrix} 1 & 0 & 0 & 0 \\ 1 & 2 & 0 & 0 \\ 2 & 1 & 3 & 0 \\ 3 & 2 & 1 & 4 \end{pmatrix}$; (3) $\begin{pmatrix} 3 & -2 & 0 & -1 \\ 0 & 2 & 2 & 1 \\ 1 & -2 & -3 & -2 \\ 0 & 1 & 2 & 1 \end{pmatrix}$.

6. 设下列矩阵为线性方程组的增广矩阵, 哪些对应的方程组无解; 有唯一解; 有无穷多个解?

6. Suppose that the following matrices are augmented matrices associated with the linear systems. Point out which of the corresponding systems have no solution; have a unique solution and have infinitely many solutions?

(1) $\begin{pmatrix} 1 & 3 & 1 \\ 0 & 1 & -1 \\ 0 & 0 & 0 \end{pmatrix}$; (2) $\begin{pmatrix} 1 & 2 & 4 \\ 0 & 1 & 3 \\ 0 & 0 & 1 \end{pmatrix}$;

(3) $\begin{pmatrix} 1 & -2 & 2 & -3 \\ 0 & 1 & -1 & 3 \\ 0 & 0 & 1 & 0 \end{pmatrix}$; (4) $\begin{pmatrix} 1 & 4 & 2 & -2 \\ 0 & 0 & 1 & 4 \\ 0 & 0 & 0 & 1 \end{pmatrix}$; (5) $(1,0,2,3)$.

7. 求解齐次线性方程组

7. Solve the homogeneous linear system

$$\begin{cases} x_1 + 2x_2 + x_3 - x_4 = 0 \\ 3x_1 + 6x_2 - x_3 - 3x_4 = 0 \\ 5x_1 + 10x_2 + x_3 - 5x_4 = 0 \end{cases}$$

8. 求解非齐次线性方程组

8. Solve the non-homogeneous linear system

$$\begin{cases} x_1 + x_2 - 2x_3 + 3x_4 = 0 \\ 2x_1 + x_2 - 6x_3 + 4x_4 = -1 \\ 3x_1 + 2x_2 - 8x_3 + 7x_4 = -1 \\ x_1 - x_2 - 6x_3 - x_4 = -2 \end{cases}$$

9. λ 取何值时, 非齐次线性方程组

9. What values should λ take on such that the non-homogeneous linear system

$$\begin{cases} \lambda x_1 + x_2 + x_3 = 1 \\ x_1 + \lambda x_2 + x_3 = \lambda \\ x_1 + x_2 + \lambda x_3 = \lambda^2 \end{cases}$$

(1) 有唯一解; (2) 无解; (3) 有无穷多个解? 在方程组有解的情形下, 求出解.

(1) has a unique solution; (2) has no solution; (3) has infinitely many solutions? In the case that the system has solutions, find the solutions.

10. 当 a, b 取何值时, 线性方程组

10. What values should a, b take on such that the linear system

$$\begin{cases} x_1 - x_2 - 2x_3 + 3x_4 = 0 \\ x_1 - 3x_2 - 6x_3 + 2x_4 = -1 \\ x_1 + 5x_2 + 10x_3 - x_4 = b \\ 3x_1 + x_2 + ax_3 + 4x_4 = 1 \end{cases}$$

(1) 无解; (2) 有唯一解; (3) 有无穷多个解? 在方程组有解的情形下, 求出解.

(1) has no solution; (2) has a unique solution; (3) has infinitely many solutions? In the case that the system has solutions, find the solutions.

11. 证明线性方程组

11. Prove that the linear system

$$\begin{cases} x_1 - x_2 = a_1 \\ x_2 - x_3 = a_2 \\ x_3 - x_4 = a_3 \\ x_4 - x_5 = a_4 \\ -x_1 + x_5 = a_5 \end{cases}$$

有解的充要条件是 $\sum_{i=1}^{5} a_i = 0$.

has solutions if and only if $\sum_{i=1}^{5} a_i = 0$.

12. 设 $\boldsymbol{A}$ 是 n 阶可逆阵, 将 $\boldsymbol{A}$ 的第 i 行与第 j 行对换, 所得的新矩阵记为 $\boldsymbol{B}$.

12. Let $\boldsymbol{A}$ be an invertible matrix of order n. Interchanging the i-th row and the j-th row of $\boldsymbol{A}$, we obtain a new matrix, written as $\boldsymbol{B}$.

(1) 证明 $\boldsymbol{B}$ 可逆; (2) 求 $\boldsymbol{A}\,\boldsymbol{B}^{-1}$.

(1) Prove that $\boldsymbol{B}$ is invertible; (2) Find $\boldsymbol{A}\,\boldsymbol{B}^{-1}$.

补充题 3

Supplement Exercise 3

1. 求

1. Find

$$\begin{pmatrix} 1 & 0 & 0 \\ 0 & 1 & 0 \\ 0 & 2 & 1 \end{pmatrix}^{2000} \begin{pmatrix} 1 & 2 & 3 \\ 2 & 3 & 4 \\ 3 & 4 & 5 \end{pmatrix} \begin{pmatrix} 0 & 0 & 1 \\ 0 & 1 & 0 \\ 1 & 0 & 0 \end{pmatrix}^{2001}$$

2. 设 $\boldsymbol{A}$ 为 3 阶方阵. 将 $\boldsymbol{A}$ 的第 1 列与第 2 列交换得到 $\boldsymbol{B}$, 再将 $\boldsymbol{B}$ 的第 2 列加到第 3 列得到 $\boldsymbol{C}$, 求满足 $\boldsymbol{AQ} = \boldsymbol{C}$ 的可逆矩阵 $\boldsymbol{Q}$.

2. Let $\boldsymbol{A}$ be a square matrix of order 3. Interchanging the first column and the second column of $\boldsymbol{A}$ yields $\boldsymbol{B}$, and then adding the second column of $\boldsymbol{B}$ to the third column gets $\boldsymbol{C}$, find an invertible matrix $\boldsymbol{Q}$ satisfying $\boldsymbol{AQ} = \boldsymbol{C}$.

3. 设

3. Suppose that

$$\boldsymbol{B} = \begin{pmatrix} 1 & -3 \\ 2 & 2 \\ 3 & -1 \end{pmatrix} = (\boldsymbol{\beta}_1, \boldsymbol{\beta}_2), \quad \boldsymbol{A} = \begin{pmatrix} 4 & 1 & -2 \\ 2 & 2 & 1 \\ 3 & 1 & -1 \end{pmatrix}$$

解线性方程组 $\boldsymbol{Ax} = \boldsymbol{\beta}_1$, $\boldsymbol{Ax} = \boldsymbol{\beta}_2$.

Solve the linear systems $\boldsymbol{Ax} = \boldsymbol{\beta}_1, \boldsymbol{Ax} = \boldsymbol{\beta}_2$.

4. 设

4. Suppose that

$$\boldsymbol{A} = \begin{pmatrix} 1 & 2 & -1 & 3 \\ 2 & 1 & 4 & 3 \\ 0 & a & 2 & -1 \end{pmatrix}, \quad \boldsymbol{\beta} = \begin{pmatrix} 1 \\ 5 \\ -6 \end{pmatrix}$$

且线性方程组 $\boldsymbol{A}\boldsymbol{x}=\boldsymbol{\beta}$ 无解. 求 a 的值.

and that the linear system $\boldsymbol{A}\boldsymbol{x}=\boldsymbol{\beta}$ has no solutions. Find the value of a.

5. 设 $\boldsymbol{AB}=\boldsymbol{O}$, 其中

5. Suppose that $\boldsymbol{AB}=\boldsymbol{O}$, where

$$\boldsymbol{A}=\begin{pmatrix} 1 & 2 & 0 \\ 2 & 0 & -4 \\ -1 & t & 5 \\ 1 & 0 & -2 \end{pmatrix}$$

$\boldsymbol{B}$ 是 3 阶非零矩阵. 求 t 的值.

and that $\boldsymbol{B}$ is a nonzero matrix of order 3. Find the value of t.

第 4 章 Chapter 4

向量 Vectors

向量是解决数学问题常用的工具, 也是线性代数课程的主要内容之一. 早在 1843 年, 英国数学家 W.R. 哈密顿研究 “四元数” 概念的同时, 就引入了向量的概念.

Vector is a common tool for solving mathematical problems and is also one of the main contents of Linear Algebra. Earlier in 1843, W. R. Hamilton, a British mathematician, introduced the concept of vectors as he studied the concept of “quarternion”.

本章将重点介绍向量组的线性关系理论以及向量组的秩. 最后, 当线性方程组有无穷多个解时, 通过线性方程组解向量组的极大线性无关组来表示其所有解, 即通解.

In this chapter, we mainly introduce the theory of linear relations and the rank of a vector set. Finally, as a system of linear equations has infinitely many solutions, we present all the solutions, i.e., the general solution, by a maximal linearly independent subset of a solution vector set of the linear system.

➢本章内容提要

Headline of this chapter

1. 向量组的线性关系

1. Linear relations of vector sets

(1) 向量组的线性相关和线性无关;

(1) Linear dependence and linear independence of vector sets;

(2) 向量组的极大无关组与向量组的秩.

(2) Maximal linearly independent subsets and ranks of vector sets.

2. 线性方程组的通解.

2. General solutions of linear systems.

4.1 向量及其线性运算

4.1 Vectors and Linear Operations

4.1.1 向量的概念

4.1.1 Concept of Vectors

定义 1 由 n 个数 $a_1, a_2, \cdots, a_n$ 组成的有顺序数组

Definition 1 An ordered array consisted of n numbers $a_1, a_2, \cdots, a_n$ given by

$$\boldsymbol{\alpha} = (a_1, a_2, \cdots, a_n) \tag{4.1}$$

称为 n 维**行向量**. 数 $a_1, a_2, \cdots, a_n$ 称为向量 $\boldsymbol{\alpha}$ 的分量, a_j 称为向量 $\boldsymbol{\alpha}$ 的第 j 个分量 (或坐标), $j = 1, 2, \cdots, n$.

is called an n dimensional **row vector**. The numbers $a_1, a_2, \cdots, a_n$ are called the components of $\boldsymbol{\alpha}$, a_j is called the j-th component (or coordinate) of $\boldsymbol{\alpha}$, where $j = 1, 2, \cdots, n$.

$$\boldsymbol{\alpha} = \begin{pmatrix} a_1 \\ a_2 \\ \vdots \\ a_n \end{pmatrix} \quad 或 \quad \boldsymbol{\alpha} = (a_1, a_2, \cdots, a_n)^{\mathrm{T}} \tag{4.2}$$

称为 n 维**列向量**. 向量 (4.1) 和 (4.2) 只是写法上不同, 没有本质的区别.

is called an n dimensional **column vector**. The two vectors (4.1) and (4.2) are only written in different forms and have no essential difference.

令 $\mathbf{R}$ 表示实数集, 以 $a_j \in \mathbf{R}(j = 1, 2, \cdots, n)$ 为分量的向量 $\boldsymbol{\alpha} = (a_1, a_2, \cdots, a_n)$ 称为实行向量. 本章只讨论定义在 $\mathbf{R}$ 上的向量.

Let $\mathbf{R}$ denote the set of real numbers, the vector $\boldsymbol{\alpha} = (a_1, a_2, \cdots, a_n)$ consisted of the components $a_j \in \mathbf{R}$ $(j = 1, 2, \cdots, n)$ is said to be a real row vector. Throughout this chapter, we only discuss the vectors on $\mathbf{R}$.

注 (1) 在解析几何中, 我们将既有大小又有方向的量称为向量, 并将可随意平行移动的有向线段作为向量的几何形象. 引入坐标系后, 又定义了向量的坐标表示式 (三个有次序实数), 即上面定义的三维

Note (1) In Analytic Geometry, the quantity with both magnitude and direction is called a vector, and the directed line segment which may be freely parallel translation is used as the geometric figure of a vec-

向量. 因此, 当 $n \leqslant 3$ 时, n 维向量可以将有向线段作为其几何形象. 但是, 当 $n > 3$ 时, n 维向量没有直观的几何形象.

tor. After introducing a coordinate system, we also define the coordinate representation (three ordered real numbers) of a vector, i.e., the 3 dimensional vector defined above. Therefore, if $n \leqslant 3$, an n dimensional vector may use a directed line segment as its geometric figure, while if $n > 3$, an n dimensional vector will have no visualized geometric figure.

(2) n 维行向量和 n 维列向量实际上就是行矩阵和列矩阵. 因此, 向量作为一种特殊的矩阵, 它有和矩阵一样的运算及其运算规律. 另外, 二者还有密切的联系.

(2) An n dimensional row vector and an n dimensional column vector are actually a row matrix and a column matrix, respectively. Therefore, as a class of special matrices, vectors have the same operations and operation rules as matrices. In addition, they have close relations.

例如, 在下面的线性方程组中

For example, in the following linear system,

$$\begin{cases} a_{11}x_1 + a_{12}x_2 + \cdots + a_{1n}x_n = b_1 \\ a_{21}x_1 + a_{22}x_2 + \cdots + a_{2n}x_n = b_2 \\ \qquad\qquad \cdots\cdots \\ a_{m1}x_1 + a_{m2}x_2 + \cdots + a_{mn}x_n = b_m \end{cases} \tag{4.3}$$

第 i 个方程的系数和常数项对应着一个 $n+1$ 维行向量

the coefficients and the constant term of the i-th equation correspond to an $n+1$ dimensional vector, namely,

$$(a_{i1}, a_{i2}, \cdots, a_{in}, b_i) \quad (i = 1, 2, \cdots, m)$$

而该方程组的一组解 $x_1 = c_1, x_2 = c_2, \cdots, x_n = c_n$ 可表示为如下的 n 维列向量, 即

Moreover, the solution of this system, given by $x_1 = c_1, x_2 = c_2, \cdots, x_n = c_n$, can be expressed as the following n dimensional column vector, namely,

$$\begin{pmatrix} c_1 \\ c_2 \\ \vdots \\ c_n \end{pmatrix}$$

称为该方程组的**解向量**.

which is called a **solution vector** of the system.

再例如, 在一个 $m \times n$ 矩阵中,

Again for example, in an $m \times n$ matrix,

$$A = \begin{pmatrix} a_{11} & a_{12} & \cdots & a_{1n} \\ a_{21} & a_{22} & \cdots & a_{2n} \\ \vdots & \vdots & & \vdots \\ a_{m1} & a_{m2} & \cdots & a_{mn} \end{pmatrix}$$

每一行构成一个 n 维行向量, 总共有 m 个 n 维行向量, 即

each row is an n dimensional row vector, so the matrix has m row vectors of dimension n, i.e.,

$$\boldsymbol{\alpha}_i = (a_{i1}, a_{i2}, \cdots, a_{in}) \quad (i = 1, 2, \cdots, m)$$

每一列构成一个 m 维列向量, 总共有 n 个 m 维列向量

each column is also an m dimensional column vector, it has n column vectors of dimension m

$$\boldsymbol{\beta}_j = \begin{pmatrix} a_{1j} \\ a_{2j} \\ \vdots \\ a_{mj} \end{pmatrix} \quad (i = 1, 2, \cdots, n)$$

矩阵 $\boldsymbol{A}$ 可用向量表示为

The matrix $\boldsymbol{A}$ can be represented by vectors, as follows

$$\boldsymbol{A} = \begin{pmatrix} \boldsymbol{\alpha}_1 \\ \boldsymbol{\alpha}_2 \\ \vdots \\ \boldsymbol{\alpha}_m \end{pmatrix}$$

或

or

$$\boldsymbol{A} = (\boldsymbol{\beta}_1, \boldsymbol{\beta}_2, \cdots, \boldsymbol{\beta}_n)$$

这也可以看成是 $\boldsymbol{A}$ 的一种分块. 反过来, 同维数的向量可以组成一个矩阵 $\boldsymbol{A}$.

This may also be viewed as a partition of $\boldsymbol{A}$. Vice versa, the vectors with the same dimension may form a matrix.

4.1.2 向量的线性运算

向量作为一种特殊的矩阵, 它有和矩阵一样的运算及其运算规律. 下面仅就列向量进行讨论.

对于如下两个 n 维向量

4.1.2 Linear Operations of Vectors

As a class of special matrices, vectors have the same operations and operation rules as matrices. In the following part we only discuss column vectors.

For the following two n dimensional vectors

$$\boldsymbol{\alpha}=\begin{pmatrix}a_1\\a_2\\\vdots\\a_n\end{pmatrix},\quad \boldsymbol{\beta}=\begin{pmatrix}b_1\\b_2\\\vdots\\b_n\end{pmatrix}$$

若它们对应分量都相等, 即

if their components are all equal correspondingly, namely,

$$a_i=b_i\quad(i=1,2,\cdots,n)$$

则称向量 $\boldsymbol{\alpha}$ 与 $\boldsymbol{\beta}$**相等**, 记为 $\boldsymbol{\alpha}=\boldsymbol{\beta}$.

we call the vectors $\boldsymbol{\alpha}$ and $\boldsymbol{\beta}$ are **equal**, written as $\boldsymbol{\alpha}=\boldsymbol{\beta}$.

分量都是零的向量称为**零向量**, 记作 $\mathbf{0}$, 即

The vector is said to be a **zero vector** if its components are all zeros, written as $\mathbf{0}$, namely,

$$\mathbf{0}=\begin{pmatrix}0\\0\\\vdots\\0\end{pmatrix}$$

注 维数不同的零向量不相等, 如

Note The zero vectors of different dimensions are not equal, such as

$$\mathbf{0}_1=\begin{pmatrix}0\\0\\0\\0\end{pmatrix},\quad \mathbf{0}_2=\begin{pmatrix}0\\0\\0\end{pmatrix}$$

都是零向量, 但 $\mathbf{0}_1 \neq \mathbf{0}_2$.

are all zero vectors, but $\mathbf{0}_1 \neq \mathbf{0}_2$.

向量 $-\boldsymbol{\alpha}$ 称为 $\boldsymbol{\alpha}$ 的负向量, 记为

The vector $-\boldsymbol{\alpha}$ is called a **negative vector** of $\boldsymbol{\alpha}$, written as

$$-\boldsymbol{\alpha}\begin{pmatrix} -a_1 \\ -a_2 \\ \vdots \\ -a_n \end{pmatrix}$$

定义 2 设 $\boldsymbol{\alpha}, \boldsymbol{\beta}$ 为 n 维向量, 向量 $\boldsymbol{\alpha}+\boldsymbol{\beta}$ 称为 $\boldsymbol{\alpha}$ 与 $\boldsymbol{\beta}$ 的**和**, 记作

Definition 2 Let $\boldsymbol{\alpha}$ and $\boldsymbol{\beta}$ be n dimensional vectors, the vector $\boldsymbol{\alpha}+\boldsymbol{\beta}$ is called the **sum** of $\boldsymbol{\alpha}$ and $\boldsymbol{\beta}$, written as

$$\boldsymbol{\alpha}+\boldsymbol{\beta}=\begin{pmatrix} a_1 \\ a_2 \\ \vdots \\ a_n \end{pmatrix}+\begin{pmatrix} b_1 \\ b_2 \\ \vdots \\ b_n \end{pmatrix}=\begin{pmatrix} a_1+b_1 \\ a_2+b_2 \\ \vdots \\ a_n+b_n \end{pmatrix}$$

由负向量定义向量减法为

The subtraction of two vectors defined by a negative vector is

$$\boldsymbol{\alpha}-\boldsymbol{\beta}=\boldsymbol{\alpha}+(-\boldsymbol{\beta})=\begin{pmatrix} a_1-b_1 \\ a_2-b_2 \\ \vdots \\ a_n-b_n \end{pmatrix}$$

事实上, $\boldsymbol{\alpha}$ 与 $\boldsymbol{\beta}$ 的和 (或差) 就是它们的对应元素的和 (或差).

In fact, the sum (or difference) of $\boldsymbol{\alpha}$ and $\boldsymbol{\beta}$ is the sum (or difference) of their corresponding elements.

定义 3 设 $\boldsymbol{\alpha}$ 为 n 维向量, $\lambda \in \mathbf{R}$. 向量 $\lambda\boldsymbol{\alpha}$ 称为数 λ 与向量 $\boldsymbol{\alpha}$ 的**数乘**, 记作

Definition 3 Let $\boldsymbol{\alpha}$ be an n dimensional vector, $\lambda \in \mathbf{R}$. The vector $\lambda\boldsymbol{\alpha}$ is said to be a **scalar multiplication** of the vector $\boldsymbol{\alpha}$ by the number λ, written as

$$\lambda\boldsymbol{\alpha}=\begin{pmatrix} \lambda a_1 \\ \lambda a_2 \\ \vdots \\ \lambda a_n \end{pmatrix}$$

根据定义 3, 有

From Definition 3 we have

$$0\boldsymbol{\alpha}=\mathbf{0}$$
$$(-1)\boldsymbol{\alpha}=-\boldsymbol{\alpha}$$
$$\lambda\mathbf{0}=\mathbf{0}$$

如果 $\lambda\neq 0,\boldsymbol{\alpha}\neq\mathbf{0}$, 那么 $\lambda\boldsymbol{\alpha}\neq\mathbf{0}$.

If $\lambda\neq 0,\boldsymbol{\alpha}\neq\mathbf{0}$, then $\lambda\boldsymbol{\alpha}\neq\mathbf{0}$.

向量的加法及向量的数乘两种运算, 统称为向量的**线性运算**. 不难验证向量满足以下八条运算规律:

The two operations, addition and scalar multiplication of vectors, are collectively called **linear operations** of vectors. It is not difficult to show that vectors satisfy the following eight operation rules:

(1) $\boldsymbol{\alpha}+\boldsymbol{\beta}=\boldsymbol{\beta}+\boldsymbol{\alpha}$;
(2) $(\boldsymbol{\alpha}+\boldsymbol{\beta})+\boldsymbol{\gamma}=\boldsymbol{\alpha}+(\boldsymbol{\beta}+\boldsymbol{\gamma})$;
(3) $\boldsymbol{\alpha}+\mathbf{0}=\boldsymbol{\alpha}$;
(4) $\boldsymbol{\alpha}+(-\boldsymbol{\alpha})=\mathbf{0}$;
(5) $1\cdot\boldsymbol{\alpha}=\boldsymbol{\alpha}$;
(6) $\lambda(\mu\boldsymbol{\alpha})=(\lambda\mu)\boldsymbol{\alpha}$;
(7) $\lambda(\boldsymbol{\alpha}+\boldsymbol{\beta})=\lambda\boldsymbol{\alpha}+\lambda\boldsymbol{\beta}$;
(8) $(\lambda+\mu)\boldsymbol{\alpha}=\lambda\boldsymbol{\alpha}+\mu\boldsymbol{\alpha}$.

上式中 $\boldsymbol{\alpha},\boldsymbol{\beta},\boldsymbol{\gamma}$ 都是 n 维向量, $\lambda,\mu\in\mathbf{R}$.

In the above equations $\boldsymbol{\alpha},\boldsymbol{\beta},\boldsymbol{\gamma}$ are all n dimensional vectors, $\lambda,\mu\in\mathbf{R}$.

例 1 设

Example 1 Suppose that

$$\boldsymbol{\alpha}=\begin{pmatrix}1\\3\\-2\\2\end{pmatrix},\quad \boldsymbol{\beta}=\begin{pmatrix}5\\1\\-2\\0\end{pmatrix}$$

若已知 $\boldsymbol{\alpha}+2\boldsymbol{\gamma}=3\boldsymbol{\beta}$, 求向量 $\boldsymbol{\gamma}$.

If it is known that $\boldsymbol{\alpha}+2\boldsymbol{\gamma}=3\boldsymbol{\beta}$, find the vector $\boldsymbol{\gamma}$.

解 由 $\boldsymbol{\alpha}+2\gamma=3\boldsymbol{\beta}$ 可得

Solution Using $\boldsymbol{\alpha}+2\gamma=3\boldsymbol{\beta}$ yields

$$\gamma=\frac{1}{2}(3\boldsymbol{\beta}-\boldsymbol{\alpha}).$$

根据向量的定义以及运算规律得

Further, from the definition and the operation rules of vectors we have

$$\gamma = \frac{1}{2}(3\boldsymbol{\beta} - \boldsymbol{\alpha}) = \frac{1}{2}\left(\begin{pmatrix} 15 \\ 3 \\ -6 \\ 0 \end{pmatrix} - \begin{pmatrix} 1 \\ 3 \\ -2 \\ 2 \end{pmatrix}\right) = \frac{1}{2}\begin{pmatrix} 14 \\ 0 \\ -4 \\ -2 \end{pmatrix} = \begin{pmatrix} 7 \\ 0 \\ -2 \\ -1 \end{pmatrix} \qquad \blacksquare$$

4.1.3 向量组的线性组合

由若干个同维数的向量所组成的集合称为**向量组**.

定义 4 对于给定的向量组 $\{\boldsymbol{\alpha}, \boldsymbol{\alpha}_1, \boldsymbol{\alpha}_2, \cdots, \boldsymbol{\alpha}_m\}$, 如果存在数 $\lambda_1, \lambda_2, \cdots, \lambda_m$, 使得

4.1.3 Linear Combination of Vector Sets

The set consisted of several vectors of the same dimension is called a **vector set**.

Definition 4 For the given vector set $\{\boldsymbol{\alpha}, \boldsymbol{\alpha}_1, \boldsymbol{\alpha}_2, \cdots, \boldsymbol{\alpha}_m\}$, if there exist numbers $\lambda_1, \lambda_2, \cdots, \lambda_m$ such that

$$\boldsymbol{\alpha} = \lambda_1\boldsymbol{\alpha}_1 + \lambda_2\boldsymbol{\alpha}_2 + \cdots + \lambda_m\boldsymbol{\alpha}_m$$

则称向量 $\boldsymbol{\alpha}$ 是向量组 $\{\boldsymbol{\alpha}_1, \boldsymbol{\alpha}_2, \cdots, \boldsymbol{\alpha}_m\}$ 的**线性组合**, 或称 $\boldsymbol{\alpha}$ 可由 $\{\boldsymbol{\alpha}_1, \boldsymbol{\alpha}_2, \cdots, \boldsymbol{\alpha}_m\}$**线性表示**.

then it is called that $\boldsymbol{\alpha}$ is a **linear combination** of the vector set $\{\boldsymbol{\alpha}_1, \boldsymbol{\alpha}_2, \cdots, \boldsymbol{\alpha}_m\}$, or is called that $\boldsymbol{\alpha}$ can be **linearly represented** by $\{\boldsymbol{\alpha}_1, \boldsymbol{\alpha}_2, \cdots, \boldsymbol{\alpha}_m\}$.

显然, $\mathbf{0} = 0\boldsymbol{\alpha}_1 + 0\boldsymbol{\alpha}_2 + \cdots + 0\boldsymbol{\alpha}_m$, 即零向量是任何一向量组 $\{\boldsymbol{\alpha}_1, \boldsymbol{\alpha}_2, \cdots, \boldsymbol{\alpha}_m\}$ 的线性组合.

Obviously, $\mathbf{0} = 0\boldsymbol{\alpha}_1 + 0\boldsymbol{\alpha}_2 + \cdots + 0\boldsymbol{\alpha}_m$, that is to say, the zero vector is a linear combination of an arbitrary vector set $\{\boldsymbol{\alpha}_1, \boldsymbol{\alpha}_2, \cdots, \boldsymbol{\alpha}_m\}$.

对于向量组 $\{\boldsymbol{\alpha}_1, \boldsymbol{\alpha}_2, \cdots, \boldsymbol{\alpha}_m\}$, 每个向量 $\boldsymbol{\alpha}_i (i = 1, 2, \cdots, m)$ 均可由该向量组线性表示, 即

For the vector set $\{\boldsymbol{\alpha}_1, \boldsymbol{\alpha}_2, \cdots, \boldsymbol{\alpha}_m\}$, each vector $\boldsymbol{\alpha}_i (i = 1, 2, \cdots, m)$ in it can be linearly represented by this vector set, namely,

$$\boldsymbol{\alpha}_i = 0\boldsymbol{\alpha}_1 + 0\boldsymbol{\alpha}_2 + \cdots + 1\boldsymbol{\alpha}_i + \cdots + 0\boldsymbol{\alpha}_m$$

设 n 维向量组为

Suppose that an n dimensional vector set is given by

$$\boldsymbol{\varepsilon}_1 = \begin{pmatrix} 1 \\ 0 \\ \vdots \\ 0 \end{pmatrix}, \boldsymbol{\varepsilon}_2 = \begin{pmatrix} 0 \\ 1 \\ \vdots \\ 0 \end{pmatrix}, \cdots, \boldsymbol{\varepsilon}_n = \begin{pmatrix} 0 \\ 0 \\ \vdots \\ 1 \end{pmatrix}$$

通常称 $\{\varepsilon_1, \varepsilon_2, \cdots, \varepsilon_n\}$ 为 n **维单位向量组**. 令

in general, the set consisted of $\{\varepsilon_1, \varepsilon_2, \cdots, \varepsilon_n\}$ is called an n **dimensional unit vector set**. Let

$$\boldsymbol{\alpha} = \begin{pmatrix} a_1 \\ a_2 \\ \vdots \\ a_n \end{pmatrix}$$

是任意一个 n 维列向量, 由于

be an arbitrary n dimensional column vector, since

$$\boldsymbol{\alpha} = a_1\varepsilon_1 + a_2\varepsilon_2 + \cdots + a_n\varepsilon_n$$

所以, $\boldsymbol{\alpha}$ 是 $\{\varepsilon_1, \varepsilon_2, \cdots, \varepsilon_n\}$ 的线性组合, 换句话说, 任何一个 n 维向量均可由 n 维单位向量组线性表示.

$\boldsymbol{\alpha}$ is a linear combination of $\{\varepsilon_1, \varepsilon_2, \cdots, \varepsilon_n\}$, in other words, an arbitrary n dimensional vector can be linearly represented by an n dimensional unit vector set.

例 2 已知

Example 2 Known that

$$\boldsymbol{\alpha} = \begin{pmatrix} 0 \\ 4 \\ 2 \end{pmatrix}, \boldsymbol{\alpha}_1 = \begin{pmatrix} 1 \\ 2 \\ 3 \end{pmatrix}, \boldsymbol{\alpha}_2 = \begin{pmatrix} 2 \\ 3 \\ 1 \end{pmatrix}, \boldsymbol{\alpha}_3 = \begin{pmatrix} 3 \\ 1 \\ 2 \end{pmatrix}$$

证明向量 $\boldsymbol{\alpha}$ 是向量组 $\{\boldsymbol{\alpha}_1, \boldsymbol{\alpha}_2, \boldsymbol{\alpha}_3\}$ 的线性组合, 并将 $\boldsymbol{\alpha}$ 用 $\{\boldsymbol{\alpha}_1, \boldsymbol{\alpha}_2, \boldsymbol{\alpha}_3\}$ 线性表示.

Prove that the vector $\boldsymbol{\alpha}$ is a linear combination of the vector set $\{\boldsymbol{\alpha}_1, \boldsymbol{\alpha}_2, \boldsymbol{\alpha}_3\}$, and give the linear representation of $\boldsymbol{\alpha}$ by $\{\boldsymbol{\alpha}_1, \boldsymbol{\alpha}_2, \boldsymbol{\alpha}_3\}$.

解 假定 $\boldsymbol{\alpha} = \lambda_1\boldsymbol{\alpha}_1 + \lambda_2\boldsymbol{\alpha}_2 + \lambda_3\boldsymbol{\alpha}_3$, 即

Solution Suppose that $\boldsymbol{\alpha} = \lambda_1\boldsymbol{\alpha}_1 + \lambda_2\boldsymbol{\alpha}_2 + \lambda_3\boldsymbol{\alpha}_3$, namely,

$$\lambda_1\begin{pmatrix} 1 \\ 2 \\ 3 \end{pmatrix} + \lambda_2\begin{pmatrix} 2 \\ 3 \\ 1 \end{pmatrix} + \lambda_3\begin{pmatrix} 3 \\ 1 \\ 2 \end{pmatrix} = \begin{pmatrix} 0 \\ 4 \\ 2 \end{pmatrix}$$

由向量的线性运算, 得

Using linear operations of vectors, we have

$$\begin{cases} \lambda_1 + 2\lambda_2 + 3\lambda_3 = 0 \\ 2\lambda_1 + 3\lambda_2 + \lambda_3 = 4 \\ 3\lambda_1 + \lambda_2 + 2\lambda_3 = 2 \end{cases}$$

因为该线性方程组的系数行列式为

Since the coefficient determinant associated with the linear system

$$\begin{vmatrix} 1 & 2 & 3 \\ 2 & 3 & 1 \\ 3 & 1 & 2 \end{vmatrix} = -18 \neq 0$$

由克拉默法则知, 方程组有唯一解. 可以求得 $\lambda_1 = 1, \lambda_2 = 1, \lambda_3 = -1$, 于是 $\boldsymbol{\alpha}$ 可由 $\{\boldsymbol{\alpha}_1, \boldsymbol{\alpha}_2, \boldsymbol{\alpha}_3\}$ 线性表示, 且唯一表示为

From Cramer's rule we know that this system has a unique solution. It can be find that $\lambda_1 = 1, \lambda_2 = 1, \lambda_3 = -1$, which means that $\boldsymbol{\alpha}$ can be linearly represented by $\{\boldsymbol{\alpha}_1, \boldsymbol{\alpha}_2, \boldsymbol{\alpha}_3\}$ and that the unique expression is given by

$$\boldsymbol{\alpha} = \boldsymbol{\alpha}_1 + \boldsymbol{\alpha}_2 - \boldsymbol{\alpha}_3$$ ■

一般地, 一个向量 $\boldsymbol{\alpha}$ 与一组向量 $\{\boldsymbol{\alpha}_1, \boldsymbol{\alpha}_2, \cdots, \boldsymbol{\alpha}_m\}$ 的关系必为下列情形之一:

Generally, the relations between a vector $\boldsymbol{\alpha}$ and a vector set $\{\boldsymbol{\alpha}_1, \boldsymbol{\alpha}_2, \cdots, \boldsymbol{\alpha}_m\}$ must be one of the following cases:

(1) $\boldsymbol{\alpha}$ 不能由 $\{\boldsymbol{\alpha}_1, \boldsymbol{\alpha}_2, \cdots, \boldsymbol{\alpha}_m\}$ 线性表示;

(1) $\boldsymbol{\alpha}$ can not be linearly represented by $\{\boldsymbol{\alpha}_1, \boldsymbol{\alpha}_2, \cdots, \boldsymbol{\alpha}_m\}$;

(2) $\boldsymbol{\alpha}$ 可由 $\{\boldsymbol{\alpha}_1, \boldsymbol{\alpha}_2, \cdots, \boldsymbol{\alpha}_m\}$ 线性表示, 且表示唯一;

(2) $\boldsymbol{\alpha}$ can be linearly represented by $\{\boldsymbol{\alpha}_1, \boldsymbol{\alpha}_2, \cdots, \boldsymbol{\alpha}_m\}$ and the expression is unique;

(3) $\boldsymbol{\alpha}$ 可由 $\{\boldsymbol{\alpha}_1, \boldsymbol{\alpha}_2, \cdots, \boldsymbol{\alpha}_m\}$ 线性表示, 但表示不唯一.

(3) $\boldsymbol{\alpha}$ can be linearly represented by $\{\boldsymbol{\alpha}_1, \boldsymbol{\alpha}_2, \cdots, \boldsymbol{\alpha}_m\}$, but the expression is not unique.

例如,

For instance, from

$$\begin{pmatrix} 0 \\ 0 \end{pmatrix} = \begin{pmatrix} 1 \\ -1 \end{pmatrix} + \begin{pmatrix} -1 \\ 1 \end{pmatrix} = 0\begin{pmatrix} 1 \\ -1 \end{pmatrix} + 0\begin{pmatrix} -1 \\ 1 \end{pmatrix}$$

表示不唯一.

it shows that the expressions are not unique.

对于给定的线性方程组

For the given linear system

$$\begin{cases} a_{11}x_1 + \ a_{12}x_2 + \cdots + a_{1n}x_n = b_1 \\ a_{21}x_1 + \ a_{21}x_2 + \cdots + a_{2n}x_n = b_2 \\ \qquad\qquad \cdots\cdots \\ a_{m1}x_1 + a_{m2}x_2 + \cdots + a_{mn}x_n = b_m \end{cases}$$

若以 $\boldsymbol{\alpha}_j$ 表示第 j 个未知量 x_j 的系数构成的 m 维列向量, 即

If $\boldsymbol{\alpha}_j$ denotes an m dimensional column vector consisted of the coefficients of the j-th unknown x_j, namely,

$$\boldsymbol{\alpha}_j = \begin{pmatrix} a_{1j} \\ a_{2j} \\ \vdots \\ a_{mj} \end{pmatrix} \quad (j = 1, 2, \cdots, n)$$

且

and

$$\boldsymbol{\beta} = \begin{pmatrix} b_1 \\ b_2 \\ \vdots \\ b_m \end{pmatrix}$$

则该方程组的向量组合为

then the vector combination of this system is

$$x_1\boldsymbol{\alpha}_1 + x_2\boldsymbol{\alpha}_2 + \cdots + x_n\boldsymbol{\alpha}_n = \boldsymbol{\beta}$$

于是 $\boldsymbol{\beta}$ 能否由向量 $\{\boldsymbol{\alpha}_1, \boldsymbol{\alpha}_2, \cdots, \boldsymbol{\alpha}_n\}$ 线性表示就转化为线性方程组有没有解的问题. 因此, 有下面的定理.

This means that whether $\boldsymbol{\beta}$ can be linearly represented by $\{\boldsymbol{\alpha}_1, \boldsymbol{\alpha}_2, \cdots, \boldsymbol{\alpha}_n\}$ is described as the problem that whether the linear system has solutions. Therefore, we have the following theorem.

定理 1 列向量 $\boldsymbol{\beta}$ 能由列向量组 $\{\boldsymbol{\alpha}_1, \boldsymbol{\alpha}_2, \cdots, \boldsymbol{\alpha}_n\}$ 线性表示, 即

Theorem 1 The column vector $\boldsymbol{\beta}$ can be linearly represented by the column vector set $\{\boldsymbol{\alpha}_1, \boldsymbol{\alpha}_2, \cdots, \boldsymbol{\alpha}_n\}$, namely,

$$x_1\boldsymbol{\alpha}_1 + x_2\boldsymbol{\alpha}_2 + \cdots + x_n\boldsymbol{\alpha}_n = \boldsymbol{\beta}$$

$$\Leftrightarrow \begin{cases} a_{11}x_1 + \ a_{12}x_2 + \cdots + \ a_{1n}x_n = b_1 \\ a_{21}x_1 + \ a_{21}x_2 + \cdots + \ a_{2n}x_n = b_2 \\ \qquad\qquad \cdots\cdots \\ a_{m1}x_1 + a_{m2}x_2 + \cdots + a_{mn}x_n = b_m \end{cases}$$

有解.

has solutions.

进一步有如下定理.

Further, we have the following theorem.

定理 2 (1) $\boldsymbol{\beta}$ 不能由向量组 $\{\boldsymbol{\alpha}_1,\boldsymbol{\alpha}_2,\cdots,\boldsymbol{\alpha}_n\}$ 线性表示的充分必要条件是线性方程组 $x_1\boldsymbol{\alpha}_1+x_2\boldsymbol{\alpha}_2+\cdots+x_n\boldsymbol{\alpha}_n=\boldsymbol{\beta}$ 无解;

Theorem 2 (1) $\boldsymbol{\beta}$ can not be linearly represented by $\{\boldsymbol{\alpha}_1,\boldsymbol{\alpha}_2,\cdots,\boldsymbol{\alpha}_n\}$ if and only if the linear system $x_1\boldsymbol{\alpha}_1+x_2\boldsymbol{\alpha}_2+\cdots+x_n\boldsymbol{\alpha}_n=\boldsymbol{\beta}$ has no solution;

(2) $\boldsymbol{\beta}$ 能由向量组 $\{\boldsymbol{\alpha}_1,\boldsymbol{\alpha}_2,\cdots,\boldsymbol{\alpha}_n\}$ 唯一线性表示的充分必要条件是线性方程组 $x_1\boldsymbol{\alpha}_1+x_2\boldsymbol{\alpha}_2+\cdots+x_n\boldsymbol{\alpha}_n=\boldsymbol{\beta}$ 有唯一解;

(2) $\boldsymbol{\beta}$ can be linearly represented by $\{\boldsymbol{\alpha}_1,\boldsymbol{\alpha}_2,\cdots,\boldsymbol{\alpha}_n\}$ and the expression is unique if and only if the linear system $x_1\boldsymbol{\alpha}_1+x_2\boldsymbol{\alpha}_2+\cdots+x_n\boldsymbol{\alpha}_n=\boldsymbol{\beta}$ has a unique solution;

(3) $\boldsymbol{\beta}$ 能由向量组 $\{\boldsymbol{\alpha}_1,\boldsymbol{\alpha}_2,\cdots,\boldsymbol{\alpha}_n\}$ 线性表示且表示式不唯一的充分必要条件是线性方程组 $x_1\boldsymbol{\alpha}_1+x_2\boldsymbol{\alpha}_2+\cdots+x_n\boldsymbol{\alpha}_n=\boldsymbol{\beta}$ 有无穷多个解.

(3) $\boldsymbol{\beta}$ can be linearly represented by $\{\boldsymbol{\alpha}_1,\boldsymbol{\alpha}_2,\cdots,\boldsymbol{\alpha}_n\}$ and the expression is not unique if and only if the linear system $x_1\boldsymbol{\alpha}_1+x_2\boldsymbol{\alpha}_2+\cdots+x_n\boldsymbol{\alpha}_n=\boldsymbol{\beta}$ has infinitely many solutions.

定义 5 如果向量组 $\{\boldsymbol{\alpha}_1,\boldsymbol{\alpha}_2,\cdots,\boldsymbol{\alpha}_s\}$ 中每一个向量 $\boldsymbol{\alpha}_i(i=1,2,\cdots,s)$ 都可以由向量组 $\{\boldsymbol{\beta}_1,\boldsymbol{\beta}_2,\cdots,\boldsymbol{\beta}_r\}$ 线性表示, 则称 $\{\boldsymbol{\alpha}_1,\boldsymbol{\alpha}_2,\cdots,\boldsymbol{\alpha}_s\}$ 可以由 $\{\boldsymbol{\beta}_1,\boldsymbol{\beta}_2,\cdots,\boldsymbol{\beta}_r\}$ 线性表示. 如果两个向量组 $\{\boldsymbol{\alpha}_1,\boldsymbol{\alpha}_2,\cdots,\boldsymbol{\alpha}_s\}$ 与 $\{\boldsymbol{\beta}_1,\boldsymbol{\beta}_2,\cdots,\boldsymbol{\beta}_r\}$ 可以互相线性表示, 则称这两个向量组**等价**.

Definition 5 If each vector $\boldsymbol{\alpha}_i(i=1,2,\cdots,s)$ in the vector set $\{\boldsymbol{\alpha}_1,\boldsymbol{\alpha}_2,\cdots,\boldsymbol{\alpha}_s\}$ can be linearly represented by the vector set $\{\boldsymbol{\beta}_1,\boldsymbol{\beta}_2,\cdots,\boldsymbol{\beta}_r\}$, then $\{\boldsymbol{\alpha}_1,\boldsymbol{\alpha}_2,\cdots,\boldsymbol{\alpha}_s\}$ is said to be linearly represented by $\{\boldsymbol{\beta}_1,\boldsymbol{\beta}_2,\cdots,\boldsymbol{\beta}_r\}$. If the two vector sets $\{\boldsymbol{\alpha}_1,\boldsymbol{\alpha}_2,\cdots,\boldsymbol{\alpha}_s\}$ and $\{\boldsymbol{\beta}_1,\boldsymbol{\beta}_2,\cdots,\boldsymbol{\beta}_r\}$ can be linearly represented by each other, then they are said to be **equivalent**.

注 向量组等价与矩阵等价是完全不同的两个概念.

Note Equivalence of vector sets and equivalence of matrices are two entirely different concepts.

不难证明向量组的等价具有如下性质:

It is not difficult to show that equivalence of vector sets has the following properties.

(1) **反身性** 向量组 $\{\boldsymbol{\alpha}_1,\boldsymbol{\alpha}_2,\cdots,\boldsymbol{\alpha}_s\}$ 与它自己等价;

(1) **Reflexivity** The vector set $\{\boldsymbol{\alpha}_1,\boldsymbol{\alpha}_2,\cdots,\boldsymbol{\alpha}_s\}$ is equivalent to itself;

(2) **对称性** 如果向量组 $\{\boldsymbol{\alpha}_1,\boldsymbol{\alpha}_2,\cdots,\boldsymbol{\alpha}_s\}$ 与 $\{\boldsymbol{\beta}_1,\boldsymbol{\beta}_2,\cdots,\boldsymbol{\beta}_r\}$ 等价, 那么向量组 $\{\boldsymbol{\beta}_1,\boldsymbol{\beta}_2,\cdots,\boldsymbol{\beta}_r\}$ 也与 $\{\boldsymbol{\alpha}_1,\boldsymbol{\alpha}_2,\cdots,\boldsymbol{\alpha}_s\}$ 等价;

(2) **Symmetry** If the vector set $\{\boldsymbol{\alpha}_1,\boldsymbol{\alpha}_2,\cdots,\boldsymbol{\alpha}_s\}$ is equivalent to $\{\boldsymbol{\beta}_1,\boldsymbol{\beta}_2,\cdots,\boldsymbol{\beta}_r\}$, then the vector set $\{\boldsymbol{\beta}_1,\boldsymbol{\beta}_2,\cdots,\boldsymbol{\beta}_r\}$ is equivalent to $\{\boldsymbol{\alpha}_1,\boldsymbol{\alpha}_2,\cdots,\boldsymbol{\alpha}_s\}$;

(3) **传递性** 如果向量组 $\{\boldsymbol{\alpha}_1,\boldsymbol{\alpha}_2,\cdots,\boldsymbol{\alpha}_s\}$ 与 $\{\boldsymbol{\beta}_1,\boldsymbol{\beta}_2,\cdots,\boldsymbol{\beta}_r\}$ 等价, 而向量组 $\{\boldsymbol{\beta}_1,\boldsymbol{\beta}_2,\cdots,\boldsymbol{\beta}_r\}$ 又与 $\{\boldsymbol{\gamma}_1,\boldsymbol{\gamma}_2,\cdots,\boldsymbol{\gamma}_p\}$ 等价, 那么向量组 $\{\boldsymbol{\alpha}_1,\boldsymbol{\alpha}_2,\cdots,\boldsymbol{\alpha}_s\}$ 与 $\{\boldsymbol{\gamma}_1,\boldsymbol{\gamma}_2,\cdots,\boldsymbol{\gamma}_p\}$ 等价.

(3) **Transitivity** If the vector set $\{\boldsymbol{\alpha}_1,\boldsymbol{\alpha}_2,\cdots,\boldsymbol{\alpha}_s\}$ is equivalent to $\{\boldsymbol{\beta}_1,\boldsymbol{\beta}_2,\cdots,\boldsymbol{\beta}_r\}$, and if the vector set $\{\boldsymbol{\beta}_1,\boldsymbol{\beta}_2,\cdots,\boldsymbol{\beta}_r\}$ is equivalent to $\{\boldsymbol{\gamma}_1,\boldsymbol{\gamma}_2,\cdots,\boldsymbol{\gamma}_p\}$, then the vector set $\{\boldsymbol{\alpha}_1,\boldsymbol{\alpha}_2,\cdots,\boldsymbol{\alpha}_s\}$ is equivalent to $\{\boldsymbol{\gamma}_1,\boldsymbol{\gamma}_2,\cdots,\boldsymbol{\gamma}_p\}$.

例 3 已知向量组对应的矩阵为

Example 3 Known that the matrices associated with the vector sets are

$$(\boldsymbol{\alpha}_1,\boldsymbol{\alpha}_2)=\begin{pmatrix}2&3\\0&-2\\-1&1\\3&-1\end{pmatrix},\quad (\boldsymbol{\beta}_1,\boldsymbol{\beta}_2)=\begin{pmatrix}-5&4\\6&-4\\-5&3\\9&-5\end{pmatrix}$$

证明 $\{\boldsymbol{\alpha}_1,\boldsymbol{\alpha}_2\}$ 与 $\{\boldsymbol{\beta}_1,\boldsymbol{\beta}_2\}$ 等价.

Prove that $(\boldsymbol{\alpha}_1,\boldsymbol{\alpha}_2)$ is equivalent to $(\boldsymbol{\beta}_1,\boldsymbol{\beta}_2)$.

证 要证等价性, 即证存在 2 阶方阵 $\boldsymbol{X},\boldsymbol{Y}$, 使

Proof To prove the equivalence is to show that there exist $\boldsymbol{X},\boldsymbol{Y}$, which are square matrices of order 2, such that

$$(\boldsymbol{\alpha}_1,\boldsymbol{\alpha}_2)\boldsymbol{X}=(\boldsymbol{\beta}_1,\boldsymbol{\beta}_2),\quad (\boldsymbol{\beta}_1,\boldsymbol{\beta}_2)\boldsymbol{Y}=(\boldsymbol{\alpha}_1,\boldsymbol{\alpha}_2)$$

求矩阵 $\boldsymbol{X}=\begin{pmatrix}x_1&y_1\\x_2&y_2\end{pmatrix}$ 相当于求解两个非齐次线性方程组

Finding the matrix $\boldsymbol{X}=\begin{pmatrix}x_1&y_1\\x_2&y_2\end{pmatrix}$ is equivalent to solving two non-homogeneous linear systems, namely,

$$x_1\boldsymbol{\alpha}_1+x_2\boldsymbol{\alpha}_2=\boldsymbol{\beta}_1,\quad y_1\boldsymbol{\alpha}_1+y_2\boldsymbol{\alpha}_2=\boldsymbol{\beta}_2$$

对增广矩阵 $(\boldsymbol{\alpha}_1,\boldsymbol{\alpha}_2,\boldsymbol{\beta}_1,\boldsymbol{\beta}_2)$ 施以初等行变换

Performing elementary row operations on the augmented matrix $(\boldsymbol{\alpha}_1,\boldsymbol{\alpha}_2,\boldsymbol{\beta}_1,\boldsymbol{\beta}_2)$ yields

$$(\boldsymbol{\alpha}_1,\boldsymbol{\alpha}_2,\boldsymbol{\beta}_1,\boldsymbol{\beta}_2)=\begin{pmatrix}2&3&-5&4\\0&-2&6&-4\\-1&1&-5&3\\3&-1&9&-5\end{pmatrix}\xrightarrow[r_3+2r_1,r_4+3r_1]{r_1\leftrightarrow r_3}\begin{pmatrix}-1&1&-5&3\\0&-2&6&-4\\0&5&-15&10\\0&2&-6&4\end{pmatrix}$$

$$\xrightarrow[r_3-5r_2,r_4-2r_2]{-1/2r_2}\begin{pmatrix}-1&1&-5&3\\0&1&-3&2\\0&0&0&0\\0&0&0&0\end{pmatrix}\xrightarrow{r_1-r_2,-r_1}\begin{pmatrix}1&0&2&-1\\0&1&-3&2\\0&0&0&0\\0&0&0&0\end{pmatrix}$$

因此, 有

Thus, we have

$$\boldsymbol{X}=\begin{pmatrix}2&-1\\-3&2\end{pmatrix}$$

由 $|\boldsymbol{X}|=1\neq 0$ 知, $\boldsymbol{X}$ 可逆. 取 $\boldsymbol{Y}=\boldsymbol{X}^{-1}$ 即为所求. 因此向量组 $\{\boldsymbol{\alpha}_1,\boldsymbol{\alpha}_2\}$ 与 $\{\boldsymbol{\beta}_1,\boldsymbol{\beta}_2\}$ 等价. ■

From $|\boldsymbol{X}|=1\neq 0$, we see that $\boldsymbol{X}$ is invertible. Let $\boldsymbol{Y}=\boldsymbol{X}^{-1}$, which is our desired. Therefore the vector sets $\{\boldsymbol{\alpha}_1,\boldsymbol{\alpha}_2\}$ and $\{\boldsymbol{\beta}_1,\boldsymbol{\beta}_2\}$ are equivalent. ■

思 考 题

Question

不同维数的向量能进行线性运算吗?

Can vectors of different dimensions perform linear operations?

4.2 向量组的线性相关性

4.2 Linear Dependence of Vector Sets

引例 考虑下面的线性方程组

Citing example Consider the following linear system

$$\begin{cases}-x_1+3x_2+x_3=0\\2x_1+\ \ x_2\qquad\ \ =0\\x_1+4x_2+x_3=0\end{cases}$$

哪个是多余的?

Which equation is needless?

不难看出, 方程组的系数矩阵为

It is not difficult to see that the coefficient matrix of the system is

$$\boldsymbol{A}=\begin{pmatrix}-1&3&1\\2&1&0\\1&4&1\end{pmatrix}$$

显然，行向量组 $\boldsymbol{\beta}_1=(-1,3,1),\boldsymbol{\beta}_2=(2,1,0),\boldsymbol{\beta}_3=(1,4,1)$ 有下面关系

Obviously, the row vector set, given by $\boldsymbol{\beta}_1=(-1,3,1),\boldsymbol{\beta}_2=(2,1,0),\boldsymbol{\beta}_3=(1,4,1)$, has the following relation

$$\boldsymbol{\beta}_1+\boldsymbol{\beta}_2=\boldsymbol{\beta}_3$$

即第三个方程是多余.

that is to say, the third equation is needless.

一般地，若向量 $\boldsymbol{\beta}$ 能由向量组 $\{\boldsymbol{\alpha}_1,\boldsymbol{\alpha}_2,\cdots,\boldsymbol{\alpha}_s\}$ 线性表出，那么我们掌握了 $\{\boldsymbol{\alpha}_1,\boldsymbol{\alpha}_2,\cdots,\boldsymbol{\alpha}_s\}$ 的信息也就掌握了 $\boldsymbol{\beta}$ 的信息. 换句话说，一旦丢了 $\boldsymbol{\beta}$ 的信息，也可以通过 $\{\boldsymbol{\alpha}_1,\boldsymbol{\alpha}_2,\cdots,\boldsymbol{\alpha}_s\}$ 将它找回来. 因此我们在研究一个向量组时，总是关心这个向量组是否存在这种关系. 若向量组中有一个向量能用其他向量线性表出，则暂时将它丢掉也不会影响我们的讨论. 于是, 如果一个向量组存在这种关系, 就称这个向量组是线性相关的，否则称这个向量组是线性无关的. 为此, 我们给出如下定义.

In general, if the vector $\boldsymbol{\beta}$ can be linearly represented by the vector set $\{\boldsymbol{\alpha}_1,\boldsymbol{\alpha}_2,\cdots,\boldsymbol{\alpha}_s\}$, then the information about $\{\boldsymbol{\alpha}_1,\boldsymbol{\alpha}_2,\cdots,\boldsymbol{\alpha}_s\}$ that we know is equivalent to the information about $\boldsymbol{\beta}$. In other words, once we lose the information about $\boldsymbol{\beta}$, we can find it by $\{\boldsymbol{\alpha}_1,\boldsymbol{\alpha}_2,\cdots,\boldsymbol{\alpha}_s\}$. Therefore, as we study a vector set, we always focus on whether or not there exists this kind of relation. If there is a vector in this vector set can be linearly represented by other vectors, then it can not effect our discussion as this vector is dropped for the moment. Consequently, if a vector set exists this relation, then it is said to be linearly dependent; otherwise, it is said to be linearly independent. For this purpose, we give the following definition.

定义 1 对于给定的 n 维向量组

Definition 1 For the given n dimensional vector set

$$\{\boldsymbol{\alpha}_1,\boldsymbol{\alpha}_2,\cdots,\boldsymbol{\alpha}_m\}$$

如果存在不全为零的数 $k_1,k_2,\cdots,k_m$，使

if there exist numbers $k_1,k_2,\cdots,k_m$, which are not all zeros, such that

$$k_1\boldsymbol{\alpha}_1+k_2\boldsymbol{\alpha}_2+\cdots+k_m\boldsymbol{\alpha}_m=\mathbf{0} \tag{4.4}$$

则称向量组 $\{\boldsymbol{\alpha}_1,\boldsymbol{\alpha}_2,\cdots,\boldsymbol{\alpha}_m\}$ **线性相关**，否则称为**线性无关**.

then the vector set $\{\boldsymbol{\alpha}_1,\boldsymbol{\alpha}_2,\cdots,\boldsymbol{\alpha}_m\}$ is said to be **linearly dependent**; otherwise, it is said to be **linearly independent**.

注 (1) (4.4) 可以看成一个齐次线性方程组.

Note (1) (4.4) may be viewed as a homogeneous linear system.

(2) 两个三维向量 $\boldsymbol{\alpha}_1, \boldsymbol{\alpha}_2$ 相关 $\Leftrightarrow$ $\boldsymbol{\alpha}_1 = k\boldsymbol{\alpha}_2$, 即 $\boldsymbol{\alpha}_1, \boldsymbol{\alpha}_2$ 共线; 三个三维向量 $\boldsymbol{\alpha}_1, \boldsymbol{\alpha}_2, \boldsymbol{\alpha}_3$ 相关 $\Leftrightarrow$ 存在不全为零的数 k_1, k_2, k_3 使 $k_1\boldsymbol{\alpha}_1 + k_2\boldsymbol{\alpha}_2 + k_3\boldsymbol{\alpha}_3 = \mathbf{0}$, 即 $\boldsymbol{\alpha}_1, \boldsymbol{\alpha}_2, \boldsymbol{\alpha}_3$ 共面.

(2) Two vectors $\boldsymbol{\alpha}_1, \boldsymbol{\alpha}_2$ of dimension 3 are linearly dependent$\Leftrightarrow$ $\boldsymbol{\alpha}_1 = k\boldsymbol{\alpha}_2$, namely, $\boldsymbol{\alpha}_1, \boldsymbol{\alpha}_2$ are collinear; Three vectors $\boldsymbol{\alpha}_1, \boldsymbol{\alpha}_2, \boldsymbol{\alpha}_3$ of dimension 3 are linearly dependent $\Leftrightarrow$ there exist numbers k_1, k_2, k_3, which are not all zeros, such that $k_1\boldsymbol{\alpha}_1 + k_2\boldsymbol{\alpha}_2 + k_3\boldsymbol{\alpha}_3 = \mathbf{0}$, namely, $\boldsymbol{\alpha}_1, \boldsymbol{\alpha}_2, \boldsymbol{\alpha}_3$ are coplaner.

由定义 1 不难得到: 向量组 $\{\boldsymbol{\alpha}_1, \boldsymbol{\alpha}_2, \cdots, \boldsymbol{\alpha}_m\}$ 线性无关 $\Leftrightarrow$ 若 $k_1\boldsymbol{\alpha}_1 + k_2\boldsymbol{\alpha}_2 + \cdots + k_m\boldsymbol{\alpha}_m = \mathbf{0}$, 则 $k_1 = k_2 = \cdots = k_m = 0$.

From Definition 1 we easily know that: the vector set $\{\boldsymbol{\alpha}_1, \boldsymbol{\alpha}_2, \cdots, \boldsymbol{\alpha}_m\}$ is linearly independent $\Leftrightarrow$ if $k_1\boldsymbol{\alpha}_1 + k_2\boldsymbol{\alpha}_2 + \cdots + k_m\boldsymbol{\alpha}_m = \mathbf{0}$, then $k_1 = k_2 = \cdots = k_m = 0$.

定理 1 向量组 $\{\boldsymbol{\alpha}_1, \boldsymbol{\alpha}_2, \cdots, \boldsymbol{\alpha}_m\}(m \geqslant 2)$ 线性相关的充分必要条件是向量组中至少有一个向量可由其余 $m-1$ 个向量线性表示.

Theorem 1 The vector set $\{\boldsymbol{\alpha}_1, \boldsymbol{\alpha}_2, \cdots, \boldsymbol{\alpha}_m\}(m \geqslant 2)$ is linearly dependent if and only if there exists at least a vector in the vector set can be linearly represented by the other $m-1$ vectors.

证 **充分性** 设向量组中的一个向量, 例如, $\boldsymbol{\alpha}_m$ 能由其余 $m-1$ 个向量线性表示, 即有数 $\lambda_1, \lambda_2, \cdots, \lambda_{m-1}$, 使得

Proof **Sufficiency** Suppose that a vector, for instance, $\boldsymbol{\alpha}_m$, can be linearly represented by the other $m-1$ vectors, namely, there exist numbers $\lambda_1, \lambda_2, \cdots, \lambda_{m-1}$ such that

$$\boldsymbol{\alpha}_m = \lambda_1\boldsymbol{\alpha}_1 + \lambda_2\boldsymbol{\alpha}_2 + \cdots + \lambda_{m-1}\boldsymbol{\alpha}_{m-1}$$

即

namely,

$$\lambda_1\boldsymbol{\alpha}_1 + \lambda_2\boldsymbol{\alpha}_2 + \cdots + \lambda_{m-1}\boldsymbol{\alpha}_{m-1} + (-1)\boldsymbol{\alpha}_m = \mathbf{0}$$

因为 $\lambda_1, \lambda_2, \cdots, \lambda_{m-1}, -1$ 不全为零, 所以向量组 $\{\boldsymbol{\alpha}_1, \boldsymbol{\alpha}_2, \cdots, \boldsymbol{\alpha}_m\}$ 线性相关.

Since $\lambda_1, \lambda_2, \cdots, \lambda_{m-1}, -1$ are not all zeros, it means that the vector set $\{\boldsymbol{\alpha}_1, \boldsymbol{\alpha}_2, \cdots, \boldsymbol{\alpha}_m\}$ is linearly dependent.

必要性 设向量组 $\{\boldsymbol{\alpha}_1, \boldsymbol{\alpha}_2, \cdots, \boldsymbol{\alpha}_m\}$ 线性相关, 即存在 m 个不全为零的数 $k_1, k_2, \cdots, k_m$ 使得

Necessity Suppose that the vector set $\{\boldsymbol{\alpha}_1, \boldsymbol{\alpha}_2, \cdots, \boldsymbol{\alpha}_m\}$ is linearly dependent, namely, there exist m numbers $k_1, k_2, \cdots, k_m$, which are not all zeros, such that

$$k_1\boldsymbol{\alpha}_1 + k_2\boldsymbol{\alpha}_2 + \cdots + k_m\boldsymbol{\alpha}_m = \mathbf{0}$$

因为 $k_1, k_2, \cdots, k_m$ 中至少有一个不为零, 不妨设 $k_m \neq 0$, 于是

Since there is at least one nonzero number in $k_1, k_2, \cdots, k_m$, not loss of generality, suppose that $k_m \neq 0$, we have

$$\boldsymbol{\alpha}_m = \left(-\frac{k_1}{k_m}\right)\boldsymbol{\alpha}_1 + \left(-\frac{k_2}{k_m}\right)\boldsymbol{\alpha}_2 + \cdots + \left(-\frac{k_{m-1}}{k_m}\right)\boldsymbol{\alpha}_{m-1}$$

即 $\boldsymbol{\alpha}_m$ 能由其余 $m-1$ 个向量线性表示.▲

It means that $\boldsymbol{\alpha}_m$ can be linearly represented by the other $m-1$ vectors. ▲

定理 2 设向量组 $\{\boldsymbol{\alpha}_1, \boldsymbol{\alpha}_2, \cdots, \boldsymbol{\alpha}_m\}$ 线性无关, 而向量组 $\{\boldsymbol{\alpha}_1, \boldsymbol{\alpha}_2, \cdots, \boldsymbol{\alpha}_m, \boldsymbol{\beta}\}$ 线性相关, 则 $\boldsymbol{\beta}$ 能由 $\{\boldsymbol{\alpha}_1, \boldsymbol{\alpha}_2, \cdots, \boldsymbol{\alpha}_m\}$ 线性表示, 且表示式是唯一的.

Theorem 2 Suppose that the vector set $\{\boldsymbol{\alpha}_1, \boldsymbol{\alpha}_2, \cdots, \boldsymbol{\alpha}_m\}$ is linearly independent, but the vector set $\{\boldsymbol{\alpha}_1, \boldsymbol{\alpha}_2, \cdots, \boldsymbol{\alpha}_m, \boldsymbol{\beta}\}$ is linearly dependent, then $\boldsymbol{\beta}$ can be linearly represented by $\{\boldsymbol{\alpha}_1, \boldsymbol{\alpha}_2, \cdots, \boldsymbol{\alpha}_m\}$ and the expression is unique.

证 因为向量组 $\{\boldsymbol{\alpha}_1, \boldsymbol{\alpha}_2, \cdots, \boldsymbol{\alpha}_m, \boldsymbol{\beta}\}$ 线性相关, 所以存在不全为零的 $m+1$ 个数 $k_1, k_2, \cdots, k_m, k$, 使得

Proof Since the vector set $\{\boldsymbol{\alpha}_1, \boldsymbol{\alpha}_2, \cdots, \boldsymbol{\alpha}_m, \boldsymbol{\beta}\}$ is linearly dependent, there exist $m+1$ numbers $k_1, k_2, \cdots, k_m, k$, which are not all zeros, such that

$$k_1\boldsymbol{\alpha}_1 + k_2\boldsymbol{\alpha}_2 + \cdots + k_m\boldsymbol{\alpha}_m + k\boldsymbol{\beta} = \mathbf{0}$$

如果 $k=0$, 则 $k_1, k_2, \cdots, k_m$ 不全为零, 且有

If $k=0$, then $k_1, k_2, \cdots, k_m$ are not all zeros and

$$k_1\boldsymbol{\alpha}_1 + k_2\boldsymbol{\alpha}_2 + \cdots + k_m\boldsymbol{\alpha}_m = \mathbf{0}$$

于是向量组 $\{\boldsymbol{\alpha}_1, \boldsymbol{\alpha}_2, \cdots, \boldsymbol{\alpha}_m\}$ 线性相关, 这与已知条件矛盾. 因此 $k \neq 0$ 且

This shows that the vector set $\{\boldsymbol{\alpha}_1, \boldsymbol{\alpha}_2, \cdots, \boldsymbol{\alpha}_m\}$ is linearly dependent, which contradicts to the known condition, thereby, $k \neq 0$ and

$$\boldsymbol{\beta} = -\frac{k_1}{k}\boldsymbol{\alpha}_1 - \frac{k_2}{k}\boldsymbol{\alpha}_2 - \cdots - \frac{k_m}{k}\boldsymbol{\alpha}_m$$

再证表示式是唯一的. 设 $\boldsymbol{\beta}$ 有两个表示式

Next we prove the expression is unique. Suppose that $\boldsymbol{\beta}$ has two expressions

$$\boldsymbol{\beta} = \lambda_1\boldsymbol{\alpha}_1 + \cdots + \lambda_m\boldsymbol{\alpha}_m$$

及

and

$$\boldsymbol{\beta} = \mu_1\boldsymbol{\alpha}_1 + \cdots + \mu_m\boldsymbol{\alpha}_m$$

两式相减得

Subtracting the two equations leads to

$$(\lambda_1 - \mu_1)\boldsymbol{\alpha}_1 + \cdots + (\lambda_1 - \mu_1)\boldsymbol{\alpha}_m = \mathbf{0}$$

因为向量组 $\{\boldsymbol{\alpha}_1, \boldsymbol{\alpha}_2, \cdots, \boldsymbol{\alpha}_m\}$ 线性无关, 所以, $\lambda_i - \mu_i = 0$, 因此 $\lambda_i = \mu_i (i = 1, 2, \cdots, m)$. ▲

Since the vector set $\{\boldsymbol{\alpha}_1, \boldsymbol{\alpha}_2, \cdots, \boldsymbol{\alpha}_m\}$ are linearly independent, we have $\lambda_i - \mu_i = 0$, i.e., $\lambda_i = \mu_i (i = 1, 2, \cdots, m)$. ▲

根据本节定义 3 容易证明如下定理成立.

From Definition 3 in this section, it is easy to show that the following theorem is valid.

定理 3 列向量组 $\{\boldsymbol{\alpha}_1, \boldsymbol{\alpha}_2, \cdots, \boldsymbol{\alpha}_n\}$ 线性相 (无) 关 $\Leftrightarrow$ 有 (没有) 不全零的数 $x_1, x_2, \cdots, x_n$ 使得 $x_1\boldsymbol{\alpha}_1 + x_2\boldsymbol{\alpha}_2 + \cdots + x_n\boldsymbol{\alpha}_n = \mathbf{0} \Leftrightarrow$ 方程组

Theorem 3 The column vector set $\{\boldsymbol{\alpha}_1, \boldsymbol{\alpha}_2, \cdots, \boldsymbol{\alpha}_n\}$ is linearly dependent (independent) $\Leftrightarrow$ there exist (does not exist) n numbers $x_1, x_2, \cdots, x_n$, which are not all zeros, such that $x_1\boldsymbol{\alpha}_1 + x_2\boldsymbol{\alpha}_2 + \cdots + x_n\boldsymbol{\alpha}_n = \mathbf{0} \Leftrightarrow$ the system

$$\begin{cases} a_{11}x_1 + a_{12}x_2 + \cdots + a_{1n}x_n = 0 \\ a_{21}x_1 + a_{21}x_2 + \cdots + a_{2n}x_n = 0 \\ \cdots\cdots \\ a_{m1}x_1 + a_{m2}x_2 + \cdots + a_{mn}x_n = 0 \end{cases} \tag{4.5}$$

有非零解 (只有零解).

has non-zero solutions (has only a zero solution).

注 若向量组 $\{\boldsymbol{\alpha}_1, \boldsymbol{\alpha}_2, \cdots, \boldsymbol{\alpha}_m\}$ 线性无关, 则当且仅当 $k_1 = k_2 = \cdots = k_m = 0$ 时, 有

Note If the vector set $\{\boldsymbol{\alpha}_1, \boldsymbol{\alpha}_2, \cdots, \boldsymbol{\alpha}_m\}$ is linearly independent, then if and only if $k_1 = k_2 = \cdots = k_m = 0$, we have that

$$k_1\boldsymbol{\alpha}_1 + k_2\boldsymbol{\alpha}_2 + \cdots + k_m\boldsymbol{\alpha}_m = \mathbf{0}$$

才成立. 例如, 对于 $\boldsymbol{\alpha}_1 = \begin{pmatrix} 1 \\ 0 \end{pmatrix}, \boldsymbol{\alpha}_2 = \begin{pmatrix} 2 \\ 0 \end{pmatrix}$, 当 $k_1 = k_2 = 0$ 时, 有 $k_1\boldsymbol{\alpha}_1 + k_2\boldsymbol{\alpha}_2 = 0$ 成立, 但不能由此推出向量 $\boldsymbol{\alpha}_1$,

is just valid. For example, for $\boldsymbol{\alpha}_1 = \begin{pmatrix} 1 \\ 0 \end{pmatrix}, \boldsymbol{\alpha}_2 = \begin{pmatrix} 2 \\ 0 \end{pmatrix}$, as $k_1 = k_2 = 0$, the equation $k_1\boldsymbol{\alpha}_1 + k_2\boldsymbol{\alpha}_2 = \mathbf{0}$ is valid, however, it can not be induced that α_1, α_2 are linearly

$\boldsymbol{\alpha}_2$ 线性无关. 事实上, 取 $k_1 = 2, k_2 = -1$ 时, $k_1\boldsymbol{\alpha}_1 + k_2\boldsymbol{\alpha}_2 = \mathbf{0}$ 也成立, 所以 $\boldsymbol{\alpha}_1, \boldsymbol{\alpha}_2$ 线性相关.

independent. In fact, let $k_1 = 2, k_2 = -1$, the equation $k_1\boldsymbol{\alpha}_1 + k_2\boldsymbol{\alpha}_2 = \mathbf{0}$ is also valid, so $\boldsymbol{\alpha}_1, \boldsymbol{\alpha}_2$ are linearly dependent.

一个向量组中的向量只可能存在两种线性关系的一种: 或是线性相关, 或是线性无关. 由本节定义 1 和定理 2 可以得到以下结论:

The vectors in a vector set may have only one of the two linear relations, namely, either linearly dependent or linearly independent. From Definition 1 and Theorem 2 in this section we have the following results:

(1) 只有一个向量 $\boldsymbol{\alpha}$ 构成的向量组, 若 $\boldsymbol{\alpha} \neq \mathbf{0}$ 则线性无关; $\boldsymbol{\alpha} = \mathbf{0}$ 则线性相关.

(1) For a vector set containing only one vector $\boldsymbol{\alpha}$, if $\boldsymbol{\alpha} \neq \mathbf{0}$, it is linearly independent; while if $\boldsymbol{\alpha} = \mathbf{0}$, it is linearly dependent.

(2) 由两个向量 $\boldsymbol{\alpha}, \boldsymbol{\beta}$ 构成的向量组, 向量组线性相关的充要条件是它们的对应分量成比例.

(2) For a vector set containing two vectors $\boldsymbol{\alpha}, \boldsymbol{\beta}$, it is linearly dependent if and only if their corresponding components are proportional.

(3) 含有零向量的向量组必线性相关.

(3) A vector set containing a zero vector must be linearly dependent.

(4) 向量组和其部分组的关系:

(4) Relation between a vector set and its part set:

设 $\{\boldsymbol{\alpha}_1, \boldsymbol{\alpha}_2, \cdots, \boldsymbol{\alpha}_r\}$ 是向量组 $\{\boldsymbol{\alpha}_1, \boldsymbol{\alpha}_2, \cdots, \boldsymbol{\alpha}_n\}(r < n)$ 的部分向量组, 则有

Let $\{\boldsymbol{\alpha}_1, \boldsymbol{\alpha}_2, \cdots, \boldsymbol{\alpha}_r\}$ be a part set of the vector set $\{\boldsymbol{\alpha}_1, \boldsymbol{\alpha}_2, \cdots, \boldsymbol{\alpha}_n\}(r < n)$, we have

$$\text{部分向量组线性相关} \underset{\not\Leftarrow}{\Rightarrow} \text{整体向量组线性相关}$$

$$\begin{gathered}\text{A part vector set is linearly dependent}\\ \Rightarrow\\ \not\Leftarrow\\ \text{the whole vector set is linearly dependent}\end{gathered}$$

$$\text{整体向量组线性无关} \underset{\not\Leftarrow}{\Rightarrow} \text{部分向量组线性无关}$$

$$\begin{gathered}\text{A whole vector set is linearly independent}\\ \Rightarrow\\ \not\Leftarrow\\ \text{the part vector set is linearly independent}\end{gathered}$$

(5) 延伸向量组和缩短向量组的关系:

(5) Relation between an extended vector set and a shortened vector set:

称向量组 | The vector set

$$\boldsymbol{\alpha}_1=\begin{pmatrix}a_{11}\\a_{21}\\\vdots\\a_{m1}\end{pmatrix},\boldsymbol{\alpha}_2=\begin{pmatrix}a_{12}\\a_{22}\\\vdots\\a_{m2}\end{pmatrix},\cdots,\boldsymbol{\alpha}_n=\begin{pmatrix}a_{1n}\\a_{2n}\\\vdots\\a_{mn}\end{pmatrix}$$

为缩短向量组. | is called a shortened vector set.

称向量组 | The vector set

$$\boldsymbol{\alpha}_1'=\begin{pmatrix}a_{11}\\a_{21}\\\vdots\\a_{m1}\\a_{m+1,1}\end{pmatrix},\boldsymbol{\alpha}_2'=\begin{pmatrix}a_{12}\\a_{22}\\\vdots\\a_{m2}\\a_{m+1,2}\end{pmatrix},\cdots,\boldsymbol{\alpha}_n'=\begin{pmatrix}a_{1n}\\a_{2n}\\\vdots\\a_{mn}\\a_{m+1,n}\end{pmatrix}$$

为延伸向量组. | is called a extended vector set.

则有 | Then we have

延伸向量组线性相关 $\underset{\not\Leftarrow}{\Rightarrow}$ 缩短向量组线性相关 | An extended vector set is linearly dependent $\underset{\not\Leftarrow}{\Rightarrow}$ the shortened vector set is linearly dependent

缩短向量组线性无关 $\underset{\not\Leftarrow}{\Rightarrow}$ 延伸向量组线性无关 | A shortened vector set is linearly independent $\underset{\not\Leftarrow}{\Rightarrow}$ the extended vector set is linearly independent

判别向量组的线性相 (无) 关的方法: | Methods for deciding linear dependence (independence) of vector sets:

设 $\boldsymbol{\beta}_1,\boldsymbol{\beta}_2,\cdots,\boldsymbol{\beta}_n$ 为 n 个 m 维列向量, 由这些向量组成的矩阵为 $\boldsymbol{A}_{m\times n}=(\boldsymbol{\beta}_1,\boldsymbol{\beta}_2,\cdots,\boldsymbol{\beta}_n)$, 对应的齐次线性方程组为 | Let $\boldsymbol{\beta}_1,\boldsymbol{\beta}_2,\cdots,\boldsymbol{\beta}_n$ be n column vectors of dimension m. The matrix consisted of these vectors is $\boldsymbol{A}_{m\times n}=(\boldsymbol{\beta}_1,\boldsymbol{\beta}_2,\cdots,\boldsymbol{\beta}_n)$ and the corresponding system of homogeneous linear equations is given by

$$x_1\boldsymbol{\beta}_1+x_2\boldsymbol{\beta}_2+\cdots+x_n\boldsymbol{\beta}_n=\mathbf{0} \tag{4.6}$$

列向量组中向量的维数m

$\Updownarrow$

对应系数矩阵的行数m

$\Updownarrow$

对应方程组中方程的个数m

dimension m of vectors in a column vector set

$\Updownarrow$

row number m of the corresponding coefficient matrix

$\Updownarrow$

equation number m of the corresponding system

列向量组中向量的个数n

$\Updownarrow$

对应系数矩阵的列数n

$\Updownarrow$

对应方程组中未知数的个数n

vector number n in a column vector set

$\Updownarrow$

column number n of the corresponding coefficient matrix

$\Updownarrow$

unknown number n of the corresponding system

根据线性方程组的解的 3 种情况 (对应向量组的 3 种情况) 和本节定理 3, 得到如下 3 种判别方法:

From the three cases of solutions of a linear system (corresponding to the three cases of relations of a vector set) and Theorem 3 in this section, we have the following three discriminate methods:

(1) $m > n$.

用初等行变换将 $\boldsymbol{A}_{m\times n}$ 化为行阶梯形, 就可以知道对应的线性方程组 (4.6) 有非零解 $\Leftrightarrow R(\boldsymbol{A}_{m\times n}) < n$(此时向量组线性相关); 只有零解 $\Leftrightarrow R(\boldsymbol{A}_{m\times n}) = n$(此时向量组线性无关).

(1) $m > n$.

Reduce $\boldsymbol{A}_{m\times n}$ to a row echelon form by elementary row operations, and then we know that the corresponding system (4.6) has non-zero solutions $\Leftrightarrow R(\boldsymbol{A}_{m\times n}) < n$ (in this case, the vector set is linearly dependent); and that it has only a zero solution $\Leftrightarrow R(\boldsymbol{A}_{m\times n}) = n$ (in this case, the vector set is linearly independent).

(2) $m < n$.

因为 $R(\boldsymbol{A}_{m\times n}) \leqslant m < n$, 此时对应的方程组 (4.6) 一定有非零解, 所以向量组一定线性相关.

(2) $m < n$.

Since $R(\boldsymbol{A}_{m\times n}) \leqslant m < n$, in this case, the corresponding system (4.6) must have nonzero solutions, this means that the vector set must be linearly dependent.

例如, 由 5 个四维列向量构成向量组一定线性相关.

For example, a vector set consisted of 5 column vectors of dimension 4 must be linearly dependent.

(3) $m=n$.

(3) $m=n$.

由克拉默法则, 有

From Gramar's rule, we have

行列式 $|(\boldsymbol{\beta}_1,\boldsymbol{\beta}_2,\cdots,\boldsymbol{\beta}_n)|=0\Rightarrow$ 对应的方程组 (4.6) 有非零解 $\Rightarrow$ 向量组一定线性相关.

The determinant $|(\boldsymbol{\beta}_1,\boldsymbol{\beta}_2,\cdots,\boldsymbol{\beta}_n)|=0\Rightarrow$ the corresponding system (4.6) must have nonzero solutions $\Rightarrow$ the vector set must be linearly dependent.

行列式 $|(\boldsymbol{\beta}_1,\boldsymbol{\beta}_2,\cdots,\boldsymbol{\beta}_n)|\neq 0\Rightarrow$ 对应的方程组 (4.6) 只有零解 $\Rightarrow$ 向量组一定线性无关.

The determinant $|(\boldsymbol{\beta}_1,\boldsymbol{\beta}_2,\cdots,\boldsymbol{\beta}_n)|\neq 0\Rightarrow$ the corresponding system (4.6) has only a zero solutions $\Rightarrow$ the vector set must be linearly independent.

例 1 讨论 n 维单位向量组 $\{\boldsymbol{\varepsilon}_1,\boldsymbol{\varepsilon}_2,\cdots,\boldsymbol{\varepsilon}_n\}$ 的线性相关性

Example 1 Discuss the linear dependence of the n dimensional unit vector set $\{\boldsymbol{\varepsilon}_1,\boldsymbol{\varepsilon}_2,\cdots,\boldsymbol{\varepsilon}_n\}$

解 设有 n 个数 $k_1,k_2,\cdots,k_n$ 使

Solution Suppose that there exist n numbers $k_1,k_2,\cdots,k_n$ such that

$$k_1\boldsymbol{\varepsilon}_1+k_2\boldsymbol{\varepsilon}_2+\cdots+k_n\boldsymbol{\varepsilon}_n=\mathbf{0}$$

所以有

So we have

$$\begin{pmatrix}k_1\\k_2\\\vdots\\k_n\end{pmatrix}=\begin{pmatrix}0\\0\\\vdots\\0\end{pmatrix}$$

即 $k_1=k_2=\cdots=k_n=0$. 所以向量组 $\{\boldsymbol{\varepsilon}_1,\boldsymbol{\varepsilon}_3,\cdots,\boldsymbol{\varepsilon}_n\}$ 线性无关. ■

namely, $k_1=k_2=\cdots=k_n=0$. This shows that the vector set $\{\boldsymbol{\varepsilon}_1,\boldsymbol{\varepsilon}_3,\cdots,\boldsymbol{\varepsilon}_n\}$ is linearly independent. ■

例 2 判断下面向量组的线性相关性

Example 2 Decide the linear dependence of the following vector set

$$\boldsymbol{\alpha}_1=\begin{pmatrix}-1\\3\\1\end{pmatrix},\quad\boldsymbol{\alpha}_2=\begin{pmatrix}2\\1\\0\end{pmatrix},\quad\boldsymbol{\alpha}_3=\begin{pmatrix}1\\4\\1\end{pmatrix}$$

解法 1 设有数 k_1, k_2, k_3 使

Solution 1 Suppose that there exist numbers k_1, k_2, k_3 such that

$$k_1\boldsymbol{\alpha}_1 + k_2\boldsymbol{\alpha}_2 + k_3\boldsymbol{\alpha}_3 = \mathbf{0}$$

即

namely,

$$k_1\begin{pmatrix}-1\\3\\1\end{pmatrix}+k_2\begin{pmatrix}2\\1\\0\end{pmatrix}+k_3\begin{pmatrix}1\\4\\1\end{pmatrix}=\begin{pmatrix}0\\0\\0\end{pmatrix}$$

于是有

Thus, we have

$$\begin{pmatrix}-k_1+2k_2+k_3\\3k_1+k_2+4k_3\\k_1+k_3\end{pmatrix}=\begin{pmatrix}0\\0\\0\end{pmatrix}$$

由向量相等的定义, 得

Using the definition of equal vectors yields

$$\begin{cases}-k_1+2k_2+\ \ k_3=0\\3k_1+\ \ k_2+4k_3=0\\k_1\qquad\quad+\ \ k_3=0\end{cases}$$

由于

Since

$$\begin{vmatrix}-1&2&1\\3&1&4\\1&0&1\end{vmatrix}=0$$

由克拉默法可知方程组有非零解, 从而向量组 $\{\boldsymbol{\alpha}_1, \boldsymbol{\alpha}_2, \boldsymbol{\alpha}_3\}$ 线性相关.

from Gramar's rule we know that this system has non-zero solutions, therefore, the vector set $\{\boldsymbol{\alpha}_1, \boldsymbol{\alpha}_2, \boldsymbol{\alpha}_3\}$ is linearly dependent.

解法 2 因为

Solution 2 Since

$$\begin{vmatrix}-1&2&1\\3&1&4\\1&0&1\end{vmatrix}=0$$

直接用判别方法 3, 可知量组 $\{\boldsymbol{\alpha}_1, \boldsymbol{\alpha}_2, \boldsymbol{\alpha}_3\}$ 线性相关. ■

using the discriminate method 3 directly, we know that the vector set $\{\boldsymbol{\alpha}_1, \boldsymbol{\alpha}_2, \boldsymbol{\alpha}_3\}$ is linearly dependent. ■

例 3 设向量组 $\{\boldsymbol{\alpha}_1, \boldsymbol{\alpha}_2, \boldsymbol{\alpha}_3\}$ 线性无关, $\boldsymbol{\beta}_1 = \boldsymbol{\alpha}_1 + \boldsymbol{\alpha}_2, \boldsymbol{\beta}_2 = \boldsymbol{\alpha}_2 + \boldsymbol{\alpha}_3, \boldsymbol{\beta}_3 = \boldsymbol{\alpha}_3 + \boldsymbol{\alpha}_1$. 证明: $\{\boldsymbol{\beta}_1, \boldsymbol{\beta}_2, \boldsymbol{\beta}_3\}$ 也线性无关.

Example 3 Suppose that the vector set $\{\boldsymbol{\alpha}_1, \boldsymbol{\alpha}_2, \boldsymbol{\alpha}_3\}$ is linearly independent and that $\boldsymbol{\beta}_1 = \boldsymbol{\alpha}_1 + \boldsymbol{\alpha}_2$, $\boldsymbol{\beta}_2 = \boldsymbol{\alpha}_2 + \boldsymbol{\alpha}_3$, $\boldsymbol{\beta}_3 = \boldsymbol{\alpha}_3 + \boldsymbol{\alpha}_1$. Prove that the vector set $\{\boldsymbol{\beta}_1, \boldsymbol{\beta}_2, \boldsymbol{\beta}_3\}$ is linearly independent.

证 设有数 k_1, k_2, k_3, 使

Proof Suppose that there exist numbers k_1, k_2, k_3 such that

$$k_1\boldsymbol{\beta}_1 + k_2\boldsymbol{\beta}_2 + k_3\boldsymbol{\beta}_3 = \mathbf{0}$$

即

namely,

$$k_1(\boldsymbol{\alpha}_1 + \boldsymbol{\alpha}_2) + k_2(\boldsymbol{\alpha}_2 + \boldsymbol{\alpha}_3) + k_3(\boldsymbol{\alpha}_3 + \boldsymbol{\alpha}_1) = \mathbf{0}$$

于是有

Thus, we have

$$(k_1 + k_3)\boldsymbol{\alpha}_1 + (k_1 + k_2)\boldsymbol{\alpha}_2 + (k_2 + k_3)\boldsymbol{\alpha}_3 = \mathbf{0}$$

由于向量组 $\{\boldsymbol{\alpha}_1, \boldsymbol{\alpha}_2, \boldsymbol{\alpha}_3\}$ 线性无关, 有

Since the vector set $\{\boldsymbol{\alpha}_1, \boldsymbol{\alpha}_2, \boldsymbol{\alpha}_3\}$ is linearly independent, we have

$$\begin{cases} k_1 + k_3 = 0 \\ k_1 + k_2 = 0 \\ k_2 + k_3 = 0 \end{cases}$$

又由

also, from

$$\begin{vmatrix} 1 & 0 & 1 \\ 1 & 1 & 0 \\ 0 & 1 & 1 \end{vmatrix} = 2 \neq 0$$

知方程组只有零解, 即 $k_1 = k_2 = k_3 = 0$, 从而向量组 $\{\boldsymbol{\beta}_1, \boldsymbol{\beta}_2, \boldsymbol{\beta}_3\}$ 线性无关. ■

we know that this system has only a zero solution, i.e., $k_1 = k_2 = k_3 = 0$. Thus, the vector set $\{\boldsymbol{\beta}_1, \boldsymbol{\beta}_2, \boldsymbol{\beta}_3\}$ is linearly independent. ■

例 4 确定 c 值, 使向量组

Example 4 Decide the value of c such that the vector set

$$\boldsymbol{\alpha}_1 = \begin{pmatrix} 1 \\ -2 \\ 4 \end{pmatrix}, \quad \boldsymbol{\alpha}_2 = \begin{pmatrix} 0 \\ 1 \\ 2 \end{pmatrix}, \quad \boldsymbol{\alpha}_3 = \begin{pmatrix} -2 \\ 3 \\ c \end{pmatrix}$$

线性相关.

is linearly dependent.

解 要使 $\{\boldsymbol{\alpha}_1, \boldsymbol{\alpha}_2, \boldsymbol{\alpha}_3\}$ 线性相关, 只要行列式满足

Solution To make $\{\boldsymbol{\alpha}_1, \boldsymbol{\alpha}_2, \boldsymbol{\alpha}_3\}$ be linearly dependent, it only requires that the determinant satisfies

$$\begin{vmatrix} 1 & -2 & 4 \\ 0 & 1 & 2 \\ -2 & 3 & c \end{vmatrix} = 10 + c = 0$$

所以 $c = -10$. ■

so we have $c = -10$. ■

例 5 判断下面向量组的线性相关性.

Example 5 Decide the linear dependence of the following vector set.

$$\boldsymbol{\alpha}_1 = \begin{pmatrix} 1 \\ 2 \\ -1 \\ 5 \end{pmatrix}, \quad \boldsymbol{\alpha}_2 = \begin{pmatrix} 2 \\ -1 \\ 1 \\ 1 \end{pmatrix}, \quad \boldsymbol{\alpha}_3 = \begin{pmatrix} 4 \\ 3 \\ -1 \\ 11 \end{pmatrix}$$

解 对矩阵 $(\boldsymbol{\alpha}_1, \boldsymbol{\alpha}_2, \boldsymbol{\alpha}_3)$ 施以初等行变换, 将其化为行阶梯形矩阵

Solution Performing elementary row operations on the matrix $(\boldsymbol{\alpha}_1, \boldsymbol{\alpha}_2, \boldsymbol{\alpha}_3)$ yields a row echelon matrix

$$\begin{pmatrix} 1 & 2 & 4 \\ 2 & -1 & 3 \\ -1 & 1 & -1 \\ 5 & 1 & 11 \end{pmatrix} \to \begin{pmatrix} 1 & 2 & 4 \\ 0 & 1 & 1 \\ 0 & 0 & 0 \\ 0 & 0 & 0 \end{pmatrix}$$

因为 $R(\boldsymbol{\alpha}_1, \boldsymbol{\alpha}_2, \boldsymbol{\alpha}_3) = 2 < 3$, 所以向量组 $\{\boldsymbol{\alpha}_1, \boldsymbol{\alpha}_2, \boldsymbol{\alpha}_3\}$ 线性相关. ■

Since $R(\boldsymbol{\alpha}_1, \boldsymbol{\alpha}_2, \boldsymbol{\alpha}_3) = 2 < 3$, the vector set $\{\boldsymbol{\alpha}_1, \boldsymbol{\alpha}_2, \boldsymbol{\alpha}_3\}$ is linearly dependent. ■

例 6 已知向量组

Example 6 Known that the vector set, given by

$$\alpha_1 = \begin{pmatrix} 1 \\ 1 \\ 2 \\ 1 \end{pmatrix}, \quad \alpha_2 = \begin{pmatrix} 1 \\ 0 \\ 0 \\ 2 \end{pmatrix}, \quad \alpha_3 = \begin{pmatrix} -1 \\ -4 \\ -8 \\ k \end{pmatrix}$$

线性相关, 求 k.

is linearly dependent. Find the value of k.

解法 1 用初等变换讨论.

Solution 1 Discussion by elementary operations.

$$A = \begin{pmatrix} 1 & 1 & -1 \\ 1 & 0 & -4 \\ 2 & 0 & -8 \\ 1 & 2 & k \end{pmatrix} \to \begin{pmatrix} 1 & 1 & -1 \\ 0 & -1 & -3 \\ 0 & -2 & -6 \\ 0 & 1 & k+1 \end{pmatrix} \to \begin{pmatrix} 1 & 1 & -1 \\ 0 & -1 & -3 \\ 0 & 0 & k-2 \\ 0 & 0 & 0 \end{pmatrix}$$

可见, 当 $k=2$ 时, $R(A)=2<3$, 从而向量组 $\{\alpha_1, \alpha_2, \alpha_3\}$ 线性相关.

It is clear that $R(A)=2<3$ if $k=2$, consequently, the vector set $\{\alpha_1, \alpha_2, \alpha_3\}$ is linearly dependent.

解法 2 用行列式讨论.

Solution 2 Discussion by determinant.

由 $\alpha_1, \alpha_2, \alpha_3$ 的第 1,2,4 个分量构成的向量组为

The vector set obtained by taking the first, the second and the fourth components in $\alpha_1, \alpha_2, \alpha_3$ is given by

$$\alpha_1' = \begin{pmatrix} 1 \\ 1 \\ 1 \end{pmatrix}, \quad \alpha_2' = \begin{pmatrix} 1 \\ 0 \\ 2 \end{pmatrix}, \quad \alpha_3' = \begin{pmatrix} -1 \\ -4 \\ k \end{pmatrix}$$

该向量组一定线性相关(否则, 延伸向量组 $\{\alpha_1, \alpha_2, \alpha_3\}$ 线性无关, 与已知矛盾)则有

The vector set must be linearly dependent (otherwise, the extended vector set $\{\alpha_1, \alpha_2, \alpha_3\}$ is linearly independent, which contradicts to the known condition), this means that

$$|(\alpha_1', \alpha_2', \alpha_3')| = \begin{vmatrix} 1 & 1 & -1 \\ 1 & 0 & -4 \\ 1 & 2 & k \end{vmatrix} = 2-k=0$$

所以 $k=2$. ■

So $k=2$. ■

例 7 设向量组 $\{\boldsymbol{a}_1,\boldsymbol{a}_2,\boldsymbol{a}_3\}$ 线性相关, 向量组 $\{\boldsymbol{a}_2,\boldsymbol{a}_3,\boldsymbol{a}_4\}$ 线性无关. 证明

Example 7 Suppose that the vector set $\{\boldsymbol{a}_1,\boldsymbol{a}_2,\boldsymbol{a}_3\}$ is linearly dependent and that the vector set $\{\boldsymbol{a}_2,\boldsymbol{a}_3,\boldsymbol{a}_4\}$ is linearly independent. Prove that

(1) $\boldsymbol{a}_1$ 能由 $\{\boldsymbol{a}_2,\boldsymbol{a}_3\}$ 线性表示;

(1) $\boldsymbol{a}_1$ can be linearly represented by $\{\boldsymbol{a}_2,\boldsymbol{a}_3\}$;

(2) $\boldsymbol{a}_4$ 不能由 $\{\boldsymbol{a}_1,\boldsymbol{a}_2,\boldsymbol{a}_3\}$ 线性表示.

(2) $\boldsymbol{a}_4$ can not be linearly represented by $\{\boldsymbol{a}_1,\boldsymbol{a}_2,\boldsymbol{a}_3\}$.

证 (1) 因向量组 $\{\boldsymbol{a}_2,\boldsymbol{a}_3,\boldsymbol{a}_4\}$ 线性无关, 故部分组 $\{\boldsymbol{a}_2,\boldsymbol{a}_3\}$ 线性无关, 而向量组 $\{\boldsymbol{a}_1,\boldsymbol{a}_2,\boldsymbol{a}_3\}$ 线性相关, 从而 $\boldsymbol{a}_1$ 能由 $\{\boldsymbol{a}_2,\boldsymbol{a}_3\}$ 线性表示.

Proof (1) Since the vector set $\{\boldsymbol{a}_2,\boldsymbol{a}_3,\boldsymbol{a}_4\}$ is linearly independent, it means that the subset $\{\boldsymbol{a}_2,\boldsymbol{a}_3\}$ is linearly independent, moreover, since the vector set $\{\boldsymbol{a}_1,\boldsymbol{a}_2,\boldsymbol{a}_3\}$ is linearly dependent, from Theorem 2 we know that $\boldsymbol{a}_1$ can be linearly represented by $\{\boldsymbol{a}_2,\boldsymbol{a}_3\}$.

(2) **反证法**. 假设 $\boldsymbol{a}_4$ 能由 $\{\boldsymbol{a}_1,\boldsymbol{a}_2,\boldsymbol{a}_3\}$ 线性表示. 由 (1) 知, $\boldsymbol{a}_1$ 能由 $\{\boldsymbol{a}_2,\boldsymbol{a}_3\}$ 线性表示, 因此 $\boldsymbol{a}_4$ 能由 $\{\boldsymbol{a}_2,\boldsymbol{a}_3\}$ 表示, 这与 $\{\boldsymbol{a}_2,\boldsymbol{a}_3,\boldsymbol{a}_4\}$ 线性无关矛盾. ■

(2) **Proof by contradiction**. Suppose that $\boldsymbol{a}_4$ can be linearly represented by $\{\boldsymbol{a}_1,\boldsymbol{a}_2,\boldsymbol{a}_3\}$. From (1) we know that $\boldsymbol{a}_1$ can be linearly represented by $\{\boldsymbol{a}_2,\boldsymbol{a}_3\}$, so $\boldsymbol{a}_4$ can be linearly represented by $\{\boldsymbol{a}_2,\boldsymbol{a}_3\}$, which contradicts to the fact that the vector set $\{\boldsymbol{a}_2,\boldsymbol{a}_3,\boldsymbol{a}_4\}$ is linearly independent. ■

思 考 题

Questions

1. 若向量组 $\{\boldsymbol{a}_1,\boldsymbol{a}_2,\boldsymbol{a}_3\}$ 中任何两个向量都线性无关, $\{\boldsymbol{a}_1,\boldsymbol{a}_2,\boldsymbol{a}_3\}$ 是否一定线性无关?

1. If any two vectors in the vector set $\{\boldsymbol{a}_1,\boldsymbol{a}_2,\boldsymbol{a}_3\}$ are linearly independent, whether $\{\boldsymbol{a}_1,\boldsymbol{a}_2,\boldsymbol{a}_3\}$ must be linearly independent?

2. 若向量组 $\{\boldsymbol{a}_1,\boldsymbol{a}_2,\boldsymbol{a}_3\}$ 线性相关, 则 $\boldsymbol{a}_1$ 能否由 $\{\boldsymbol{a}_2,\boldsymbol{a}_3\}$ 线性表示?

2. If the vector set $\{\boldsymbol{a}_1,\boldsymbol{a}_2,\boldsymbol{a}_3\}$ is linearly dependent, whether $\boldsymbol{a}_1$ can be linearly represented by $\{\boldsymbol{a}_2,\boldsymbol{a}_3\}$?

3. 若向量组 $\{a_1, a_2, \cdots, a_m\}$ 线性无关, a_{m+1} 不能由 $\{a_1, a_2, \cdots, a_m\}$ 线性表示, 判断向量组 $\{a_1, a_2, \cdots, a_m, a_{m+1}\}$ 是线性相关还是线性无关?

3. If the vector set $\{a_1, a_2, \cdots, a_m\}$ is linearly independent, a_{m+1} can not be linearly represented by $\{a_1, a_2, \cdots, a_m\}$, decide that the vector set $\{a_1, a_2, \cdots, a_m, a_{m+1}\}$ is either linearly dependent or linearly independent?

4. 两个矩阵等价和两个向量组等价有什么区别和联系?

4. What are the difference and relation between equivalence of two matrices and equivalence of two vector sets?

4.3 向量组的极大无关组与向量组的秩

4.3 Maximal Independent Subsets and Ranks of Vector Sets

4.3.1 向量组的极大无关组

4.3.1 Maximal Independent Subsets of Vector Sets

定义 1 设有向量组 T(可以由有限多个向量组成, 也可以由无限多个向量组成).

Definition1 Suppose that T is a vector set which may be consisted of finite vectors or infinitely many vectors.

(1) 如果 T 的一个部分向量组 $\{\alpha_1, \alpha_2, \cdots, \alpha_r\}$ 线性无关;

(1) If a vector subset $\{\alpha_1, \alpha_2, \cdots, \alpha_r\}$ of T is linear independent;

(2) 在 T 的其余向量中任取一个向量 β(如果有), 由 $r+1$ 个向量组成的部分向量组 $\{\alpha_1, \alpha_2, \cdots, \alpha_r, \beta\}$ 都线性相关.

(2) For any vector β in the rest vectors of T (if there exists), the vector subset consisted of $r+1$ vectors $\{\alpha_1, \alpha_2, \cdots, \alpha_r, \beta\}$ is linearly dependent.

则称部分组 $\{\alpha_1, \alpha_2, \cdots, \alpha_r\}$ 是 T 的一个**极大线性无关向量组**, 简称**极大无关组**.

Then the subset $\{\alpha_1, \alpha_2, \cdots, \alpha_r\}$ is called a **maximal linearly independent vector subset** of T, for short, a **maximal independent subset**.

特别地, 若向量组 $\{\alpha_1, \alpha_2, \cdots, \alpha_n\}$ 是线性无关的, 则它的极大线性无关向量组是它本身.

In particular, if the vector set $\{\alpha_1, \alpha_2, \cdots, \alpha_n\}$ is linearly independent, then the maximal linearly independent vector subset is itself.

由 4.2 节定理 2 知, 上面定义中的条件 (2) 也可叙述为: T 中任一向量可由 $\{\boldsymbol{\alpha}_1,\boldsymbol{\alpha}_2,\cdots,\boldsymbol{\alpha}_r\}$ 线性表示. 换句话说, 部分向量组 $\{\boldsymbol{\alpha}_1,\boldsymbol{\alpha}_2,\cdots,\boldsymbol{\alpha}_r\}$ 又可以由整体向量组 T 线性表示. 因此有如下性质:

From Theorem 2 in Section 4.2 we know that Condition (2) in the above definition can also be stated as: Any vector in T can be linearly represented by $\{\boldsymbol{\alpha}_1,\boldsymbol{\alpha}_2,\cdots,\boldsymbol{\alpha}_r\}$. In other words, the vector subset $\{\boldsymbol{\alpha}_1,\boldsymbol{\alpha}_2,\cdots,\boldsymbol{\alpha}_r\}$ can also be linearly represented by the whole vector set T. Consequently, we have the following properties:

性质 1 一个向量组与它的任一极大无关组等价.

Property 1 A vector set is equivalent to any one of its maximal independent vector subset.

性质 2 一个向量组的任意两个极大无关组等价.

Property 2 Any two maximal independent vector subset of a vector set are equivalent.

例 1 求向量组的极大无关组.

Example 1 Find a maximal independent subset of the vector set

$$\boldsymbol{\alpha}_1=\begin{pmatrix}1\\2\\4\end{pmatrix},\quad \boldsymbol{\alpha}_2=\begin{pmatrix}-1\\2\\0\end{pmatrix},\quad \boldsymbol{\alpha}_3=\begin{pmatrix}0\\4\\4\end{pmatrix}$$

解 因 $\boldsymbol{\alpha}_1\neq\mathbf{0}$, 故部分组 $\boldsymbol{\alpha}_1$ 线性无关. 又 $\boldsymbol{\alpha}_1$ 和 $\boldsymbol{\alpha}_2$ 对应的分量不成比例, 所以部分组 $\{\boldsymbol{\alpha}_1,\boldsymbol{\alpha}_2\}$ 线性无关. 易见, $\boldsymbol{\alpha}_3=\boldsymbol{\alpha}_1+\boldsymbol{\alpha}_2$, 故 $\{\boldsymbol{\alpha}_1,\boldsymbol{\alpha}_2,\boldsymbol{\alpha}_3\}$ 线性相关. 从而部分组 $\{\boldsymbol{\alpha}_1,\boldsymbol{\alpha}_2\}$ 是 $\{\boldsymbol{\alpha}_1,\boldsymbol{\alpha}_2,\boldsymbol{\alpha}_3\}$ 的极大无关组. 同理可验证 $\{\boldsymbol{\alpha}_1,\boldsymbol{\alpha}_3\}$ 和 $\{\boldsymbol{\alpha}_2,\boldsymbol{\alpha}_3\}$ 也是 $\{\boldsymbol{\alpha}_1,\boldsymbol{\alpha}_2,\boldsymbol{\alpha}_3\}$ 的极大无关组. ■

Solution Since $\boldsymbol{\alpha}_1\neq\mathbf{0}$, the subset $\boldsymbol{\alpha}_1$ is linearly independent. The corresponding components in $\boldsymbol{\alpha}_1$ and $\boldsymbol{\alpha}_2$ are not proportional, so the subset $\{\boldsymbol{\alpha}_1,\boldsymbol{\alpha}_2\}$ is linearly independent. Obviously, $\boldsymbol{\alpha}_3=\boldsymbol{\alpha}_1+\boldsymbol{\alpha}_2$, this means that the vector set $\{\boldsymbol{\alpha}_1,\boldsymbol{\alpha}_2,\boldsymbol{\alpha}_3\}$ is linearly dependent. Therefore, the subset $\{\boldsymbol{\alpha}_1,\boldsymbol{\alpha}_2\}$ is a maximal independent subset of $\{\boldsymbol{\alpha}_1,\boldsymbol{\alpha}_2,\boldsymbol{\alpha}_3\}$. In a similar way, it is shown that the subsets $\{\boldsymbol{\alpha}_1,\boldsymbol{\alpha}_3\}$ and $\{\boldsymbol{\alpha}_2,\boldsymbol{\alpha}_3\}$ are also maximal independent subsets of $\{\boldsymbol{\alpha}_1,\boldsymbol{\alpha}_2,\boldsymbol{\alpha}_3\}$. ■

引例 设三个向量 $\boldsymbol{\alpha}_1,\boldsymbol{\alpha}_2,\boldsymbol{\alpha}_3$ 由两个向量 $\boldsymbol{\beta}_1,\boldsymbol{\beta}_2$ 线性表示为

Citing example Suppose that the linear representation of the three vectors $\boldsymbol{\alpha}_1,\boldsymbol{\alpha}_2,\boldsymbol{\alpha}_3$ by the two vectors is $\boldsymbol{\beta}_1,\boldsymbol{\beta}_2$, as follows,

$$\begin{cases} \boldsymbol{\alpha}_1 = \boldsymbol{\beta}_1 + \boldsymbol{\beta}_2 \\ \boldsymbol{\alpha}_2 = 2\boldsymbol{\beta}_1 + 3\boldsymbol{\beta}_2 \\ \boldsymbol{\alpha}_3 = \boldsymbol{\beta}_1 + 2\boldsymbol{\beta}_2 \end{cases}$$

显然有

Obviously, we have

$$\boldsymbol{\alpha}_1 - \boldsymbol{\alpha}_2 + \boldsymbol{\alpha}_3 = \mathbf{0}$$

即向量组 $\{\boldsymbol{\alpha}_1, \boldsymbol{\alpha}_2, \boldsymbol{\alpha}_3\}$ 线性相关.

namely, the vector set $\{\boldsymbol{\alpha}_1, \boldsymbol{\alpha}_2, \boldsymbol{\alpha}_3\}$ is linearly dependent.

一般地, 有下面的定理:

In general, we have the following theorem:

定理 1 若向量组 $\{\boldsymbol{\alpha}_1, \boldsymbol{\alpha}_2, \cdots, \boldsymbol{\alpha}_r\}$ 能由向量组 $\{\boldsymbol{\beta}_1, \boldsymbol{\beta}_2, \cdots, \boldsymbol{\beta}_s\}$ 线性表示, 且 $r > s$, 则向量组 $\{\boldsymbol{\alpha}_1, \boldsymbol{\alpha}_2, \cdots, \boldsymbol{\alpha}_r\}$ 一定线性相关.

Theorem 1 If the vector set $\{\boldsymbol{\alpha}_1, \boldsymbol{\alpha}_2, \cdots, \boldsymbol{\alpha}_r\}$ can be linearly represented by the vector set $\{\boldsymbol{\beta}_1, \boldsymbol{\beta}_2, \cdots, \boldsymbol{\beta}_s\}$, and if $r > s$, then the vector set $\{\boldsymbol{\alpha}_1, \boldsymbol{\alpha}_2, \cdots, \boldsymbol{\alpha}_r\}$ must be linearly dependent.

证 因向量组 $\{\boldsymbol{\alpha}_1, \boldsymbol{\alpha}_2, \cdots, \boldsymbol{\alpha}_r\}$ 能由向量组 $\{\boldsymbol{\beta}_1, \boldsymbol{\beta}_2, \cdots, \boldsymbol{\beta}_s\}$ 线性表示, 故有

Proof Since the vector set $\{\boldsymbol{\alpha}_1, \boldsymbol{\alpha}_2, \cdots, \boldsymbol{\alpha}_r\}$ can be linearly represented by the vector set $\{\boldsymbol{\beta}_1, \boldsymbol{\beta}_2, \cdots, \boldsymbol{\beta}_s\}$, we have

$$\begin{cases} \boldsymbol{\alpha}_1 = a_{11}\boldsymbol{\beta}_1 + a_{21}\boldsymbol{\beta}_2 + \cdots + a_{s1}\boldsymbol{\beta}_s \\ \boldsymbol{\alpha}_2 = a_{12}\boldsymbol{\beta}_1 + a_{22}\boldsymbol{\beta}_2 + \cdots + a_{s2}\boldsymbol{\beta}_s \\ \quad\cdots\cdots \\ \boldsymbol{\alpha}_r = a_{1r}\boldsymbol{\beta}_1 + a_{2r}\boldsymbol{\beta}_2 + \cdots + a_{sr}\boldsymbol{\beta}_s \end{cases}$$

即

namely,

$$\begin{aligned} (\boldsymbol{\alpha}_1, \boldsymbol{\alpha}_2, \cdots, \boldsymbol{\alpha}_r)_{1\times r} &= (\boldsymbol{\beta}_1, \boldsymbol{\beta}_2, \cdots, \boldsymbol{\beta}_s)_{1\times s} \begin{pmatrix} a_{11} & a_{12} & \cdots & a_{1r} \\ a_{21} & a_{22} & \cdots & a_{2r} \\ \vdots & \vdots & & \vdots \\ a_{s1} & a_{s2} & \cdots & a_{sr} \end{pmatrix}_{s\times r} \\ &= (\boldsymbol{\beta}_1, \boldsymbol{\beta}_2, \cdots, \boldsymbol{\beta}_s)_{1\times s}\boldsymbol{A}_{s\times r} \end{aligned}$$

若 $r > s$, 则齐次线性方程组 $\boldsymbol{A}_{s\times r}\boldsymbol{x} = \mathbf{0}$ 必有非零解, 记为

If $r > s$, then the homogeneous linear system $\boldsymbol{A}_{s\times r}\boldsymbol{x} = \mathbf{0}$ must have non-zero solutions, written as

$$x = \begin{pmatrix} x_1 \\ x_2 \\ \vdots \\ x_r \end{pmatrix}$$

即存在不全零数 $x_1, x_2, \cdots, x_r$ 使

namely, there exist numbers $x_1, x_2, \cdots, x_r$, which are not all zeros, such that

$$A_{s\times r}x = \begin{pmatrix} a_{11} & a_{12} & \cdots & a_{1r} \\ a_{21} & a_{22} & \cdots & a_{2r} \\ \vdots & \vdots & & \vdots \\ a_{s1} & a_{s2} & \cdots & a_{sr} \end{pmatrix}_{s\times r} \begin{pmatrix} x_1 \\ x_2 \\ \vdots \\ x_r \end{pmatrix} = \mathbf{0}$$

进而得到

and then we obtain

$$\begin{aligned} &(\alpha_1, \alpha_2, \cdots, \alpha_r)_{1\times r} \begin{pmatrix} x_1 \\ x_2 \\ \vdots \\ x_r \end{pmatrix} \\ &= (\beta_1, \beta_2, \cdots, \beta_s)_{1\times s} \begin{pmatrix} a_{11} & a_{12} & \cdots & a_{1r} \\ a_{21} & a_{22} & \cdots & a_{2r} \\ \vdots & \vdots & & \vdots \\ a_{s1} & a_{s2} & \cdots & a_{sr} \end{pmatrix}_{s\times r} \begin{pmatrix} x_1 \\ x_2 \\ \vdots \\ x_r \end{pmatrix} \\ &= \mathbf{0} \end{aligned}$$

也就是说, 存在不全零数 $x_1, x_2, \cdots, x_r$ 使

in other words, there exist numbers $x_1, x_2, \cdots, x_r$, which are not all zeros, such that

$$x_1\alpha_1 + x_2\alpha_2 + \cdots + x_r\alpha_r = \mathbf{0}$$

因此向量组 $\{\alpha_1, \alpha_2, \cdots, \alpha_r\}$ 线性相关. ▲

This means that the vector set $\{\alpha_1, \alpha_2, \cdots, \alpha_r\}$ is linearly dependent. ▲

推论 1 若向量组 $\{\alpha_1, \alpha_2, \cdots, \alpha_r\}$ 线性无关, 且 $\{\alpha_1, \alpha_2, \cdots, \alpha_r\}$ 能由 $\{\beta_1, \beta_2, \cdots, \beta_s\}$ 线性表示, 则 $r \leqslant s$.

Corollary 1 If the vector set $\{\alpha_1, \alpha_2, \cdots, \alpha_r\}$ is linearly independent and if $\{\alpha_1, \alpha_2, \cdots, \alpha_r\}$ can be linearly represented by $\{\beta_1, \beta_2, \cdots, \beta_s\}$, then $r \leqslant s$.

推论 2 一个向量组的所有极大线性无关组所含向量的个数都相同.

Corollary 2 All maximal linearly independent vector subsets of a vector set have the same number of vectors.

定义 2 向量组 T 的极大无关组所含向量个数 (唯一) 称为 T 的**秩**, 记为 $R(T)$.

Definition 2 The number of vectors contained in a maximal independent subset of the vector set T is called the **rank** of T, written as $R(T)$.

规定只含零向量的向量组的秩等于零.

It is specified that the rank of a vector set only containing zero vectors is equal to zero.

由等价的对称性、传递性和定理 1 的推论 1 可得如下推论.

Using the symmetry, transitivity of equivalence and Corollary 1 of Theorem 1, we obtain the following corollaries.

推论 3 对于两个给定的向量组 (Ⅰ) 和 (Ⅱ), 如果 (Ⅰ) 可以由 (Ⅱ) 线性表示, 则 (Ⅰ) 的秩不大于 (Ⅱ) 的秩, 即 $R(\text{Ⅰ})\leqslant R(\text{Ⅱ})$.

Corollary 3 For two given vector sets (Ⅰ) and (Ⅱ), if (Ⅰ) can be linearly represented by (Ⅱ), then the rank of (Ⅰ) is not greater than the rank of (Ⅱ), i.e., $R(\text{Ⅰ})\leqslant R(\text{Ⅱ})$.

推论 4 等价的向量组有相等的秩.

Corollary 4 Arbitrary equivalent vector sets have the same rank.

对于任意一个向量组 $\{\boldsymbol{\alpha}_1,\boldsymbol{\alpha}_2,\cdots,\boldsymbol{\alpha}_m\}$, 由定义 2 知 $R(\boldsymbol{\alpha}_1,\boldsymbol{\alpha}_2,\cdots,\boldsymbol{\alpha}_m)\leqslant m$, 于是有如下定理.

For an arbitrary vector set $\{\boldsymbol{\alpha}_1,\boldsymbol{\alpha}_2,\cdots,\boldsymbol{\alpha}_m\}$, using Definition 2 we know that $R(\boldsymbol{\alpha}_1,\boldsymbol{\alpha}_2,\cdots,\boldsymbol{\alpha}_m)\leqslant m$, this leads to the following theorem.

定理 2 向量组 $\{\boldsymbol{\alpha}_1,\boldsymbol{\alpha}_2,\cdots,\boldsymbol{\alpha}_m\}$ 线性无关 (或线性相关) 的充分必要条件是 $R(\boldsymbol{\alpha}_1,\boldsymbol{\alpha}_2,\cdots,\boldsymbol{\alpha}_m)=m$(或 $<m$).

Theorem 2 The vector set $\{\boldsymbol{\alpha}_1,\boldsymbol{\alpha}_2,\cdots,\boldsymbol{\alpha}_m\}$ is linearly independent (or linearly dependent) if and only if $R(\boldsymbol{\alpha}_1,\boldsymbol{\alpha}_2,\cdots,\boldsymbol{\alpha}_m)=m$ (or $<m$).

这就是说, 可以通过求一个向量组的秩来判定该向量组的线性相关性.

That is to say, we can decide the linear dependence of a vector set by finding the rank of the vector set.

4.3.2 向量组的秩与矩阵的秩之间的关系

4.3.2 Relations of Ranks between of Vector Sets and Matrices

设 n 个 m 维列向量组 $\{\boldsymbol{\alpha}_1, \boldsymbol{\alpha}_2, \cdots, \boldsymbol{\alpha}_n\}$, 则对应的 $m \times n$ 矩阵为

Let $\{\boldsymbol{\alpha}_1, \boldsymbol{\alpha}_2, \cdots, \boldsymbol{\alpha}_n\}$ be a vector set consisted of n vectors of dimension m. Then the corresponding $m \times n$ matrix is given by

$$\boldsymbol{A} = (\boldsymbol{\alpha}_1, \boldsymbol{\alpha}_2, \cdots, \boldsymbol{\alpha}_n)$$

反过来, 若将一个 $m \times n$ 矩阵 $\boldsymbol{A}$ 的每一列作为一个向量, 则可以得到 n 个 m 维列向量组 $\{\boldsymbol{\alpha}_1, \boldsymbol{\alpha}_2, \cdots, \boldsymbol{\alpha}_n\}$. 因此, 讨论一个向量组的线性相关性及秩的问题, 可以通过矩阵的初等变换来完成.

Conversely, if each column of an $m \times n$ matrix $\boldsymbol{A}$ is treated as a vector, then we obtain a vector set consisted of n vectors of dimension m, given by $\{\boldsymbol{\alpha}_1, \boldsymbol{\alpha}_2, \cdots, \boldsymbol{\alpha}_n\}$. Therefore, discussing the problems of linear dependence and rank of a vector set can be achieved by elementary operations on matrices.

定理 3 对矩阵实施的初等行变换不改变矩阵的列向量组的线性相关性和线性组合关系, **即初等变换是同关系变换**.

Theorem 3 Elementary row operations performing on a matrix do not change the linear dependence and the relation of linear combination of the column vector set of the matrix. That is to say, **elementary operations are operations of invariance relations**.

证 设矩阵 $\boldsymbol{A}$ 经 l 次初等行变换化为矩阵 $\boldsymbol{B}$, 即

Proof Suppose that the matrix $\boldsymbol{B}$ is obtained by performing elementary row operations l times on $\boldsymbol{A}$, namely,

$$\boldsymbol{A} = (\boldsymbol{\alpha}_1, \boldsymbol{\alpha}_2, \cdots, \boldsymbol{\alpha}_n) \xrightarrow[\text{elementary row operations}]{\text{初等行变换}} \boldsymbol{B} = (\boldsymbol{\beta}_1, \boldsymbol{\beta}_2, \cdots, \boldsymbol{\beta}_n)$$

则存在初等矩阵 $\boldsymbol{P}_1, \boldsymbol{P}_2, \cdots, \boldsymbol{P}_l$, 使

then there exist elementary matrices $\boldsymbol{P}_1, \boldsymbol{P}_2, \cdots, \boldsymbol{P}_l$ such that

$$\boldsymbol{P}_1\boldsymbol{P}_2\cdots\boldsymbol{P}_l\boldsymbol{A} = \boldsymbol{B}$$

令 $\boldsymbol{P} = \boldsymbol{P}_1\boldsymbol{P}_2\cdots\boldsymbol{P}_l$. 则 $\boldsymbol{P}$ 可逆, 并且

Let $\boldsymbol{P} = \boldsymbol{P}_1\boldsymbol{P}_2\cdots\boldsymbol{P}$. Then $\boldsymbol{P}$ is invertible, and

$$\boldsymbol{PA} = \boldsymbol{B}$$

因此有

Therefore,

$$P\alpha_i = \beta_i \quad (i = 1, 2, \cdots, n)$$

(1) 若存在数 $k_1, k_2, \cdots, k_n$, 使

(1) If there exist numbers $k_1, k_2, \cdots, k_n$ such that

$$k_1\alpha_1 + k_2\alpha_2 + \cdots + k_n\alpha_n = 0$$

上式两边左乘 P, 得

Premultiplying both sides of the above equation by P, we obtain

$$k_1P\alpha_1 + k_2P\alpha_2 + \cdots + k_nP\alpha_n = 0$$

即

namely,

$$k_1\beta_1 + k_2\beta_2 + \cdots + k_n\beta_n = 0$$

所以向量组 $\{\alpha_1, \alpha_2, \cdots, \alpha_n\}$ 与向量组 $\{\beta_1, \beta_2, \cdots, \beta_n\}$ 同时线性相关 (或无关), 也就是说, 对 A 实施的初等行变换不改变 A 的列向量组的线性相关性.

Thus, the vector sets $\{\alpha_1, \alpha_2, \cdots, \alpha_n\}$ and $\{\beta_1, \beta_2, \cdots, \beta_n\}$ are linearly dependent (or independent) simultaneously, that is to say, elementary row operations performing on A do not change the linear dependence of the column vector set of A.

(2) 若 A 的列向量 $\{\alpha_1, \alpha_2, \cdots, \alpha_n\}$ 之间存在某种线性组合关系, 不妨设 α_n 能由 $\{\alpha_1, \alpha_2, \cdots, \alpha_{n-1}\}$ 线性表示, 即存在一组数 $\lambda_1, \lambda_1, \cdots, \lambda_{n-1}$ 使

(2) If the column vector set $\{\alpha_1, \alpha_2, \cdots, \alpha_n\}$ of A have a certain relation of linear combination, not loss of generality, suppose that α_n can be linearly represented by $\{\alpha_1, \alpha_2, \cdots, \alpha_{n-1}\}$, i.e., there exist numbers $\lambda_1, \lambda_1, \cdots, \lambda_{n-1}$ such that

$$\alpha_n = \lambda_1\alpha_1 + \lambda_2\alpha_2 + \cdots + \lambda_{n-1}\alpha_{n-1}$$

进而有

and then

$$\lambda_1\alpha_1 + \lambda_2\alpha_2 + \cdots + \lambda_{n-1}\alpha_{n-1} + (-1)\alpha_n = 0$$

即

namely,

$$(\alpha_1, \alpha_2, \cdots, \alpha_{n-1}, \alpha_n)\begin{pmatrix} \lambda_1 \\ \lambda_2 \\ \vdots \\ \lambda_{n-1} \\ -1 \end{pmatrix} = 0$$

上式两端左乘以 $\boldsymbol{P}$, 得

Premultiplying both sides of the above equation by $\boldsymbol{P}$ leads to

$$\boldsymbol{P}(\boldsymbol{\alpha}_1,\boldsymbol{\alpha}_2,\cdots,\boldsymbol{\alpha}_{n-1},\boldsymbol{\alpha}_n)\begin{pmatrix}\lambda_1\\ \lambda_2\\ \vdots\\ \lambda_{n-1}\\ -1\end{pmatrix}=\mathbf{0}$$

即

namely,

$$(\boldsymbol{\beta}_1,\boldsymbol{\beta}_2,\cdots,\boldsymbol{\beta}_{n-1},\boldsymbol{\beta}_n)\begin{pmatrix}\lambda_1\\ \lambda_2\\ \vdots\\ \lambda_{n-1}\\ -1\end{pmatrix}=\mathbf{0}$$

所以有

so we have

$$\lambda_1\boldsymbol{\beta}_1+\cdots+\lambda_{n-1}\boldsymbol{\beta}_{n-1}-\boldsymbol{\beta}_n=\mathbf{0}$$

即

namely,

$$\boldsymbol{\beta}_n=\lambda_1\boldsymbol{\beta}_1+\lambda_2\boldsymbol{\beta}_2+\cdots+\lambda_{n-1}\boldsymbol{\beta}_{n-1}$$

从而 $\boldsymbol{B}$ 的列向量间也具有与 $\boldsymbol{A}$ 同样的线性组合关系. ▲

Therefore, the relation of linear combination of the column vectors of $\boldsymbol{B}$ is the same as that of $\boldsymbol{A}$. ▲

例 2 讨论下面向量组的线性相关性及线性组合关系.

Example 2 Discuss the linear dependence and the relation of linear combination of the following vector set.

$$\boldsymbol{\alpha}_1=\begin{pmatrix}1\\0\\1\\0\end{pmatrix},\quad \boldsymbol{\alpha}_2=\begin{pmatrix}-2\\1\\3\\-7\end{pmatrix},\quad \boldsymbol{\alpha}_3=\begin{pmatrix}3\\-1\\0\\3\end{pmatrix},\quad \boldsymbol{\alpha}_4=\begin{pmatrix}-4\\1\\-3\\1\end{pmatrix}$$

解 令 $\boldsymbol{\alpha}_1,\boldsymbol{\alpha}_2,\boldsymbol{\alpha}_3,\boldsymbol{\alpha}_4$ 为矩阵 $\boldsymbol{A}$ 的列向量. 经初等行变换将 $\boldsymbol{A}$ 化为行最简形, 即

Solution Let $\boldsymbol{\alpha}_1,\boldsymbol{\alpha}_2,\boldsymbol{\alpha}_3,\boldsymbol{\alpha}_4$ be column vectors of the matrix $\boldsymbol{A}$. Performing elementary row operations on $\boldsymbol{A}$ yields the row-reduced form, namely,

$$A=\begin{pmatrix}1&-2&3&-4\\0&1&-1&1\\1&3&0&-3\\0&-7&3&1\end{pmatrix}\xrightarrow{r_3-r_1}\begin{pmatrix}1&-2&3&-4\\0&1&-1&1\\0&5&-3&1\\0&-7&3&1\end{pmatrix}$$

$$\xrightarrow[r_4+7r_2]{r_3-5r_2}\begin{pmatrix}1&-2&3&-4\\0&1&-1&1\\0&0&2&-4\\0&0&-4&8\end{pmatrix}\xrightarrow[\frac{1}{2}r_3]{r_4+2r_3}\begin{pmatrix}1&-2&3&-4\\0&1&-1&1\\0&0&1&-2\\0&0&0&0\end{pmatrix}$$

$$\xrightarrow[r_1-3r_3]{r_2+r_3}\begin{pmatrix}1&-2&0&2\\0&1&0&-1\\0&0&1&-2\\0&0&0&0\end{pmatrix}\xrightarrow{r_1+2r_2}\begin{pmatrix}1&0&0&0\\0&1&0&-1\\0&0&1&-2\\0&0&0&0\end{pmatrix}$$

$$=(\boldsymbol{\beta}_1,\boldsymbol{\beta}_2,\boldsymbol{\beta}_3,\boldsymbol{\beta}_4)=\boldsymbol{B}$$

作为 $\boldsymbol{A}$ 的行最简形, $\boldsymbol{B}$ 的列向量之间的线性关系为

As a row-reduced form of $\boldsymbol{A}$, the linear relation among the column vectors of $\boldsymbol{B}$ is

$$\boldsymbol{\beta}_4=0\boldsymbol{\beta}_1-\boldsymbol{\beta}_2-2\boldsymbol{\beta}_3$$

所以 $\boldsymbol{B}$ 的列向量组线性相关. 由定理 3 知, $\boldsymbol{A}$ 的列向量组 $\{\boldsymbol{\alpha}_1,\boldsymbol{\alpha}_2,\boldsymbol{\alpha}_3,\boldsymbol{\alpha}_4\}$ 也线性相关, 且

So the column vector set of $\boldsymbol{B}$ is linearly dependent. From Theorem 3 we know that the column vector set $\{\boldsymbol{\alpha}_1,\boldsymbol{\alpha}_2,\boldsymbol{\alpha}_3,\boldsymbol{\alpha}_4\}$ of $\boldsymbol{A}$ is also linearly dependent, and

$$\boldsymbol{\alpha}_4=0\boldsymbol{\alpha}_1-\boldsymbol{\alpha}_2-2\boldsymbol{\alpha}_3.$$ ■

定理 4 矩阵 $\boldsymbol{A}$ 的秩等于 $\boldsymbol{A}$ 的列向量组的秩.

Theorem 4 The rank of a matrix $\boldsymbol{A}$ is equal to the rank of the column vector set of $\boldsymbol{A}$.

证 设 $R(\boldsymbol{A})=r$, 对 $\boldsymbol{A}$ 实施初等行变换化为行最简形, 如下:

Proof Suppose that $R(\boldsymbol{A})=r$. Performing elementary row operations on $\boldsymbol{A}$ yields the row-reduced form, as follows,

$$\boldsymbol{A}=\begin{pmatrix}a_{11}&a_{12}&\cdots&a_{1n}\\a_{21}&a_{22}&\cdots&a_{2n}\\\vdots&\vdots&&\vdots\\a_{m1}&a_{m2}&\cdots&a_{mn}\end{pmatrix}=(\boldsymbol{\alpha}_1,\boldsymbol{\alpha}_2,\cdots,\boldsymbol{\alpha}_n)$$

$$\longrightarrow B=\begin{pmatrix}1&0&\cdots&0&c_{1r+1}&\cdots&c_{1n}\\0&1&\cdots&0&c_{2r+1}&\cdots&c_{2n}\\\vdots&\vdots&&\vdots&\vdots&&\vdots\\0&0&\cdots&1&c_{rr+1}&\cdots&c_{m}\\0&0&\cdots&0&0&\cdots&0\\\vdots&\vdots&&\vdots&\vdots&&\vdots\\0&0&\cdots&0&0&\cdots&0\end{pmatrix}=(\boldsymbol{\varepsilon}_1,\boldsymbol{\varepsilon}_2,\cdots,\boldsymbol{\varepsilon}_r,\boldsymbol{\beta}_{r+1},\cdots,\boldsymbol{\beta}_n)$$

则 $R(\boldsymbol{A})=R(\boldsymbol{B})=r$. 由于 $\boldsymbol{B}$ 中的前 r 个列向量 $\boldsymbol{\varepsilon}_1,\boldsymbol{\varepsilon}_2,\cdots,\boldsymbol{\varepsilon}_r$ 线性无关, 而 $\boldsymbol{\beta}_{r+1},\cdots,\boldsymbol{\beta}_n$ 中的每个向量至多只有前 r 维分量非零, 所以 $\boldsymbol{\beta}_i(i=r+1,\cdots,n)$ 是 $\boldsymbol{\varepsilon}_1,\boldsymbol{\varepsilon}_2,\cdots,\boldsymbol{\varepsilon}_r$ 的线性组合. 故 $\boldsymbol{\varepsilon}_1,\boldsymbol{\varepsilon}_2,\cdots,\boldsymbol{\varepsilon}_r$ 为 $\boldsymbol{B}$ 的列向量组的极大无关组, 则

then $R(\boldsymbol{A})=R(\boldsymbol{B})=r$. Since the subset of the first r column vectors $\boldsymbol{\varepsilon}_1,\boldsymbol{\varepsilon}_2,\cdots,\boldsymbol{\varepsilon}_r$ of $\boldsymbol{B}$ is linearly independent, and at most the first r components of each vector in $\boldsymbol{\beta}_{r+1},\cdots,\boldsymbol{\beta}_n$ are nonzero, so $\boldsymbol{\beta}_i(i=r+1,\cdots,n)$ is the linear combination of $\boldsymbol{\varepsilon}_1,\boldsymbol{\varepsilon}_2,\cdots,\boldsymbol{\varepsilon}_r$. This means that the vector subset $\boldsymbol{\varepsilon}_1,\boldsymbol{\varepsilon}_2,\cdots,\boldsymbol{\varepsilon}_r$ is a maximal independent vector subset of the column vector set of $\boldsymbol{B}$, then

$$R(\boldsymbol{\varepsilon}_1,\boldsymbol{\varepsilon}_2,\cdots,\boldsymbol{\varepsilon}_r,\boldsymbol{\beta}_{r+1},\cdots,\boldsymbol{\beta}_n)=r$$

由本节定理 3 知, $\boldsymbol{A}$ 的列向量组的秩与 $\boldsymbol{B}$ 的列向量组的秩相等, 即

From Theorem 3 in this section we know that the rank of the column vector set of $\boldsymbol{A}$ and that of $\boldsymbol{B}$ are equal, namely

$$R(\boldsymbol{\alpha}_1,\boldsymbol{\alpha}_2,\cdots,\boldsymbol{\alpha}_n)=R(\boldsymbol{A})=r.$$

故 $\boldsymbol{A}$ 的秩等于 $\boldsymbol{A}$ 的列向组的秩. ▲

So the rank of $\boldsymbol{A}$ is equal to the rank of the column vector set of $\boldsymbol{A}$. ▲

由于 $R(\boldsymbol{A})=R(\boldsymbol{A}^{\mathrm{T}})$, 将上述证明用在 $\boldsymbol{A}^{\mathrm{T}}$ 上, 可得 $\boldsymbol{A}$ 的秩等于 $\boldsymbol{A}$ 的行向量组的秩. 因此

Since $R(\boldsymbol{A})=R(\boldsymbol{A}^{\mathrm{T}})$, using the above proof on $\boldsymbol{A}^{\mathrm{T}}$ yields that the rank of $\boldsymbol{A}$ is equal to the rank of the row vector set of $\boldsymbol{A}$. Therefore,

矩阵$\boldsymbol{A}$的秩矩阵 = $\boldsymbol{A}$列向量组的秩矩阵
= $\boldsymbol{A}$行向量组的秩

The rank of a matrix the rank $\boldsymbol{A}$
= of the column vector set of $\boldsymbol{A}$ the rank
= of the row vector set of $\boldsymbol{A}$

推论 5 令 $\boldsymbol{A}$ 为 $m\times n$ 矩阵. 若 $R(\boldsymbol{A})=r$, 则有

(1) 当 $r=m$ 时, $\boldsymbol{A}$ 的行向量组线性无关; 当 $r<m$ 时, $\boldsymbol{A}$ 的行向量组线性相关.

(2) 当 $r=n$ 时, $\boldsymbol{A}$ 的列向量组线性无关; 当 $r<n$ 时, $\boldsymbol{A}$ 的列向量组线性相关.

Corollary 5 Let $\boldsymbol{A}$ be an $m\times n$ matrix. If $R(\boldsymbol{A})=r$, then

(1) As $r=m$, the row vector set of $\boldsymbol{A}$ is linearly independent; while as $r<m$, the row vector set of $\boldsymbol{A}$ is linearly dependent.

(2) As $r=n$, the column vector set of $\boldsymbol{A}$ is linearly independent; while as $r<n$, the column vector set of $\boldsymbol{A}$ is linearly dependent.

定理 5 对下面给定的 n 元齐次线性方程组

Theorem 5 For the following given homogeneous linear system in n unknowns

$$\begin{cases} a_{11}x_1+a_{12}x_2+\cdots+a_{1n}x_n=0\\ a_{21}x_1+a_{22}x_2+\cdots+a_{2n}x_n=0\\ \cdots\cdots\\ a_{n1}x_1+a_{n2}x_2+\cdots+a_{nn}x_n=0 \end{cases} \tag{4.7}$$

如果系数行列式等于零, 则齐次线性方程组 (4.7) 有非零解.

if the coefficient determinant is equal to zero, then homogeneous linear system (4.7) has nonzero solutions.

证 设向量组为

Proof Let the vector set be

$$\boldsymbol{\alpha}_j=\begin{pmatrix} a_{1j}\\ a_{2j}\\ \vdots\\ a_{nj}\end{pmatrix}\quad (j=1,2,\cdots,n)$$

则系数矩阵为

then the coefficient matrix is given by

$$\boldsymbol{A}=\begin{pmatrix} a_{11} & a_{12} & \cdots & a_{1n}\\ a_{21} & a_{22} & \cdots & a_{2n}\\ \vdots & \vdots & & \vdots\\ a_{n1} & a_{n2} & \cdots & a_{nn}\end{pmatrix}=(\boldsymbol{\alpha}_1,\boldsymbol{\alpha}_2,\cdots,\boldsymbol{\alpha}_n)$$

则方程组对应的向量组的线性组合形式为

the linear combination form of the vector set associated with the system is

$$x_1\boldsymbol{\alpha}_1+x_2\boldsymbol{\alpha}_2+\cdots+x_n\boldsymbol{\alpha}_n=\boldsymbol{0} \tag{4.8}$$

若方程组 (4.7) 的系数行列式等于零, 即 $|\boldsymbol{A}| = 0$, 则 $R(\boldsymbol{A}) < n$, 从而 $\boldsymbol{A}$ 的列向量组线性相关, 即存在不全为零的 n 个数 $x_1, x_2, \cdots, x_n$ 使式 (4.8) 成立. 因此, (4.7) 有非零解. 这表明, 系数行列式等于零也是 (4.7) 有非零解的充分条件. 至此 1.4 节的定理 $2'$ 便得到证明. ▲

If the determinant of the coefficient determinant of System (4.7) is equal to zero, i.e., $|\boldsymbol{A}| = 0$, then we have $R(\boldsymbol{A}) < n$, so the column vector set of $\boldsymbol{A}$ is linearly dependent, namely, there exist n numbers $x_1, x_2, \cdots, x_n$, which are non all zeros, such that Equation (4.8) is valid. Therefore, (4.7) has nonzero solutions. This illustrates that the coefficient determinant is equal to zero is a sufficient condition that (4.7) has nonzero solutions. Till now, the proof of Theorem $2'$ in Section 1.4 is completed. ▲

综合以上分析, 可以利用对矩阵的初等变换确定一个向量组的线性相关性及线性组合关系. 具体做法如下:

Combining the above analyses, we can use elementary operations on matrices to determine the linear dependence and the relation of linear combination of a vector set. The concrete method is as follows:

以 $\boldsymbol{\alpha}_1, \boldsymbol{\alpha}_2, \cdots, \boldsymbol{\alpha}_m$ 作为矩阵 $\boldsymbol{A}$ 的列向量, 对 $\boldsymbol{A}$ 进行初等行变换化为行阶梯形, 即

$$\boldsymbol{A} = (\boldsymbol{\alpha}_1, \boldsymbol{\alpha}_2, \cdots, \boldsymbol{\alpha}_m) \longrightarrow \text{行阶梯形矩阵}$$

Using $\boldsymbol{\alpha}_1, \boldsymbol{\alpha}_2, \cdots, \boldsymbol{\alpha}_m$ as the column vectors of a matrix $\boldsymbol{A}$, and then performing elementary row operations on $\boldsymbol{A}$, we obtain a row echelon form, namely,

$$\boldsymbol{A} = (\boldsymbol{\alpha}_1, \boldsymbol{\alpha}_2, \cdots, \boldsymbol{\alpha}_m) \longrightarrow \text{row echelon matrix}$$

根据阶梯形矩阵, ① 求出 $\boldsymbol{A}$ 的列向量组 $\{\boldsymbol{\alpha}_1, \boldsymbol{\alpha}_2, \cdots, \boldsymbol{\alpha}_m\}$ 的秩; ② 确定向量组的线性相关性; ③ 找出向量组的极大无关组. 若向量组线性相关, 可对 $\boldsymbol{A}$ 的阶梯形继续作初等行变换化为行最简形, 确定其余向量用极大无关组的线性表示式.

From the row echelon matrix we can ① find the rank of the column vector set of $\{\boldsymbol{\alpha}_1, \boldsymbol{\alpha}_2, \cdots, \boldsymbol{\alpha}_m\}$ of $\boldsymbol{A}$; ② decide the linear dependence of the vector set; ③ find a maximal independent subset of the vector set. If the vector set is linearly dependent, we can perform elementary row operations on $\boldsymbol{A}$ continuously to reduce it to a row-reduced form, and then determine the linear representation of other vectors by the maximal independent subset.

例 3 已知向量组

Example 3 Known that the vector set is given by

$$\boldsymbol{\alpha}_1=\begin{pmatrix}1\\1\\3\\1\end{pmatrix},\quad \boldsymbol{\alpha}_2=\begin{pmatrix}-1\\1\\-1\\3\end{pmatrix},\quad \boldsymbol{\alpha}_3=\begin{pmatrix}5\\-2\\8\\9\end{pmatrix},\quad \boldsymbol{\alpha}_4=\begin{pmatrix}-1\\3\\1\\7\end{pmatrix}$$

求向量组的秩和一个极大无关组, 并将其余向量用该极大无关组线性表示.

Find the rank and a maximal independent subset of the vector set, and represent the other vectors linearly by the maximal independent subset.

解 将 $\boldsymbol{\alpha}_1,\boldsymbol{\alpha}_2,\boldsymbol{\alpha}_3,\boldsymbol{\alpha}_4$ 作为列向量构造矩阵 $\boldsymbol{A}$. 对 $\boldsymbol{A}$ 作初等行变换得

Solution Use $\boldsymbol{\alpha}_1,\boldsymbol{\alpha}_2,\boldsymbol{\alpha}_3,\boldsymbol{\alpha}_4$ as column vectors to construct a matrix $\boldsymbol{A}$. Performing elementary row operations on $\boldsymbol{A}$ yields

$$\boldsymbol{A}=\begin{pmatrix}1&-1&5&-1\\1&1&-2&3\\3&-1&8&1\\1&3&-9&7\end{pmatrix}\xrightarrow[r_4-r_1]{\substack{r_2-r_1\\r_3-3r_1}}\begin{pmatrix}1&-1&5&-1\\0&2&-7&4\\0&2&-7&4\\0&4&-14&8\end{pmatrix}$$

$$\xrightarrow[r_4-2r_2]{r_3-r_2}\begin{pmatrix}1&-1&5&-1\\0&2&-7&4\\0&0&0&0\\0&0&0&0\end{pmatrix}$$

由 $\boldsymbol{A}$ 的阶梯形知 $R(\boldsymbol{\alpha}_1,\boldsymbol{\alpha}_2,\boldsymbol{\alpha}_3,\boldsymbol{\alpha}_4)$=2. 它的一个极大无关组为 $\{\boldsymbol{\alpha}_1,\boldsymbol{\alpha}_2\}$(或 $\{\boldsymbol{\alpha}_1,\boldsymbol{\alpha}_3\}$ 或 $\{\boldsymbol{\alpha}_1,\boldsymbol{\alpha}_4\}$).

From the echelon form of $\boldsymbol{A}$ we know that $R(\boldsymbol{\alpha}_1,\boldsymbol{\alpha}_2,\boldsymbol{\alpha}_3,\boldsymbol{\alpha}_4)$=2, it implies that one of the maximal independent subsets is $\{\boldsymbol{\alpha}_1,\boldsymbol{\alpha}_2\}$ (or $\{\boldsymbol{\alpha}_1,\boldsymbol{\alpha}_3\}$ or $\{\boldsymbol{\alpha}_1,\boldsymbol{\alpha}_4\}$).

进一步对 $\boldsymbol{A}$ 作初等行变换化为如下的行最简形

Further, performing elementary row operations on $\boldsymbol{A}$ yields the following row-reduced form

$$\begin{pmatrix}1&-1&5&-1\\0&2&-7&4\\0&0&0&0\\0&0&0&0\end{pmatrix}\longrightarrow\begin{pmatrix}1&0&\dfrac{3}{2}&1\\0&1&-\dfrac{7}{2}&2\\0&0&0&0\\0&0&0&0\end{pmatrix}=(\boldsymbol{\beta}_1,\boldsymbol{\beta}_2,\boldsymbol{\beta}_3,\boldsymbol{\beta}_4)=\boldsymbol{B}$$

显然,

Obviously,

$$\boldsymbol{\beta}_3=\frac{3}{2}\boldsymbol{\beta}_1-\frac{7}{2}\boldsymbol{\beta}_2, \boldsymbol{\beta}_4=\boldsymbol{\beta}_1+2\boldsymbol{\beta}_2$$

由定理 3 知, 初等行变换不改变矩阵列向量组的线性组合关系, 所以

From Theorem 3 we know that elementary row operations do not change the relation of linear combination of a vector set, thus

$$\boldsymbol{\alpha}_3=\frac{3}{2}\boldsymbol{\alpha}_1-\frac{7}{2}\boldsymbol{\alpha}_2, \quad \boldsymbol{\alpha}_4=\boldsymbol{\alpha}_1+2\boldsymbol{\alpha}_2$$

或由 $\boldsymbol{A}$ 的行最简形可知, 非齐次线性方程组

In another way, from the row-reduced form of $\boldsymbol{A}$ we know that the solutions of the non-homogeneous linear system

$$\boldsymbol{\alpha}_3=x_1\boldsymbol{\alpha}_1+x_2\boldsymbol{\alpha}_2, \quad \boldsymbol{\alpha}_4=y_1\boldsymbol{\alpha}_1+y_2\boldsymbol{\alpha}_2$$

的解分别为

are respectively given by

$$x_1=\frac{3}{2}, \quad x_2=-\frac{7}{2}; \quad y_1=1, \quad y_2=2$$

因此,

therefore,

$$\boldsymbol{\alpha}_3=\frac{3}{2}\boldsymbol{\alpha}_1-\frac{7}{2}\boldsymbol{\alpha}_2, \quad \boldsymbol{\alpha}_4=\boldsymbol{\alpha}_1+2\boldsymbol{\alpha}_2$$

■

例 4 已知向量组为

Example 4 Known that the vector set is

$$\boldsymbol{\alpha}_1=\begin{pmatrix}1\\1\\1\end{pmatrix}, \quad \boldsymbol{\alpha}_2=\begin{pmatrix}1\\2\\4\end{pmatrix}, \quad \boldsymbol{\alpha}_3=\begin{pmatrix}1\\3\\t\end{pmatrix}$$

(1) t 为何值时, 向量组 $\{\boldsymbol{\alpha}_1,\boldsymbol{\alpha}_2,\boldsymbol{\alpha}_3\}$ 线性相关?

(1) What value does t take on such that the vector set $\{\boldsymbol{\alpha}_1,\boldsymbol{\alpha}_2,\boldsymbol{\alpha}_3\}$ is linearly dependent?

(2) 若向量组 $\{\boldsymbol{\alpha}_1,\boldsymbol{\alpha}_2,\boldsymbol{\alpha}_3\}$ 线性相关, 求该向量组的一个极大无关组, 并将其余向量用极大无关组线性表示.

(2) If the vector set $\{\boldsymbol{\alpha}_1,\boldsymbol{\alpha}_2,\boldsymbol{\alpha}_3\}$ is linearly dependent, find a maximal independent subset of the vector set and represent other vectors by the maximal independent subset.

解 以 $\boldsymbol{\alpha}_1, \boldsymbol{\alpha}_2, \boldsymbol{\alpha}_3$ 为列向量得矩阵 $\boldsymbol{A}$, 并对 $\boldsymbol{A}$ 施以行初等行变换得

Solution Using $\boldsymbol{\alpha}_1, \boldsymbol{\alpha}_2, \boldsymbol{\alpha}_3$ as column vectors we obtain a matrix $\boldsymbol{A}$, and then performing elementary row operations on $\boldsymbol{A}$ yields

$$\boldsymbol{A} = \begin{pmatrix} 1 & 1 & 1 \\ 1 & 2 & 3 \\ 1 & 4 & t \end{pmatrix} \xrightarrow[r_3 - r_1]{r_2 - r_1} \begin{pmatrix} 1 & 1 & 1 \\ 0 & 1 & 2 \\ 0 & 3 & t-1 \end{pmatrix} \xrightarrow{r_3 - 3r_2} \begin{pmatrix} 1 & 1 & 1 \\ 0 & 1 & 2 \\ 0 & 0 & t-7 \end{pmatrix}$$

(1) 易知, 当 $t = 7$ 时, $R(\boldsymbol{A}) = 2 < 3$, 所以 $\{\boldsymbol{\alpha}_1, \boldsymbol{\alpha}_2, \boldsymbol{\alpha}_3\}$ 线性相关.

(1) Obviously, as $t = 7$, we have $R(\boldsymbol{A}) = 2 < 3$, and so the vector set $\{\boldsymbol{\alpha}_1, \boldsymbol{\alpha}_2, \boldsymbol{\alpha}_3\}$ is linearly dependent.

(2) 当 $t = 7$ 时, 继续对 $\boldsymbol{A}$ 施以初等行变换

(2) As $t = 7$, performing elementary row operations on $\boldsymbol{A}$ continuously

$$\boldsymbol{A} \to \begin{pmatrix} 1 & 1 & 1 \\ 0 & 1 & 2 \\ 0 & 0 & 0 \end{pmatrix} \to \begin{pmatrix} 1 & 0 & -1 \\ 0 & 1 & 2 \\ 0 & 0 & 0 \end{pmatrix} = \boldsymbol{B}$$

矩阵 $\boldsymbol{B}$ 的列向量间的线性关系为

The linear relation among the column vectors of the matrix $\boldsymbol{B}$ is

$$\begin{pmatrix} -1 \\ 2 \\ 0 \end{pmatrix} = -\begin{pmatrix} 1 \\ 0 \\ 0 \end{pmatrix} + 2\begin{pmatrix} 0 \\ 1 \\ 0 \end{pmatrix}$$

由定理 3, $\boldsymbol{\alpha}_3 = -\boldsymbol{\alpha}_1 + 2\boldsymbol{\alpha}_2$, 而且 $\{\boldsymbol{\alpha}_1, \boldsymbol{\alpha}_2\}$ 为该向量组的一个极大无关组. ■

From Theorem 3 we know that $\boldsymbol{\alpha}_3 = -\boldsymbol{\alpha}_1 + 2\boldsymbol{\alpha}_2$, and that $\{\boldsymbol{\alpha}_1, \boldsymbol{\alpha}_2\}$ is a maximal independent subset. ■

例 5* 证明: 两个矩阵 $\boldsymbol{A}_{m\times n}$ 与 $\boldsymbol{B}_{n\times s}$ 乘积的秩不大于 $\boldsymbol{A}_{m\times n}$ 的秩和 $\boldsymbol{B}_{n\times s}$ 的秩, 即

Example 5* Prove that the rank of the product of two matrices $\boldsymbol{A}_{m\times n}$ and $\boldsymbol{B}_{n\times s}$ is not greater than the ranks of $\boldsymbol{A}_{m\times n}$ and $\boldsymbol{B}_{n\times s}$, namely,

$$R(\boldsymbol{AB}) \leqslant \min\{R(\boldsymbol{A}), R(\boldsymbol{B})\}$$

证 设

Proof Suppose that

$$\boldsymbol{A} = (a_{ij})_{m\times n} = (\boldsymbol{\alpha}_1, \boldsymbol{\alpha}_2, \cdots, \boldsymbol{\alpha}_n), \quad \boldsymbol{B} = (b_{ij})_{n\times s}$$

$$\boldsymbol{AB} = \boldsymbol{C} = (c_{ij})_{m\times s} = (\boldsymbol{\gamma}_1, \boldsymbol{\gamma}_2, \cdots, \boldsymbol{\gamma}_s)$$

即

namely,

$$(\gamma_1,\gamma_2,\cdots,\gamma_s)_{1\times s}=(\alpha_1,\alpha_2,\cdots,\alpha_n)_{1\times n}\begin{pmatrix} b_{11} & \cdots & b_{1j} & \cdots & b_{1s}\\ b_{21} & \cdots & b_{2j} & \cdots & b_{2s}\\ \vdots & & \vdots & & \vdots\\ b_{n1} & \cdots & b_{nj} & \cdots & b_{ns}\end{pmatrix}_{n\times s}$$

因此, 有

Thus, we have

$$\gamma_j=b_{1j}\alpha_1+b_{2j}\alpha_2+\cdots+b_{nj}\alpha_n \quad (j=1,2,\cdots,s)$$

即 $\boldsymbol{AB}$ 的列向量组 $\{\gamma_1,\gamma_2,\cdots,\gamma_s\}$ 可由 $\boldsymbol{A}$ 的列向量组 $\{\alpha_1,\alpha_2,\cdots,\alpha_n\}$ 线性表示. 由定理 1 的推论 3 可知 $R(\boldsymbol{AB})\leqslant R(\boldsymbol{A})$.

That is to say, the column vector set $\{\gamma_1,\gamma_2,\cdots,\gamma_s\}$ of $\boldsymbol{AB}$ can be linearly represented by the column vector set $\{\alpha_1,\alpha_2,\cdots,\alpha_n\}$ of $\boldsymbol{A}$. From Corollary 3 of Theorem 1 we know that $R(\boldsymbol{AB})\leqslant R(\boldsymbol{A})$.

同理, $\boldsymbol{AB}$ 的行向量组可由 $\boldsymbol{B}$ 的行向量组线性表示, 即 $R(\boldsymbol{AB})\leqslant R(\boldsymbol{B})$. 因此,

In a similar way, the row vector set of $\boldsymbol{AB}$ can be linearly represented by the row vector set of $\boldsymbol{B}$, it implies that $R(\boldsymbol{AB})\leqslant R(\boldsymbol{B})$. Consequently,

$$R(\boldsymbol{AB})\leqslant\min\{R(\boldsymbol{A}),R(\boldsymbol{B})\}$$

■

4.3.3* 向量空间

4.3.3* Vector Spaces

定义 3 设 $\mathbf{R}^n$ 是一个非空的 n 维向量集合, P 是一个数域. 如果 $\mathbf{R}^n$ 中的向量对于线性运算封闭 (即对任意 $\alpha,\beta\in\mathbf{R}^n$, 都有 $\alpha+\beta,k\alpha\in\mathbf{R}^n$, 其中 $k\in P$), 则称 $\mathbf{R}^n$ 是数域 P 上的**向量空间**.

Definition 3 Let $\mathbf{R}^n$ be a nonempty set of n dimensional vectors and let P be a number field. If the linear operations of vectors in $\mathbf{R}^n$ are closed (namely, for any $\alpha,\beta\in\mathbf{R}^n$, we have $\alpha+\beta,k\alpha\in\mathbf{R}^n$, where $k\in P$), then $\mathbf{R}^n$ is called a **vector space** over the number field P.

例 6 在解析几何里, 平面或空间中从坐标原点引出的一切向量的集合对向量的加法和数乘运算来说分别构成实数域上的两个向量空间.

Example 6 In Analytic Geometry, the sets consisted of all the vectors starting from the origin in a plane or a space respectively constitute two vector spaces over the real field for the operations of vector addition and scalar multiplication.

易知, 三维几何空间 $\mathbf{R}^3$ 中的向量组

Obviously, the vector set in the three dimensional geometric space $\mathbf{R}^3$ given by

$$\varepsilon_1=\begin{pmatrix}1\\0\\0\end{pmatrix},\quad \varepsilon_2=\begin{pmatrix}0\\1\\0\end{pmatrix},\quad \varepsilon_3=\begin{pmatrix}0\\0\\1\end{pmatrix}$$

是线性无关的, 并且对于任一个向量

is linearly independent, and for any vector

$$\boldsymbol{\alpha}=\begin{pmatrix}a_1\\a_2\\a_3\end{pmatrix}$$

有

we have

$$\boldsymbol{\alpha}=a_1\varepsilon_1+a_2\varepsilon_2+a_3\varepsilon_3$$

向量组 $\{\varepsilon_1,\varepsilon_2,\varepsilon_3\}$ 称为 $\mathbf{R}^3$ 的基底, 而 a_1,a_2,a_3 称为向量 $\boldsymbol{\alpha}$ 在基底 $\{\varepsilon_1,\varepsilon_2,\varepsilon_3\}$ 下的坐标. 一般地, 我们有如下的定义.

The vector set $\{\varepsilon_1,\varepsilon_2,\varepsilon_3\}$ is called a baiss of $\mathbf{R}^3$ and a_1,a_2,a_3 are called the coordinates of the vector $\boldsymbol{\alpha}$ under the basis $\{\varepsilon_1,\varepsilon_2,\varepsilon_3\}$. In general, we have the following definition.

定义 4 向量空间 $\mathbf{R}^n$ 中的一个极大线性无关组 $\{\varepsilon_1,\varepsilon_2,\cdots,\varepsilon_n\}$ 称为 $\mathbf{R}^n$ 的一个**基底**, n 称为向量空间 $\mathbf{R}^n$ 的**维数**.

Definition 4 The maximal linearly independent subset in the vector space $\mathbf{R}^n$, given by $\{\varepsilon_1,\varepsilon_2,\cdots,\varepsilon_n\}$, is called a **basis** of $\mathbf{R}^n$, n is called the **dimension** of the vector space $\mathbf{R}^n$.

注 (1) $\mathbf{R}^n$ 的基底不唯一.

Note (1) The basis of $\mathbf{R}^n$ is not unique.

(2) $\mathbf{R}^n$ 中的任一向量 $\boldsymbol{\alpha}$ 均可由基底 $\{\varepsilon_1,\varepsilon_2,\cdots,\varepsilon_n\}$ 线性表示, 即

(2) Any vector $\boldsymbol{\alpha}$ in $\mathbf{R}^n$ can be linearly represented by the basis $\{\varepsilon_1,\varepsilon_2,\cdots,\varepsilon_n\}$, namely,

$$\boldsymbol{\alpha}=x_1\varepsilon_1+x_2\varepsilon_2+\cdots+x_n\varepsilon_n$$

其中 $x_1,x_2,\cdots,x_n$ 称为向量 $\boldsymbol{\alpha}$ 在基底 $\{\varepsilon_1,\varepsilon_2,\cdots,\varepsilon_n\}$ 下的**坐标**.

where $x_1,x_2,\cdots,x_n$ are called the **coordinates** of $\boldsymbol{\alpha}$ under the basis $\{\varepsilon_1,\varepsilon_2,\cdots,\varepsilon_n\}$.

例 7 给定 $\mathbf{R}^3$ 中向量

Example 7 Given the vectors in $\mathbf{R}^3$

$$\boldsymbol{\alpha}=\begin{pmatrix}1\\0\\6\end{pmatrix},\boldsymbol{\varepsilon}_1=\begin{pmatrix}1\\0\\2\end{pmatrix},\boldsymbol{\varepsilon}_2=\begin{pmatrix}0\\1\\-1\end{pmatrix},\boldsymbol{\varepsilon}_3=\begin{pmatrix}1\\1\\3\end{pmatrix}$$

求 $\boldsymbol{\alpha}$ 在基底 $\{\boldsymbol{\varepsilon}_1,\boldsymbol{\varepsilon}_2,\boldsymbol{\varepsilon}_3\}$ 下的坐标向量.

Find the coordinate vector of $\boldsymbol{\alpha}$ under the basis $\{\boldsymbol{\varepsilon}_1,\boldsymbol{\varepsilon}_2,\boldsymbol{\varepsilon}_3\}$.

解 令 $\boldsymbol{A}=(\boldsymbol{\varepsilon}_1,\boldsymbol{\varepsilon}_2,\boldsymbol{\varepsilon}_3)$. 下面求解线性方程组 $\boldsymbol{A}\boldsymbol{x}=\boldsymbol{\alpha}$. 对方程组的增广矩阵实施初等行变换, 得到行最简形

Solution Let $\boldsymbol{A}=(\boldsymbol{\varepsilon}_1,\boldsymbol{\varepsilon}_2,\boldsymbol{\varepsilon}_3)$. In the following part we solve the linear system $\boldsymbol{A}\boldsymbol{x}=\boldsymbol{\alpha}$. Performing elementary row operations on the augmented matrix of the system leads to the row-reduced form

$$\boldsymbol{A}=\begin{pmatrix}1&0&1&1\\0&1&1&0\\2&-1&3&6\end{pmatrix}\xrightarrow{r_3-2r_1}\begin{pmatrix}1&0&1&1\\0&1&1&0\\0&-1&1&4\end{pmatrix}$$
$$\xrightarrow{r_3+r_2}\begin{pmatrix}1&0&1&1\\0&1&1&0\\0&0&2&4\end{pmatrix}\xrightarrow{r_3\times\frac{1}{2}}\begin{pmatrix}1&0&1&1\\0&1&1&0\\0&0&1&2\end{pmatrix}$$
$$\xrightarrow{r_1-r_3,r_2-r_3}\begin{pmatrix}1&0&0&-1\\0&1&0&-2\\0&0&1&2\end{pmatrix}$$

所以, 向量 $\boldsymbol{\alpha}$ 在基底 $\{\boldsymbol{\varepsilon}_1,\boldsymbol{\varepsilon}_2,\boldsymbol{\varepsilon}_3\}$ 下的坐标向量为

Therefore, the coordinate vector of the vector $\boldsymbol{\alpha}$ under the basis $\{\boldsymbol{\varepsilon}_1,\boldsymbol{\varepsilon}_2,\boldsymbol{\varepsilon}_3\}$ in $\mathbf{R}^3$ is

$$\begin{pmatrix}-1\\-2\\2\end{pmatrix}.$$

对于同一向量在不同基底下的坐标, 它们之间存在某种内在的联系. 设 $x_1,x_2,\cdots,x_n$ 和 $y_1,y_2,\cdots,y_n$ 分别是向量 $\boldsymbol{\alpha}$ 在基底 $\{\boldsymbol{\alpha}_1,\boldsymbol{\alpha}_2,\cdots,\boldsymbol{\alpha}_n\}$ 和 $\{\boldsymbol{\beta}_1,\boldsymbol{\beta}_2,\cdots,\boldsymbol{\beta}_n\}$ 下的坐标, 首先将 $\{\boldsymbol{\alpha}_1,\boldsymbol{\alpha}_2,\cdots,\boldsymbol{\alpha}_n\}$ 用 $\{\boldsymbol{\beta}_1,\boldsymbol{\beta}_2,\cdots,\boldsymbol{\beta}_n\}$ 线性表示

For coordinates of a vector under different bases, they have a certain interior relation. Let $x_1,x_2,\cdots,x_n$ and $y_1,y_2,\cdots,y_n$ be the coordinates of the vector $\boldsymbol{\alpha}$ under the bases $\{\boldsymbol{\alpha}_1,\boldsymbol{\alpha}_2,\cdots,\boldsymbol{\alpha}_n\}$ and $\{\boldsymbol{\beta}_1,\boldsymbol{\beta}_2,\cdots,\boldsymbol{\beta}_n\}$. Firstly, represent $\{\boldsymbol{\alpha}_1,\boldsymbol{\alpha}_2,\cdots,\boldsymbol{\alpha}_n\}$ by $\{\boldsymbol{\beta}_1,\boldsymbol{\beta}_2,\cdots,\boldsymbol{\beta}_n\}$ linearly,

$$\begin{cases} \boldsymbol{\alpha}_1 = a_{11}\boldsymbol{\beta}_1 + a_{21}\boldsymbol{\beta}_2 + \cdots + a_{n1}\boldsymbol{\beta}_n \\ \boldsymbol{\alpha}_2 = a_{12}\boldsymbol{\beta}_1 + a_{22}\boldsymbol{\beta}_2 + \cdots + a_{n2}\boldsymbol{\beta}_n \\ \cdots\cdots \\ \boldsymbol{\alpha}_n = a_{1n}\boldsymbol{\beta}_1 + a_{2n}\boldsymbol{\beta}_2 + \cdots + a_{nn}\boldsymbol{\beta}_n \end{cases}$$

即

namely,

$$(\boldsymbol{\alpha}_1, \boldsymbol{\alpha}_2, \cdots, \boldsymbol{\alpha}_n) = (\boldsymbol{\beta}_1, \boldsymbol{\beta}_2, \cdots, \boldsymbol{\beta}_n) \begin{pmatrix} a_{11} & a_{12} & \cdots & a_{1n} \\ a_{21} & a_{22} & \cdots & a_{2n} \\ \vdots & \vdots & & \vdots \\ a_{n1} & a_{n2} & \cdots & a_{nn} \end{pmatrix}$$

令

Let

$$\boldsymbol{P} = \begin{pmatrix} a_{11} & a_{12} & \cdots & a_{1n} \\ a_{21} & a_{22} & \cdots & a_{2n} \\ \vdots & \vdots & & \vdots \\ a_{n1} & a_{n2} & \cdots & a_{nn} \end{pmatrix}$$

矩阵 $\boldsymbol{P}$ 称为由基底 $\{\boldsymbol{\beta}_1, \boldsymbol{\beta}_2, \cdots, \boldsymbol{\beta}_n\}$ 到基底 $\{\boldsymbol{\alpha}_1, \boldsymbol{\alpha}_2, \cdots, \boldsymbol{\alpha}_n\}$ 的**过渡矩阵**. 显然, $|\boldsymbol{P}| \neq 0$. 表达式

The matrix $\boldsymbol{P}$ is called a **transition matrix** from the basis $\{\boldsymbol{\beta}_1, \boldsymbol{\beta}_2, \cdots, \boldsymbol{\beta}_n\}$ to the basis $\{\boldsymbol{\alpha}_1, \boldsymbol{\alpha}_2, \cdots, \boldsymbol{\alpha}_n\}$. Obviously, $|\boldsymbol{P}| \neq 0$. The expression

$$\boldsymbol{\alpha} = x_1\boldsymbol{\alpha}_1 + x_2\boldsymbol{\alpha}_2 + \cdots + x_n\boldsymbol{\alpha}_n = y_1\boldsymbol{\beta}_1 + y_2\boldsymbol{\beta}_2 + \cdots + y_n\boldsymbol{\beta}_n$$

可写成如下的矩阵形式

may be written in the following matrix form

$$\boldsymbol{\alpha} = (\boldsymbol{\alpha}_1, \boldsymbol{\alpha}_2, \cdots, \boldsymbol{\alpha}_n) \begin{pmatrix} x_1 \\ x_2 \\ \vdots \\ x_n \end{pmatrix} = (\boldsymbol{\beta}_1, \boldsymbol{\beta}_2, \cdots, \boldsymbol{\beta}_n) \begin{pmatrix} y_1 \\ y_2 \\ \vdots \\ y_n \end{pmatrix}$$

由以上两式得

From the above two equations we get

$$\boldsymbol{\alpha} = (\boldsymbol{\alpha}_1, \boldsymbol{\alpha}_2, \cdots, \boldsymbol{\alpha}_n) \begin{pmatrix} x_1 \\ x_2 \\ \vdots \\ x_n \end{pmatrix} = (\boldsymbol{\beta}_1, \boldsymbol{\beta}_2, \cdots, \boldsymbol{\beta}_n)\boldsymbol{P} \begin{pmatrix} x_1 \\ x_2 \\ \vdots \\ x_n \end{pmatrix}$$

$$= (\boldsymbol{\beta}_1, \boldsymbol{\beta}_2, \cdots, \boldsymbol{\beta}_n) \begin{pmatrix} y_1 \\ y_2 \\ \vdots \\ y_n \end{pmatrix}$$

因为一个向量在同一基底下的坐标是唯一的, 所以

Since the coordinates of a vector under the same basis are unique, we have

$$\boldsymbol{P} \begin{pmatrix} x_1 \\ x_2 \\ \vdots \\ x_n \end{pmatrix} = \begin{pmatrix} y_1 \\ y_2 \\ \vdots \\ y_n \end{pmatrix}$$

或

or

$$\begin{pmatrix} x_1 \\ x_2 \\ \vdots \\ x_n \end{pmatrix} = \boldsymbol{P}^{-1} \begin{pmatrix} y_1 \\ y_2 \\ \vdots \\ y_n \end{pmatrix}$$

上式给出了向量 $\boldsymbol{\alpha}$ 在不同基底下的坐标之间的关系.

The above equations give the relation of coordinates of the vector $\boldsymbol{\alpha}$ under different bases.

例 8 已知 $\mathbf{R}^3$ 中两个向量组分别为

Example 8 Known that the two vector sets in $\mathbf{R}^3$ are respectively given by

$$\boldsymbol{\alpha}_1 = \begin{pmatrix} 1 \\ 0 \\ 1 \end{pmatrix}, \quad \boldsymbol{\alpha}_2 = \begin{pmatrix} 0 \\ 1 \\ -1 \end{pmatrix}, \quad \boldsymbol{\alpha}_3 = \begin{pmatrix} 1 \\ 2 \\ 0 \end{pmatrix}$$

和

and

$$\boldsymbol{\beta}_1 = \begin{pmatrix} 1 \\ 1 \\ 1 \end{pmatrix}, \quad \boldsymbol{\beta}_2 = \begin{pmatrix} 0 \\ 1 \\ 1 \end{pmatrix}, \quad \boldsymbol{\beta}_3 = \begin{pmatrix} 0 \\ 0 \\ 1 \end{pmatrix}$$

(1) 求从基底 $\{\boldsymbol{\beta}_1, \boldsymbol{\beta}_2, \boldsymbol{\beta}_3\}$ 到基底 $\{\boldsymbol{\alpha}_1, \boldsymbol{\alpha}_2, \boldsymbol{\alpha}_3\}$ 的过渡矩阵 $\boldsymbol{P}$;

(1) Find the transition matrix $\boldsymbol{P}$ from the basis $\{\boldsymbol{\beta}_1, \boldsymbol{\beta}_2, \boldsymbol{\beta}_3\}$ to the basis $\{\boldsymbol{\alpha}_1, \boldsymbol{\alpha}_2, \boldsymbol{\alpha}_3\}$;

(2) 设向量 $\boldsymbol{\alpha}$ 在基底 $\{\boldsymbol{\beta}_1,\boldsymbol{\beta}_2,\boldsymbol{\beta}_3\}$ 下的坐标向量为

(2) Suppose that the coordinate vector of $\boldsymbol{\alpha}$ under the basis $\{\boldsymbol{\beta}_1,\boldsymbol{\beta}_2,\boldsymbol{\beta}_3\}$ is

$$\boldsymbol{y}=\begin{pmatrix}1\\-2\\-1\end{pmatrix}$$

求向量 $\boldsymbol{\alpha}$ 在基底 $\{\boldsymbol{\alpha}_1,\boldsymbol{\alpha}_2,\boldsymbol{\alpha}_3\}$ 下的坐标向量 $\boldsymbol{x}$.

find the coordinate vector $\boldsymbol{x}$ under the basis $\{\boldsymbol{\alpha}_1,\boldsymbol{\alpha}_2,\boldsymbol{\alpha}_3\}$.

解 (1) 令 $\boldsymbol{B}=(\boldsymbol{\alpha}_1,\boldsymbol{\alpha}_2,\boldsymbol{\alpha}_3)$, $\boldsymbol{A}=(\boldsymbol{\beta}_1,\boldsymbol{\beta}_2,\boldsymbol{\beta}_3)$, 则

Solution (1) Let $\boldsymbol{B}=(\boldsymbol{\alpha}_1,\boldsymbol{\alpha}_2,\boldsymbol{\alpha}_3)$ and $\boldsymbol{A}=(\boldsymbol{\beta}_1,\boldsymbol{\beta}_2,\boldsymbol{\beta}_3)$, then

$$(\boldsymbol{\alpha}_1,\boldsymbol{\alpha}_2,\boldsymbol{\alpha}_3)=(\boldsymbol{\beta}_1,\boldsymbol{\beta}_2,\boldsymbol{\beta}_3)\boldsymbol{P}$$

即

namely,

$$\boldsymbol{B}=\boldsymbol{A}\boldsymbol{P}$$

所以

so

$$\boldsymbol{P}=\boldsymbol{A}^{-1}\boldsymbol{B}$$

$$(\boldsymbol{A},\boldsymbol{B})=\begin{pmatrix}1&0&0&1&0&1\\1&1&0&0&1&2\\1&1&1&1&-1&0\end{pmatrix}\longrightarrow\begin{pmatrix}1&0&0&1&0&1\\0&1&0&-1&1&1\\0&0&1&1&-2&-2\end{pmatrix}$$

$$\boldsymbol{P}=\begin{pmatrix}1&0&1\\-1&1&1\\1&-2&-2\end{pmatrix}$$

(2) 由

(2) Since

$$(\boldsymbol{P},\boldsymbol{y})=\begin{pmatrix}1&0&1&1\\-1&1&1&-2\\1&-2&-2&-1\end{pmatrix}\longrightarrow\begin{pmatrix}1&0&0&5\\0&1&0&7\\0&0&1&-4\end{pmatrix}$$

得

we obtain

$$\begin{pmatrix}x_1\\x_2\\x_3\end{pmatrix}=\boldsymbol{P}^{-1}\begin{pmatrix}1\\-2\\-1\end{pmatrix}=\begin{pmatrix}5\\7\\-4\end{pmatrix}$$ ■

思 考 题

1. 秩相等的向量组是否等价?

2. 若向量组 (I) 能由向量组 (II) 线性表示, (II) 是否能由 (I) 线性表示?

Questions

1. Are vector sets of the same rank equivalent?

2. If the vector set (I) can be linearly represented by the vector set (II), can (II) be linearly represented by (I)?

4.4 线性方程组解的结构

4.4 Solution Structures of Linear Systems

4.4.1 齐次线性方程组解的结构

4.4.1 Solution Structures of Homogeneous Linear Systems

对于 n 元齐次线性方程组

For a homogeneous linear system in n unknowns, given by

$$\boldsymbol{A}\boldsymbol{x} = \mathbf{0} \tag{4.9}$$

其中

where

$$\boldsymbol{x} = \begin{pmatrix} x_1 \\ x_2 \\ \vdots \\ x_n \end{pmatrix}, \quad \mathbf{0} = \begin{pmatrix} 0 \\ 0 \\ \vdots \\ 0 \end{pmatrix}$$

对应的系数矩阵为

the corresponding coefficient matrix is

$$\boldsymbol{A} = \begin{pmatrix} a_{11} & a_{12} & \cdots & a_{1n} \\ a_{21} & a_{22} & \cdots & a_{2n} \\ \vdots & \vdots & & \vdots \\ a_{m1} & a_{m2} & \cdots & a_{mn} \end{pmatrix}$$

显然, $\boldsymbol{x} = \mathbf{0}$ 是 $\boldsymbol{A}\boldsymbol{x} = \mathbf{0}$ 的解, 称为方程组 (4.9) 的**零解**. 因此, (4.9) 的全部解向量构成的集合 S 非空.

Obviously, $\boldsymbol{x} = \mathbf{0}$ is a solution of $\boldsymbol{A}\boldsymbol{x} = \mathbf{0}$ and it is called a **zero solution** of System (4.9). Therefore, the set S consisted of the whole solution vectors of (4.9) is non-empty.

问题 由 3.3 节定理 2 知, 若 $R(\boldsymbol{A}) = r < n$, 则方程组 $\boldsymbol{Ax} = \boldsymbol{0}$ 有无穷多个非零解. 那么如何求出这些解?

Problem From Theorem 2 in Section 3.3 we know that if $R(\boldsymbol{A}) = r < n$, then the system $\boldsymbol{Ax} = \boldsymbol{0}$ has infinitely many nonzero solutions. How to find these solutions?

为了求出这些解, 令 $\boldsymbol{X}$ 为所有解向量组成的集合. 由 4.3 节极大无关组的性质 1 知, 我们只需求出这个解向量组 $\boldsymbol{X}$ 的一个极大无关组, 就可以得到所有非零解, 也就是**通解**. 为此我们研究 $\boldsymbol{Ax} = \boldsymbol{0}$ 的解的结构.

To do this, let $\boldsymbol{X}$ be a set of all solution vectors. From Property 1 in Section 4.3 on maximal independent subsets, we only require finding a maximal independent subset of $\boldsymbol{X}$, and then we can obtain all nonzero solutions, namely, the **general solution**. For this purpose, we study the solution structures of $\boldsymbol{Ax} = \boldsymbol{0}$.

先讨论非零解的性质. 容易验证如下性质成立:

Firstly, we discuss the properties of nonzero solutions. It is easy to show that the following properties are valid:

性质 1 若向量 $\boldsymbol{\xi}_1, \boldsymbol{\xi}_2$ 是方程组 (4.9) 的解, 则 $\boldsymbol{\xi}_1 + \boldsymbol{\xi}_2$ 也是 (4.9) 的解.

Property 1 If the vectors $\boldsymbol{\xi}_1, \boldsymbol{\xi}_2$ are solutions of System (4.9), so is $\boldsymbol{\xi}_1 + \boldsymbol{\xi}_2$.

性质 2 若向量 $\boldsymbol{\xi}$ 是方程组 (4.9) 的解, k 为任意实数, 则 $k\boldsymbol{\xi}$ 也为 (4.9) 的解.

Property 2 If the vector $\boldsymbol{\xi}$ is a solution of System (4.9), k is an arbitrary real number, then $k\boldsymbol{\xi}$ is also a solution of (4.9).

由性质 1 和性质 2 可知, 若 $\boldsymbol{\xi}_1, \boldsymbol{\xi}_2, \cdots, \boldsymbol{\xi}_{n-r}$ 为 (4.9) 的解, 则线性组合

From Properties 1 and 2 we know that if $\boldsymbol{\xi}_1, \boldsymbol{\xi}_2, \cdots, \boldsymbol{\xi}_{n-r}$ are solutions of (4.9), then the following linear combination

$$\boldsymbol{x} = k_1\boldsymbol{\xi}_1 + k_2\boldsymbol{\xi}_2 + \cdots + k_{n-r}\boldsymbol{\xi}_{n-r}$$

也是 (4.9) 的解. 又若向量组 $\{\boldsymbol{\xi}_1, \boldsymbol{\xi}_2, \cdots, \boldsymbol{\xi}_{n-r}\}$ 为 (4.9) 解向量组 $\boldsymbol{X}$ 的极大线性无关组, 则称之为方程组 (4.9) 的**基础解系**, 对应的通解为

is also a solution of (4.9). Also if the vector set $\{\boldsymbol{\xi}_1, \boldsymbol{\xi}_2, \cdots, \boldsymbol{\xi}_{n-r}\}$ is a maximal linearly independent subset of the solution vector set $\boldsymbol{X}$ of (4.9), then it is called a **system of fundamental solutions** of System (4.9) and the corresponding general solution is as follows,

$$\boldsymbol{X} = \left\{k_1\boldsymbol{\xi}_1 + k_2\boldsymbol{\xi}_2 + \cdots + k_{n-r}\boldsymbol{\xi}_{n-r} | k_1, k_2, \cdots, k_{n-r} \in \mathbf{R}\right\}$$

下面我们给出求方程组 (4.9) 的一个基础解系的方法.

In the following part we give the method for finding a system of fundamental solutions of System (4.9).

对于 n 元齐次线性方程组 $\boldsymbol{Ax}=\boldsymbol{0}$, 设其系数矩阵的秩为 $R(\boldsymbol{A})=r<n$. 对 $\boldsymbol{A}$ 施以初等行变换, 可将其化为如下的行最简形, 即

For the system of homogeneous linear equations in n unknowns, given by $\boldsymbol{Ax}=\boldsymbol{0}$, suppose that the rank of the coefficient matrix is that $R(\boldsymbol{A})=r<n$. Performing elementary row operations on $\boldsymbol{A}$ yields the following row-reduced form, namely,

$$\boldsymbol{A}=\begin{pmatrix} a_{11} & a_{12} & \cdots & a_{1n} \\ a_{21} & a_{22} & \cdots & a_{2n} \\ \vdots & \vdots & & \vdots \\ a_{m1} & a_{m2} & \cdots & a_{mn} \end{pmatrix} \longrightarrow \begin{pmatrix} 1 & 0 & \cdots & 0 & b_{1r+1} & \cdots & b_{1n} \\ 0 & 1 & \cdots & 0 & b_{2r+1} & \cdots & b_{2n} \\ \vdots & \vdots & & \vdots & \vdots & & \vdots \\ 0 & 0 & \cdots & 1 & b_{rr+1} & \cdots & b_{rn} \\ 0 & 0 & \cdots & 0 & 0 & \cdots & 0 \\ \vdots & \vdots & & \vdots & \vdots & & \vdots \\ 0 & 0 & \cdots & 0 & 0 & \cdots & 0 \end{pmatrix}$$

与 $\boldsymbol{Ax}=\boldsymbol{0}$ 同解的线性方程组为

the linear system equivalent to $\boldsymbol{Ax}=\boldsymbol{0}$ is

$$\begin{cases} x_1 \qquad\qquad + b_{1r+1}x_{r+1} + \cdots + b_{1n}x_n = 0 \\ \qquad x_2 \qquad + b_{2r+1}x_{r+1} + \cdots + b_{2n}x_n = 0 \\ \qquad\qquad \cdots\cdots \\ \qquad\qquad x_r + b_{rr+1}x_{r+1} + \cdots + b_{rn}x_n = 0 \end{cases}$$

即

namely,

$$\begin{cases} x_1 = -b_{1r+1}x_{r+1} - \cdots - b_{1n}x_n \\ x_2 = -b_{2r+1}x_{r+1} - \cdots - b_{2n}x_n \\ \qquad \cdots\cdots \\ x_r = -b_{rr+1}x_{r+1} - \cdots - b_{rn}x_n \end{cases} \tag{4.10}$$

显然, 方程组 (4.9) 与方程组 (4.10) 同解.

Obviously, Systems (4.9) and (4.10) are equivalent.

在 (4.10) 中, $x_1,x_2,\cdots,x_r$ 由 $x_{r+1},\cdots,x_n$ 表示, 任给 $x_{r+1},\cdots,x_n$ 一组值, 可唯一确定 $x_1,x_2,\cdots,x_r$ 的值, 从而得到 (4.9) 的一组解. 将 $x_1,x_2,\cdots,x_r$ 称为**主变**

In (4.10), $x_1,x_2,\cdots,x_r$ are represented by $x_{r+1},\cdots,x_n$. For any given values of $x_{r+1},\cdots,x_n$, we can uniquely determine the values of $x_1,x_2,\cdots,x_r$, thus we

量, $x_{r+1},\cdots,x_n$ 称为可取任意实数**自由未知量**.

obtain a solution of (4.9). $x_1,x_2,\cdots,x_r$ are called **principal variables** and $x_{r+1},\cdots,x_n$ are called **free unknowns** which can take any real numbers.

设自由未知量 $x_{r+1}=k_1,\cdots,x_n=k_{n-r}$, 其中 $k_1,k_2,\cdots,k_{n-r}$ 为任意实数, 则与 (4.9) 同解的方程组为

Suppose that the free unknowns are $x_{r+1}=k_1,\cdots,x_n=k_{n-r}$, where k_1, $k_2,\cdots,k_{n-r}$ are arbitrary real numbers, then the system equivalent to (4.10) is

$$\begin{cases} x_1=-b_{1r+1}k_1-\cdots-b_{1n}k_{n-r} \\ \qquad\cdots\cdots \\ x_r=-b_{rr+1}k_1-\cdots-b_{rn}k_{n-r} \\ x_{r+1}= \quad k_1 \\ \qquad\cdots\cdots \\ x_n= \qquad\qquad k_{n-r} \end{cases}$$

则

then

$$\boldsymbol{x}=\begin{pmatrix} x_1\\ x_2\\ \vdots\\ x_r\\ x_{r+1}\\ x_{r+2}\\ \vdots\\ x_n \end{pmatrix}=k_1\begin{pmatrix} -b_{1r+1}\\ -b_{2r+1}\\ \vdots\\ -b_{rr+1}\\ 1\\ 0\\ \vdots\\ 0 \end{pmatrix}+k_2\begin{pmatrix} -b_{1r+2}\\ -b_{2r+2}\\ \vdots\\ -b_{rr+2}\\ 0\\ 1\\ \vdots\\ 0 \end{pmatrix}+\cdots+k_{n-r}\begin{pmatrix} -b_{1n}\\ -b_{2n}\\ \vdots\\ -b_{rn}\\ 0\\ 0\\ \vdots\\ 1 \end{pmatrix} \tag{4.11}$$

因此, (4.11) 为方程组 (4.9) 的解. 令

Therefore, (4.11) is a solution of System (4.9). Let

$$\boldsymbol{\xi}_1=\begin{pmatrix} -b_{1r+1}\\ -b_{2r+1}\\ \vdots\\ -b_{rr+1}\\ 1\\ 0\\ \vdots\\ 0 \end{pmatrix},\boldsymbol{\xi}_2=\begin{pmatrix} -b_{1r+2}\\ -b_{2r+2}\\ \vdots\\ -b_{rr+2}\\ 0\\ 1\\ \vdots\\ 0 \end{pmatrix},\cdots,\boldsymbol{\xi}_{n-r}=\begin{pmatrix} -b_{1n}\\ -b_{2n}\\ \vdots\\ -b_{rn}\\ 0\\ 0\\ \vdots\\ 1 \end{pmatrix}$$

则

then

$$\boldsymbol{x} = k_1\boldsymbol{\xi}_1 + k_2\boldsymbol{\xi}_2 + \cdots + k_{n-r}\boldsymbol{\xi}_{n-r} \tag{4.12}$$

$k_1, k_2, \cdots, k_{n-r}$ 为任意实数.

where $k_1, k_2, \cdots, k_{n-r}$ are arbitrary real numbers.

一般地, 用下面方法求 $\boldsymbol{\xi}_1, \boldsymbol{\xi}_2, \cdots, \boldsymbol{\xi}_{n-r}$ 更方便. 在方程组 (4.10) 中, 令自由未知量 $x_{r+1}, x_{r+2}, \cdots, x_n$ 构成的向量分别为如下的 $n-r$ 个向量

In general, the following method for finding $\boldsymbol{\xi}_1, \boldsymbol{\xi}_2, \cdots, \boldsymbol{\xi}_{n-r}$ is more convenient. In System (4.10), let the vector consisted of free unknowns $x_{r+1}, x_{r+2}, \cdots, x_n$ be the following $n-r$ vectors, respectively,

$$\begin{pmatrix} x_{r+1} \\ x_{r+2} \\ \vdots \\ x_n \end{pmatrix} = \begin{pmatrix} 1 \\ 0 \\ \vdots \\ 0 \end{pmatrix}, \begin{pmatrix} 0 \\ 1 \\ \vdots \\ 0 \end{pmatrix}, \cdots, \begin{pmatrix} 0 \\ 0 \\ \vdots \\ 1 \end{pmatrix}$$

则由 (4.10) 可得

then from (4.10) we have

$$\begin{pmatrix} x_1 \\ x_2 \\ \vdots \\ x_r \end{pmatrix} = \begin{pmatrix} -b_{1r+1} \\ -b_{2r+1} \\ \vdots \\ -b_{rr+1} \end{pmatrix}, \begin{pmatrix} -b_{1r+2} \\ -b_{2r+2} \\ \vdots \\ -b_{rr+2} \end{pmatrix}, \cdots, \begin{pmatrix} -b_{1n} \\ -b_{2n} \\ \vdots \\ -b_{rn} \end{pmatrix}$$

合起来便得到

Combining them yields

$$\boldsymbol{\xi}_1 = \begin{pmatrix} -b_{1r+1} \\ -b_{2r+1} \\ \vdots \\ -b_{rr+1} \\ 1 \\ 0 \\ \vdots \\ 0 \end{pmatrix}, \boldsymbol{\xi}_2 = \begin{pmatrix} -b_{1r+2} \\ -b_{2r+2} \\ \vdots \\ -b_{rr+2} \\ 0 \\ 1 \\ \vdots \\ 0 \end{pmatrix}, \cdots, \boldsymbol{\xi}_{n-r} = \begin{pmatrix} -b_{1n} \\ -b_{2n} \\ \vdots \\ -b_{rn} \\ 0 \\ 0 \\ \vdots \\ 1 \end{pmatrix}$$

显然, 向量 $\boldsymbol{\xi}_1, \boldsymbol{\xi}_2, \cdots, \boldsymbol{\xi}_{n-r}$ 是方程组 (4.9) 的解, 而且容易验证解向量组 $\{\boldsymbol{\xi}_1, \boldsymbol{\xi}_2, \cdots, \boldsymbol{\xi}_{n-r}\}$ 线性无关; 由 (4.12) 知, (4.9) 的任一解向量均可由 $\{\boldsymbol{\xi}_1, \boldsymbol{\xi}_2, \cdots, \boldsymbol{\xi}_{n-r}\}$ 线性表示. 所以, 向量组

Obviously, the vectors $\boldsymbol{\xi}_1, \boldsymbol{\xi}_2, \cdots, \boldsymbol{\xi}_{n-r}$ are solutions of System (4.9), and it is easy to show that the solution vector set $\{\boldsymbol{\xi}_1, \boldsymbol{\xi}_2, \cdots, \boldsymbol{\xi}_{n-r}\}$ is linearly independent;

$\{\boldsymbol{\xi}_1, \boldsymbol{\xi}_2, \cdots, \boldsymbol{\xi}_{n-r}\}$ 为 (4.9) 的解向量组 $\boldsymbol{X}$ 的一个极大无关组, 即基础解系. 因此方程组的通解为

and from (4.12) we see that an arbitrary solution vector of (4.9) can be linearly represented by $\{\boldsymbol{\xi}_1, \boldsymbol{\xi}_2, \cdots, \boldsymbol{\xi}_{n-r}\}$. Therefore, the vector set $\{\boldsymbol{\xi}_1, \boldsymbol{\xi}_2, \cdots, \boldsymbol{\xi}_{n-r}\}$ is a maximal independent subset of the solution set $\boldsymbol{X}$ of (4.9), namely, the system of fundamental solutions. Thus the general solution of the system is

$$\boldsymbol{X} = \{k_1\boldsymbol{\xi}_1 + k_2\boldsymbol{\xi}_2 + \cdots + k_{n-r}\boldsymbol{\xi}_{n-r} | k_1, k_2, \cdots, k_{n-r} \in \mathbf{R}\}$$

综上所述, 对于 n 元齐次线性方程组 $\boldsymbol{Ax} = \mathbf{0}$, 若 $R(\boldsymbol{A}) = n$, 则方程组只有零解; 若 $R(\boldsymbol{A}) = r < n$, 则方程组有无穷多个非零解, 其基础解系含有 $n-r$ 个向量, 如 $\boldsymbol{\xi}_1, \boldsymbol{\xi}_2, \cdots, \boldsymbol{\xi}_{n-r}$, 通解形式为

In sum, for a system of homogeneous linear equations in n unknowns, given by $\boldsymbol{Ax} = \mathbf{0}$, if $R(\boldsymbol{A}) = n$, the system has only a zero solution; while if $R(\boldsymbol{A}) = r < n$, the system has infinitely many solutions, and the system of fundamental solutions contains $n-r$ vectors, such as $\boldsymbol{\xi}_1, \boldsymbol{\xi}_2, \cdots, \boldsymbol{\xi}_{n-r}$, the form of general solutions is

$$\boldsymbol{X} = \{k_1\boldsymbol{\xi}_1 + k_2\boldsymbol{\xi}_2 + \cdots + k_{n-r}\boldsymbol{\xi}_{n-r} | k_1, k_2, \cdots, k_{n-r} \in \mathbf{R}\}$$

现在我们通过例题讲解如何求线性方程组的一个基础解系和通解.

Now we explain how to find a system of fundamental solutions and a general solution of a linear system by examples.

例 1 求齐次线性方程组的一个基础解系和通解.

Example 1 Find a system of fundamental solutions and a general solution of the homogeneous linear system.

$$\begin{cases} x_1 + 2x_2 + 2x_3 + x_4 = 0 \\ 2x_1 + x_2 - 2x_3 - 2x_4 = 0 \\ x_1 - x_2 - 4x_3 - 3x_4 = 0 \end{cases}$$

解 对系数矩阵施以初等行变换, 将其化为如下的行最简形

Solution Performing elementary row operations on the coefficient matrix, we obtain the following row-reduced form

$$\boldsymbol{A} = \begin{pmatrix} 1 & 2 & 2 & 1 \\ 2 & 1 & -2 & -2 \\ 1 & -1 & -4 & -3 \end{pmatrix} \xrightarrow{r_2-2r_1, r_3-r_1} \begin{pmatrix} 1 & 2 & 2 & 1 \\ 0 & -3 & -6 & -4 \\ 0 & -3 & -6 & -4 \end{pmatrix}$$

$$\xrightarrow{r_2-r_3,(-1/3)r_2}\begin{pmatrix}1&2&2&1\\0&1&2&\frac{4}{3}\\0&0&0&0\end{pmatrix}\xrightarrow{r_1-2r_2}\begin{pmatrix}1&0&-2&-\frac{5}{3}\\0&1&2&\frac{4}{3}\\0&0&0&0\end{pmatrix}$$

同解方程组

The equivalent system is

$$\begin{cases}x_1-2x_3-\dfrac{5}{3}x_4=0\\x_2+2x_3+\dfrac{4}{3}x_4=0\end{cases}$$

即

namely,

$$\begin{cases}x_1=2x_3+\dfrac{5}{3}x_4\\x_2=-2x_3-\dfrac{4}{3}x_4\end{cases}$$

令 x_3, x_4 为自由未知量, 分别取

Let x_3, x_4 be free unknowns, and respectively take

$$\begin{pmatrix}x_3\\x_4\end{pmatrix}=\begin{pmatrix}1\\0\end{pmatrix},\quad\begin{pmatrix}0\\1\end{pmatrix}$$

则

then

$$\begin{pmatrix}x_1\\x_2\end{pmatrix}=\begin{pmatrix}2\\-2\end{pmatrix},\quad\begin{pmatrix}\frac{5}{3}\\\frac{-4}{3}\end{pmatrix}$$

得到线性方程组的基础解系

We obtain the system of fundamental solutions of the linear system

$$\boldsymbol{\xi}_1=\begin{pmatrix}2\\-2\\1\\0\end{pmatrix},\quad\boldsymbol{\xi}_2=\begin{pmatrix}\frac{5}{3}\\-\frac{4}{3}\\0\\1\end{pmatrix}$$

于是原方程组通解为

Therefore, the general solution of the original system is

$$\boldsymbol{x}=\begin{pmatrix}x_1\\x_2\\x_3\\x_4\end{pmatrix}=k_1\begin{pmatrix}2\\-2\\1\\0\end{pmatrix}+k_2\begin{pmatrix}\frac{5}{3}\\-\frac{4}{3}\\0\\1\end{pmatrix}$$

其中 k_1, k_2 为任何实数. ■

where k_1, k_2 are arbitrary real numbers. ■

例 2 求齐次线性方程组 $\boldsymbol{Ax}=\boldsymbol{0}$ 的基础解系和通解, 其系数矩阵为

Example 2 Find a system of fundamental solutions and a general solution of the homogeneous linear system $\boldsymbol{Ax}=\boldsymbol{0}$, the coefficient matrix is

$$\boldsymbol{A}=\begin{pmatrix}1 & 1 & -2 & 1 & 3\\ 2 & -1 & 2 & 2 & 6\\ 3 & 2 & -4 & -5 & -7\end{pmatrix}$$

解 对系数矩阵 $\boldsymbol{A}$ 施以初等行变换, 将其化为如下的行最简形

Solution Performing elementary row operations on the coefficient matrix $\boldsymbol{A}$, we obtain the following row-reduced form

$$\boldsymbol{A}=\begin{pmatrix}1 & 1 & -2 & 1 & 3\\ 2 & -1 & 2 & 2 & 6\\ 3 & 2 & -4 & -5 & -7\end{pmatrix}\xrightarrow[r_3-3r_1]{r_2-2r_1}\begin{pmatrix}1 & 1 & -2 & 1 & 3\\ 0 & -3 & 6 & 0 & 0\\ 0 & -1 & 2 & -8 & -16\end{pmatrix}$$

$$\xrightarrow[r_3+r_2]{(-1/3)r_2}\begin{pmatrix}1 & 1 & -2 & 1 & 3\\ 0 & 1 & -2 & 0 & 0\\ 0 & 0 & 0 & -8 & -16\end{pmatrix}$$

$$\xrightarrow{r_1-r_2,(-1/8)r_3,r_1-r_3}\begin{pmatrix}1 & 0 & 0 & 0 & 1\\ 0 & 1 & -2 & 0 & 0\\ 0 & 0 & 0 & 1 & 2\end{pmatrix}$$

同解方程组为

The equivalent system is

$$\begin{cases}x_1+x_5=0\\ x_2-2x_3=0\\ x_4+2x_5=0\end{cases}$$

即

namely,

$$\begin{cases}x_1= & & -x_5\\ x_2=2x_3 & & \\ x_4= & & -2x_5\end{cases}$$

令 x_3, x_5 为自由未知量. 取

Let x_3, x_5 be free unknowns. Take

$$\begin{pmatrix}x_3\\ x_5\end{pmatrix}=\begin{pmatrix}1\\ 0\end{pmatrix},\quad\begin{pmatrix}0\\ 1\end{pmatrix}$$

则 | then

$$\begin{pmatrix} x_1 \\ x_2 \\ x_4 \end{pmatrix} = \begin{pmatrix} 0 \\ 2 \\ 0 \end{pmatrix}, \quad \begin{pmatrix} -1 \\ 0 \\ -2 \end{pmatrix}$$

即得到方程组的基础解系 | We obtain the system of fundamental solutions of the system

$$\boldsymbol{\xi}_1 = \begin{pmatrix} 0 \\ 2 \\ 1 \\ 0 \\ 0 \end{pmatrix}, \quad \boldsymbol{\xi}_2 = \begin{pmatrix} -1 \\ 0 \\ 0 \\ -2 \\ 1 \end{pmatrix}$$

于是原方程组的通解为 | Therefore, the general solution of the original system is

$$\boldsymbol{x} = \begin{pmatrix} x_1 \\ x_2 \\ x_3 \\ x_4 \\ x_5 \end{pmatrix} = k_1 \begin{pmatrix} 0 \\ 2 \\ 1 \\ 0 \\ 0 \end{pmatrix} + k_2 \begin{pmatrix} -1 \\ 0 \\ 0 \\ -2 \\ 1 \end{pmatrix}$$

其中 k_1, k_2 为任何实数. | where k_1, k_2 are arbitrary real numbers.

4.4.2 非齐次线性方程组解的结构 | 4.4.2 Solution Structures of Non-homogeneous Linear Systems

对于 n 元非齐次线性方程组 | For a system of non-homogeneous linear equations in n unknowns

$$\boldsymbol{Ax} = \boldsymbol{b} \tag{4.13}$$

其中 $\boldsymbol{A}$ 为 $m \times n$ 矩阵, $\boldsymbol{b} \neq \boldsymbol{0}$. | where $\boldsymbol{A}$ is an $m \times n$ matrix, $\boldsymbol{b} \neq \boldsymbol{0}$.

对应的齐次线性方程组 | The corresponding homogeneous linear system

$$\boldsymbol{Ax} = \boldsymbol{0} \tag{4.14}$$

称为 $\boldsymbol{Ax}=\boldsymbol{b}$ 的**导出组**.

is called a **derived system** of $\boldsymbol{Ax}=\boldsymbol{b}$.

关于方程组 (4.13) 的解, 具有下列性质:

For solutions of System (4.13), we have the following properties:

性质 3 若向量 $\boldsymbol{\eta}_1,\boldsymbol{\eta}_2$ 都是方程组 (4.13) 的解, 则 $\boldsymbol{\eta}_1-\boldsymbol{\eta}_2$ 是与 (4.13) 对应的齐次线性方程组 (4.14) 的解.

Property 3 If the vectors $\boldsymbol{\eta}_1,\boldsymbol{\eta}_2$ are solutions of System (4.13), then $\boldsymbol{\eta}_1-\boldsymbol{\eta}_2$ is solution of System (4.14) associated with (4.13).

证 由已知可得 $A\boldsymbol{\eta}_1=\boldsymbol{b},\boldsymbol{A\eta}_2=\boldsymbol{b}$, 所以 $\boldsymbol{A}(\boldsymbol{\eta}_1-\boldsymbol{\eta}_2)=\boldsymbol{A\eta}_1-A\boldsymbol{\eta}_2=\mathbf{0}$, 即 $\boldsymbol{\eta}_1-\boldsymbol{\eta}_2$ 为 (4.14) 的解.

Proof From the known condition we see that $\boldsymbol{A\eta}_1=\boldsymbol{b},\boldsymbol{A\eta}_2=\boldsymbol{b}$, so $\boldsymbol{A}(\boldsymbol{\eta}_1-\boldsymbol{\eta}_2)=\boldsymbol{A\eta}_1-\boldsymbol{A\eta}_2=\mathbf{0}$, namely, $\boldsymbol{\eta}_1-\boldsymbol{\eta}_2$ is a solution of (4.14).

性质 4 若 $\boldsymbol{\eta}$ 是 (4.13) 的解, $\boldsymbol{\xi}$ 是 (4.14) 的解, 则 $\boldsymbol{\eta}+\boldsymbol{\xi}$ 是 (4.13) 的解.

Property 4 If $\boldsymbol{\eta}$ is a solution of (4.13) and if $\boldsymbol{\xi}$ is a solution of (4.14), then $\boldsymbol{\eta}+\boldsymbol{\xi}$ is a solution of (4.13).

证 由已知条件可得, $\boldsymbol{A\eta}=\boldsymbol{b},\boldsymbol{A\xi}=\mathbf{0}$, 故

Proof From the known conditions we see that $\boldsymbol{A\eta}=\boldsymbol{b},\boldsymbol{A\xi}=\mathbf{0}$, so

$$\boldsymbol{A}(\boldsymbol{\eta}+\boldsymbol{\xi})=\boldsymbol{A\eta}+\boldsymbol{A\xi}=\boldsymbol{b}+\mathbf{0}=\boldsymbol{b}$$

从而 $\boldsymbol{\eta}+\boldsymbol{\xi}$ 是 (4.13) 的解.

Therefore, $\boldsymbol{\eta}+\boldsymbol{\xi}$ is a solution of (4.13).

根据以上性质, 给出非齐次线性方程组解的结构.

From the above properties, we give the solution structures of non-homogeneous linear systems.

定理 1 设非齐次线性方程组 $\boldsymbol{Ax}=\boldsymbol{b}$ 有解, $\boldsymbol{\eta}^*$ 是它的一个 (特) 解, $\boldsymbol{\xi}$ 为对应齐次线性方程组 $\boldsymbol{Ax}=\mathbf{0}$ 的通解, 则非齐次线性方程组 $\boldsymbol{Ax}=\boldsymbol{b}$ 的通解为

Theorem 1 Suppose that the non-homogeneous linear system, given by $\boldsymbol{Ax}=\boldsymbol{b}$ has solutions, and that $\boldsymbol{\eta}^*$ is its (particular) solution, and that $\boldsymbol{\xi}$ is a general solution of the corresponding homogeneous linear system $\boldsymbol{Ax}=\mathbf{0}$, then the general solution of $\boldsymbol{Ax}=\boldsymbol{b}$ is

$$\boldsymbol{x}=\boldsymbol{\eta}^*+\boldsymbol{\xi}$$

证 令 $\boldsymbol{x}$ 是 $\boldsymbol{Ax}=\boldsymbol{b}$ 的任意一个解. 令 $\boldsymbol{\xi}=\boldsymbol{x}-\boldsymbol{\eta}^*$, 由性质 3 知, $\boldsymbol{\xi}=\boldsymbol{x}-\boldsymbol{\eta}^*$ 是 $\boldsymbol{Ax}=\boldsymbol{0}$ 的解. $\boldsymbol{x}=\boldsymbol{\eta}^*+\boldsymbol{\xi}$ 是 $\boldsymbol{Ax}=\boldsymbol{b}$ 的任意一个解.

Proof Let $\boldsymbol{x}$ be an arbitrary solution of $\boldsymbol{Ax}=\boldsymbol{b}$. From Property 3 we know that $\boldsymbol{\xi}=\boldsymbol{x}-\boldsymbol{\eta}^*$ is a solution of $\boldsymbol{Ax}=\boldsymbol{0}$. Thus, $\boldsymbol{x}=\boldsymbol{\eta}^*+\boldsymbol{\xi}$ is an arbitrary solution of $\boldsymbol{Ax}=\boldsymbol{b}$.

一方面, 由性质 4 知, $\boldsymbol{x}=\boldsymbol{\eta}^*+\boldsymbol{\xi}$ 是 $\boldsymbol{Ax}=\boldsymbol{b}$ 的解;

On the one hand, from Property 4 we know that $\boldsymbol{x}=\boldsymbol{\eta}^*+\boldsymbol{\xi}$ is a solution of $\boldsymbol{Ax}=\boldsymbol{b}$;

另一方面, 若 $\boldsymbol{\xi}$ 为 $\boldsymbol{Ax}=\boldsymbol{0}$ 通解, 则 $\boldsymbol{x}=\boldsymbol{\eta}^*+\boldsymbol{\xi}$ 为 $\boldsymbol{Ax}=\boldsymbol{b}$ 的通解.

On the other hand, if $\boldsymbol{\xi}$ is a general solution of $\boldsymbol{Ax}=\boldsymbol{0}$, then $\boldsymbol{x}=\boldsymbol{\eta}^*+\boldsymbol{\xi}$ is a general solution of $\boldsymbol{Ax}=\boldsymbol{b}$.

设 $R(\boldsymbol{A})=r$, 向量组 $\{\boldsymbol{\xi}_1,\boldsymbol{\xi}_2,\cdots,\boldsymbol{\xi}_{n-r}\}$ 为 $\boldsymbol{Ax}=\boldsymbol{0}$ 的基础解系, 则 $\boldsymbol{Ax}=\boldsymbol{b}$ 的通解为

Suppose that $R(\boldsymbol{A})=r$ and that the vector set $\{\boldsymbol{\xi}_1,\boldsymbol{\xi}_2,\cdots,\boldsymbol{\xi}_{n-r}\}$ is a system of fundamental solutions of $\boldsymbol{Ax}=\boldsymbol{0}$, then the general solution of $\boldsymbol{Ax}=\boldsymbol{b}$ is

$$\boldsymbol{x}=\boldsymbol{\eta}^*+k_1\boldsymbol{\xi}_1+k_2\boldsymbol{\xi}_2+\cdots+k_{n-r}\boldsymbol{\xi}_{n-r}, k_1,k_2,\cdots,k_{n-r}\in\mathbf{R}$$ ▲

例 3 求下面非齐次线性方程组的一个基础解系和通解.

Example 3 Find a system of fundamental solutions and a general solution of the following non-homogeneous linear system.

$$\begin{cases} x_1-x_2-\ \ x_3+\ \ x_4=0\\ x_1-x_2+\ \ x_3-3x_4=1\\ x_1-x_2-2x_3+3x_4=-\dfrac{1}{2}\end{cases}$$

解 对增广矩阵施以初等行变换, 将其化成如下的行最简形

Solution Performing elementary operations on the augmented matrix yields the following row-reduced form

$$\overline{\boldsymbol{A}}=\begin{pmatrix}1&-1&-1&1&0\\1&-1&1&-3&1\\1&-1&-2&3&-\dfrac{1}{2}\end{pmatrix}\xrightarrow{r_2-r_1,r_3-r_1}\begin{pmatrix}1&-1&-1&1&0\\0&0&2&-4&1\\0&0&-1&2&-\dfrac{1}{2}\end{pmatrix}$$

$$\xrightarrow{1/2r_2,r_3+r_2}\begin{pmatrix}1&-1&-1&1&0\\0&0&1&-2&\dfrac{1}{2}\\0&0&0&0&0\end{pmatrix}\xrightarrow{r_1+r_2}\begin{pmatrix}1&-1&0&-1&\dfrac{1}{2}\\0&0&1&-2&\dfrac{1}{2}\\0&0&0&0&0\end{pmatrix}$$

易见, $R(\boldsymbol{A}) = R(\overline{\boldsymbol{A}}) = 2 < 4$, 方程组有无穷多解. 同解方程组

Obviously, $R(\boldsymbol{A}) = R(\overline{\boldsymbol{A}}) = 2 < 4$, it means that this system has infinitely many solutions. The equivalent system is

$$\begin{cases} x_1 - x_2 - x_4 = \dfrac{1}{2} \\ x_3 - 2x_4 = \dfrac{1}{2} \end{cases}$$

导出组为

The derived system is

$$\begin{cases} x_1 - x_2 - x_4 = 0 \\ x_3 - 2x_4 = 0 \end{cases}$$

令 x_2, x_4 为自由未知量. 取

Let x_2, x_4 be free unknowns. Setting

$$\begin{pmatrix} x_2 \\ x_4 \end{pmatrix} = \begin{pmatrix} 0 \\ 0 \end{pmatrix}$$

得到非齐次线性方程组的一个特解

we obtain a particular solution of the non-homogeneous linear system

$$\boldsymbol{\eta}^* = \begin{pmatrix} \dfrac{1}{2} \\ 0 \\ \dfrac{1}{2} \\ 0 \end{pmatrix}$$

令

Let

$$\begin{pmatrix} x_2 \\ x_4 \end{pmatrix} = \begin{pmatrix} 1 \\ 0 \end{pmatrix}, \begin{pmatrix} 0 \\ 1 \end{pmatrix}.$$

则有

Then we have

$$\begin{pmatrix} x_1 \\ x_3 \end{pmatrix} = \begin{pmatrix} 1 \\ 0 \end{pmatrix}, \quad \begin{pmatrix} 1 \\ 2 \end{pmatrix}$$

得到导出组的基础解系为

The system of fundamental solutions of the obtained derived system is

$$\boldsymbol{\xi}_1 = \begin{pmatrix} 1 \\ 1 \\ 0 \\ 0 \end{pmatrix}, \quad \boldsymbol{\xi}_2 = \begin{pmatrix} 1 \\ 0 \\ 2 \\ 1 \end{pmatrix}$$

于是原方程组的通解为

Therefore, the general solution of the original system is

$$\boldsymbol{x}=\begin{pmatrix}x_1\\x_2\\x_3\\x_4\end{pmatrix}=\begin{pmatrix}\frac{1}{2}\\0\\\frac{1}{2}\\0\end{pmatrix}+k_1\begin{pmatrix}1\\1\\0\\0\end{pmatrix}+k_2\begin{pmatrix}1\\0\\2\\1\end{pmatrix}$$

其中 k_1, k_2 为任何实数. ■

where k_1, k_2 are arbitrary real numbers. ■

例 4 对于一个四元非齐次线性方程组, 设其系数矩阵的秩为 3, 已知 $\boldsymbol{\eta}_1, \boldsymbol{\eta}_2, \boldsymbol{\eta}_3$ 是它的三个解向量, 且

Example 4 For a system of linear equations in 4 unknowns, suppose that the rank of the coefficient matrix is equal to 3, and that $\boldsymbol{\eta}_1, \boldsymbol{\eta}_2, \boldsymbol{\eta}_3$ are known solution vectors satisfying

$$\boldsymbol{\eta}_1=\begin{pmatrix}2\\3\\4\\5\end{pmatrix},\quad \boldsymbol{\eta}_2+\boldsymbol{\eta}_3=\begin{pmatrix}1\\2\\3\\4\end{pmatrix}$$

求该方程组的通解.

Find the general solution of the system.

解 设方程组为 $\boldsymbol{Ax}=\boldsymbol{b}$, 则 $\boldsymbol{A\eta}_1=\boldsymbol{b}, \boldsymbol{A\eta}_2=\boldsymbol{b}, \boldsymbol{A\eta}_3=\boldsymbol{b}$. 不难验证下式成立

Solution Suppose that the system is given by $\boldsymbol{Ax}=\boldsymbol{b}$, then we have $\boldsymbol{A\eta}_1=\boldsymbol{b}, \boldsymbol{A\eta}_2=\boldsymbol{b}$, $\boldsymbol{A\eta}_3=\boldsymbol{b}$. It is not difficult to show that the following equation is valid, namely,

$$\boldsymbol{A}\frac{1}{2}(\boldsymbol{\eta}_2+\boldsymbol{\eta}_3)=\frac{1}{2}\boldsymbol{A\eta}_2+\frac{1}{2}\boldsymbol{A\eta}_3=\boldsymbol{b}$$

故 $\frac{1}{2}(\boldsymbol{\eta}_2+\boldsymbol{\eta}_3)$ 也是 $\boldsymbol{Ax}=\boldsymbol{b}$ 的解.

It implies that $\frac{1}{2}(\boldsymbol{\eta}_2+\boldsymbol{\eta}_3)$ is also a solution of $\boldsymbol{Ax}=\boldsymbol{b}$.

由性质 3 知,

From Property 3 we know that

$$\boldsymbol{\eta}_1-\frac{1}{2}(\boldsymbol{\eta}_2+\boldsymbol{\eta}_3)=\begin{pmatrix}\frac{3}{2}\\2\\\frac{5}{2}\\3\end{pmatrix}$$

为对应的齐次线性方程 $\boldsymbol{Ax}=\mathbf{0}$ 的非零解. 因为 $R(\boldsymbol{A})=3$, 所以 $\boldsymbol{Ax}=\mathbf{0}$ 的基础解系只含一个向量, 故 $\boldsymbol{\eta}_1-\dfrac{1}{2}(\boldsymbol{\eta}_2+\boldsymbol{\eta}_3)$ 是 $\boldsymbol{Ax}=\mathbf{0}$ 的基础解系. 由非齐次线性方程组解的结构得 $\boldsymbol{Ax}=\boldsymbol{b}$ 的通解

is a non-zero solution of the corresponding homogeneous linear system $\boldsymbol{Ax}=\mathbf{0}$. Since $R(\boldsymbol{A})=3$, the system of fundamental solutions of $\boldsymbol{Ax}=\mathbf{0}$ has only a vector, which means that $\boldsymbol{\eta}_1-\dfrac{1}{2}(\boldsymbol{\eta}_2+\boldsymbol{\eta}_3)$ is a system of fundamental solutions of $\boldsymbol{Ax}=\mathbf{0}$. From the solution structures of non-homogeneous linear systems we know that the general solution of $\boldsymbol{Ax}=\boldsymbol{b}$ is given by

$$\boldsymbol{x}=\boldsymbol{\eta}_1+k\left[\boldsymbol{\eta}_1-\frac{1}{2}(\boldsymbol{\eta}_2+\boldsymbol{\eta}_3)\right]=\begin{pmatrix}2\\3\\4\\5\end{pmatrix}+k\begin{pmatrix}\frac{3}{2}\\2\\\frac{5}{2}\\3\end{pmatrix}$$

其中 k 为任何实数.

where k is an arbitrary real number.

例 5 λ 取何值时, 使得线性方程组

Example 5 What values do λ take on such that the linear system

$$\begin{cases}-2x_1+x_2+x_3=-2\\x_1-2x_2+x_3=\lambda\\x_1+x_2-2x_3=\lambda^2\end{cases}$$

有解? 求出它的通解.

has solutions? Find its general solutions.

解 对增广矩阵施以初等行变换, 得到

Solution Performing elementary row operations on the augmented matrix we obtain

$$\overline{\boldsymbol{A}}=\begin{pmatrix}-2&1&1&-2\\1&-2&1&\lambda\\1&1&-2&\lambda^2\end{pmatrix}\xrightarrow{r_3\leftrightarrow r_1}\begin{pmatrix}1&1&-2&\lambda^2\\1&-2&1&\lambda\\-2&1&1&-2\end{pmatrix}$$

$$\xrightarrow{r_2-r_1,r_3+2r_1}\begin{pmatrix}1&1&-2&\lambda^2\\0&-3&3&\lambda-\lambda^2\\0&3&-3&-2+2\lambda^2\end{pmatrix}$$

$$\xrightarrow{r_3+r_2}\begin{pmatrix}1&1&-2&\lambda^2\\0&-3&3&\lambda-\lambda^2\\0&0&0&-2+\lambda+\lambda^2\end{pmatrix}$$

若 $-2+\lambda+\lambda^2=0$, 即 $\lambda=1$ 或 $\lambda=-2$, 则有 $R(\boldsymbol{A})=R(\overline{\boldsymbol{A}})=2$, 此时方程组有解.

If $-2+\lambda+\lambda^2=0$, i.e., $\lambda=1$ or $\lambda=-2$, then $R(\boldsymbol{A})=R(\overline{\boldsymbol{A}})=2$, this implies the system has solutions.

当 $\lambda=1$, 有

If $\lambda=1$, we have

$$\overline{\boldsymbol{A}} \to \begin{pmatrix} 1 & 1 & -2 & 1 \\ 0 & -3 & 3 & 0 \\ 0 & 0 & 0 & 0 \end{pmatrix} \to \begin{pmatrix} 1 & 0 & -1 & 1 \\ 0 & 1 & -1 & 0 \\ 0 & 0 & 0 & 0 \end{pmatrix}$$

同解方程组为

the equivalent system is

$$\begin{cases} x_1 = x_3 + 1 \\ x_2 = x_3 \end{cases}$$

导出组为

the derived system is

$$\begin{cases} x_1 = x_3 \\ x_2 = x_3 \end{cases}$$

令 x_3 为自由未知量. 取 $x_3=0$, 得到非齐次线性方程组的一个特解

Let x_3 be a free unknown. Setting $x_3=0$, we obtain a particular solution of the non-homogeneous linear system,

$$\boldsymbol{\eta}^* = \begin{pmatrix} 1 \\ 0 \\ 0 \end{pmatrix}$$

取 $x_3=1$, 得到导出组的一个基础解系

Setting $x_3=1$, we obtain a system of fundamental solutions of the derived system

$$\boldsymbol{\xi} = \begin{pmatrix} 1 \\ 1 \\ 1 \end{pmatrix}$$

即当 $\lambda=1$ 时, 原方程组的通解为

i.e., the general solution of the original system as $\lambda=1$ is given by

$$\boldsymbol{x} = \begin{pmatrix} x_1 \\ x_2 \\ x_3 \end{pmatrix} = \begin{pmatrix} 1 \\ 0 \\ 0 \end{pmatrix} + k \begin{pmatrix} 1 \\ 1 \\ 1 \end{pmatrix}$$

其中 k 为任何实数.

where k is an arbitrary real number.

当 $\lambda=2$, 则有

If $\lambda=2$, then we have

$$\overline{\boldsymbol{A}}\to\begin{pmatrix}1&1&-2&4\\0&-3&3&-6\\0&0&0&0\end{pmatrix}\to\begin{pmatrix}1&0&-1&2\\0&1&-1&2\\0&0&0&0\end{pmatrix}$$

同解方程组为

The equivalent system is

$$\begin{cases}x_1-x_3=2\\x_2-x_3=2\end{cases}$$

导出组为

The derived system is

$$\begin{cases}x_1-x_3=0\\x_2-x_3=0\end{cases}$$

令 x_3 为自由未知量, 取 $x_3=0$, 可得到非齐次线性方程组的一个特解

Let x_3 be a free unknown. Setting $x_3=0$, we obtain a particular solution of the non-homogeneous linear system,

$$\boldsymbol{\eta}^*=\begin{pmatrix}2\\2\\0\end{pmatrix}$$

取 $x_3=1$, 得到导出组的一个基础解系

Setting $x_3=1$, we obtain a system of fundamental solutions of the derived system

$$\boldsymbol{\xi}=\begin{pmatrix}1\\1\\1\end{pmatrix}$$

即当 $\lambda=-2$ 时, 原方程的通解为

i.e., the general solution of the original system as $\lambda=-2$ is given by

$$\boldsymbol{x}=\begin{pmatrix}x_1\\x_2\\x_3\end{pmatrix}=\begin{pmatrix}2\\2\\0\end{pmatrix}+k\begin{pmatrix}1\\1\\1\end{pmatrix}$$ ■

例 6 问 a,b 取何值时, 使得如下线性方程组

Example 6 What values do a,b take on such that the following linear system

$$\begin{cases} x_1 + 2x_2 = 3 \\ 4x_1 + 7x_2 + x_3 = 10 \\ x_2 - x_3 = b \\ 2x_1 + 3x_2 + ax_3 = 4 \end{cases}$$

(1) 无解; (2) 有唯一解; (3) 有无穷多个解? 在有解的情况下, 求出唯一解或通解.

(1) has no solution; (2) has a unique solution; (3) has infinitely many solutions? In the case that this system has solutions, find the unique solution or the general solution.

解 对增广矩阵施以初等行变换, 有

Solution Performing elementary row operations on the augmented matrix, we have

$$\overline{\boldsymbol{A}} = \begin{pmatrix} 1 & 2 & 0 & 3 \\ 4 & 7 & 1 & 10 \\ 0 & 1 & -1 & b \\ 2 & 3 & a & 4 \end{pmatrix} \xrightarrow[r_4-2r_1]{r_2-4r_1,} \begin{pmatrix} 1 & 2 & 0 & 3 \\ 0 & -1 & 1 & -2 \\ 0 & 1 & -1 & b \\ 0 & -1 & a & -2 \end{pmatrix}$$

$$\xrightarrow[r_4-r_2]{r_3+r_2,} \begin{pmatrix} 1 & 2 & 0 & 3 \\ 0 & -1 & 1 & -2 \\ 0 & 0 & 0 & b-2 \\ 0 & 0 & a-1 & 0 \end{pmatrix} \xrightarrow{r_3 \leftrightarrow r_4} \begin{pmatrix} 1 & 2 & 0 & 3 \\ 0 & -1 & 1 & -2 \\ 0 & 0 & a-1 & 0 \\ 0 & 0 & 0 & b-2 \end{pmatrix}$$

(1) 当 $b \neq 2$ 时, $R(\boldsymbol{A}) < R(\overline{\boldsymbol{A}})$, 方程组无解.

(1) If $b \neq 2$, then $R(\boldsymbol{A}) < R(\overline{\boldsymbol{A}})$, this means that the system has no solution.

(2) 当 $b = 2, a \neq 1$ 时, $R(\boldsymbol{A}) = R(\overline{\boldsymbol{A}}) = 3$, 方程组有唯一解. 此时

(2) If $b = 2, a \neq 1$, then $R(\boldsymbol{A}) = R(\overline{\boldsymbol{A}}) = 3$, this implies that the system has a unique solution. In this case,

$$\overline{\boldsymbol{A}} = \begin{pmatrix} 1 & 2 & 0 & 3 \\ 0 & -1 & 1 & -2 \\ 0 & 0 & a-1 & 0 \\ 0 & 0 & 0 & 0 \end{pmatrix} \rightarrow \begin{pmatrix} 1 & 0 & 0 & -1 \\ 0 & -1 & 0 & -2 \\ 0 & 0 & 1 & 0 \\ 0 & 0 & 0 & 0 \end{pmatrix}$$

同解方程组为

the equivalent system is

$$\begin{cases} x_1 = -1 \\ x_2 = 2 \\ x_3 = 0 \end{cases}$$

因此, 方程组有唯一解

Thus, the system has a unique solution

$$\boldsymbol{x}=\begin{pmatrix}x_1\\x_2\\x_3\end{pmatrix}=\begin{pmatrix}-1\\2\\0\end{pmatrix}$$

(3) 当 $b=2, a=1$ 时, $R(\boldsymbol{A})=R(\overline{\boldsymbol{A}})=2$, 方程组有无穷多个解. 此时

(3) If $b=2, a=1$, then $R(\boldsymbol{A})=R(\overline{\boldsymbol{A}})=2$, this shows that the system has infinitely many solutions. In this case,

$$\overline{\boldsymbol{A}}=\begin{pmatrix}1&2&0&3\\0&-1&1&-2\\0&0&0&0\\0&0&0&0\end{pmatrix}\to\begin{pmatrix}1&0&2&-1\\0&1&-1&2\\0&0&0&0\\0&0&0&0\end{pmatrix}$$

同解方程组为

The equivalent system is

$$\begin{cases}x_1+2x_3=-1\\x_2-x_3=2\end{cases}$$

导出组为

The derived system is

$$\begin{cases}x_1+2x_3=0\\x_2-x_3=0\end{cases}$$

令 x_3 为自由未知量. 取 $x_3=0$, 得到非齐次线性方程组的一个特解

Let x_3 be a free unknown. Setting $x_3=0$, we obtain a particular solution of the non-homogeneous linear system,

$$\boldsymbol{\eta}^*=\begin{pmatrix}-1\\2\\0\end{pmatrix}$$

令 $x_3=1$ 得到导出组的一个基础解系

Setting $x_3=1$, we obtain a system of fundamental solutions of the derived system

$$\boldsymbol{\xi}=\begin{pmatrix}-2\\1\\1\end{pmatrix}$$

于是原方程组的通解

Thus the general solution of the original system is

$$x = \begin{pmatrix} x_1 \\ x_2 \\ x_3 \end{pmatrix} = \begin{pmatrix} -1 \\ 2 \\ 0 \end{pmatrix} + k \begin{pmatrix} -2 \\ 1 \\ 1 \end{pmatrix}$$

其中 k 为任何实数. ■

where k is an arbitrary real number. ■

注 对系数矩阵和增广矩阵作初等行变换时, 不能将参数 a 放到分母上运算. 例如, 不能作如下初等行变换: $r_2 + \dfrac{1}{1-a}r_4$.

Note As we perform elementary row operations on the coefficient matrix and the augmented matrix, the parameter a can not be put on the denominator, for example, the elementary row operation $r_2 + \dfrac{1}{1-a}r_4$ is wrong.

例 7 例 6 还可以换一种形式描述, 已知向量组为

Example 7 Example 6 may be described as another form, known that the vector set is

$$\alpha_1 = \begin{pmatrix} 1 \\ 4 \\ 0 \\ 2 \end{pmatrix}, \quad \alpha_2 = \begin{pmatrix} 2 \\ 7 \\ 1 \\ 3 \end{pmatrix}, \quad \alpha_3 = \begin{pmatrix} 0 \\ 1 \\ -1 \\ a \end{pmatrix}, \quad \beta = \begin{pmatrix} 3 \\ 10 \\ b \\ 4 \end{pmatrix}$$

(1) a, b 取何值时, β 不能由 $\alpha_1, \alpha_2, \alpha_3$ 线性表示?

(1) What values do a, b take on such that β can not be linearly represented by $\alpha_1, \alpha_2, \alpha_3$?

(2) a, b 取何值时, β 可由 $\alpha_1, \alpha_2, \alpha_3$ 唯一线性表示?

(2) What values do a, b take on such that β can be linearly represented by $\alpha_1, \alpha_2, \alpha_3$ and the expression is unique?

(3) a, b 取何值时, β 可由 $\alpha_1, \alpha_2, \alpha_3$ 线性表示但表示式不唯一? 写出此表示式.

(3) What values do a, b take on such that β can be linearly represented by $\alpha_1, \alpha_2, \alpha_3$, but the expression is not uniquely? Write out this representation.

解 由于 β 由 $\alpha_1, \alpha_2, \alpha_3$ 线性表示的形式为

Solution Since the linear representation form of β by $\alpha_1, \alpha_2, \alpha_3$ is

$$x_1\alpha_1 + x_2\alpha_2 + x_3\alpha_3 = \beta$$

显然, 它等价于

Obviously, it is equivalent to

$$(\boldsymbol{\alpha}_1, \boldsymbol{\alpha}_2, \boldsymbol{\alpha}_3)\begin{pmatrix} x_1 \\ x_2 \\ x_3 \end{pmatrix} = \boldsymbol{\beta}$$

即例 6 中的方程组. 以上三个问题变成讨论, a, b 取何值时, 方程组无解; 有唯一解; 有无穷多个解的问题. 并且由例 6 知

namely, the system in Example 6. The discussions of the above three problems turn into the problems of what values a, b take on such that the system has no solution; has a unique solution; has infinitely many solutions. From Example 6 we know that

(1) 若 $b \neq 2$, $\boldsymbol{\beta}$ 不能由 $\boldsymbol{\alpha}_1, \boldsymbol{\alpha}_2, \boldsymbol{\alpha}_3$ 线性表示.

(1) If $b \neq 2$, $\boldsymbol{\beta}$ can not be linearly represented by $\boldsymbol{\alpha}_1, \boldsymbol{\alpha}_2, \boldsymbol{\alpha}_3$.

(2) 若 $b = 2, a \neq 1$, $\boldsymbol{\beta}$ 可唯一表示为 $\boldsymbol{\beta} = -\boldsymbol{\alpha}_1 + 2\boldsymbol{\alpha}_2 + 0\boldsymbol{\alpha}_3$.

(2) If $b = 2, a \neq 1$, $\boldsymbol{\beta}$ can be linearly represented by $\boldsymbol{\alpha}_1, \boldsymbol{\alpha}_2, \boldsymbol{\alpha}_3$, and the expression is uniquely given by $\boldsymbol{\beta} = -\boldsymbol{\alpha}_1 + 2\boldsymbol{\alpha}_2 + 0\boldsymbol{\alpha}_3$.

(3) 若 $b = 2, a = 1$, $\boldsymbol{\beta}$ 由 $\boldsymbol{\alpha}_1, \boldsymbol{\alpha}_2, \boldsymbol{\alpha}_3$ 的线性表示式为

(3) If $b = 2, a = 1$, the linear representation of $\boldsymbol{\beta}$ by $\boldsymbol{\alpha}_1, \boldsymbol{\alpha}_2, \boldsymbol{\alpha}_3$ is

$$\boldsymbol{\beta} = -(2k+1)\boldsymbol{\alpha}_1 + (k+2)\boldsymbol{\alpha}_2 + k\boldsymbol{\alpha}_3$$

其中 k 为任何实数.

where k is an arbitrary real number.

例 8 证明 $R(\boldsymbol{A}^{\mathrm{T}}\boldsymbol{A}) = R(\boldsymbol{A})$.

Example 8 Prove that $R(\boldsymbol{A}^{\mathrm{T}}\boldsymbol{A}) = R(\boldsymbol{A})$.

证 设 $\boldsymbol{A}$ 为 $m \times n$ 矩阵, $\boldsymbol{x}$ 为 n 维列向量, 若 $\boldsymbol{x}$ 满足 $\boldsymbol{Ax} = \boldsymbol{0}$, 则有 $\boldsymbol{A}^{\mathrm{T}}(\boldsymbol{Ax}) = \boldsymbol{0}$, 即 $(\boldsymbol{A}^{\mathrm{T}}\boldsymbol{A})\boldsymbol{x} = \boldsymbol{0}$.

Proof Let $\boldsymbol{A}$ be an $m \times n$ matrix and let $\boldsymbol{x}$ be an n dimensional column vector. If $\boldsymbol{x}$ satisfies $\boldsymbol{Ax} = \boldsymbol{0}$, then we have $\boldsymbol{A}^{\mathrm{T}}(\boldsymbol{Ax}) = \boldsymbol{0}$, namely, $(\boldsymbol{A}^{\mathrm{T}}\boldsymbol{A})\boldsymbol{x} = \boldsymbol{0}$.

若 $\boldsymbol{x}$ 满足 $(\boldsymbol{A}^{\mathrm{T}}\boldsymbol{A})\boldsymbol{x} = \boldsymbol{0}$, 则有 $\boldsymbol{x}^{\mathrm{T}}(\boldsymbol{A}^{\mathrm{T}}\boldsymbol{A})\boldsymbol{x} = 0$, 即 $(\boldsymbol{Ax})^{\mathrm{T}}(\boldsymbol{Ax}) = 0$ 从而 $\boldsymbol{Ax} = \boldsymbol{0}$.

If $\boldsymbol{x}$ satisfies $(\boldsymbol{A}^{\mathrm{T}}\boldsymbol{A})\boldsymbol{x} = \boldsymbol{0}$, then we have $\boldsymbol{x}^{\mathrm{T}}(\boldsymbol{A}^{\mathrm{T}}\boldsymbol{A})\boldsymbol{x} = 0$, namely, $(\boldsymbol{Ax})^{\mathrm{T}}(\boldsymbol{Ax}) = 0$, it means that $\boldsymbol{Ax} = \boldsymbol{0}$.

综上, 可知方程组 $\boldsymbol{Ax}=\boldsymbol{0}$ 与 $(\boldsymbol{A}^{\mathrm{T}}\boldsymbol{A})\boldsymbol{x}=\boldsymbol{0}$ 同解, 所以 $n-R(\boldsymbol{A}^{\mathrm{T}}\boldsymbol{A})=n-R(\boldsymbol{A})$, 即

In sum, we know that the systems $\boldsymbol{Ax} = \boldsymbol{0}$ and $(\boldsymbol{A}^{\mathrm{T}}\boldsymbol{A})\boldsymbol{x} = \boldsymbol{0}$ are equivalent, and so $n - R(\boldsymbol{A}^{\mathrm{T}}\boldsymbol{A}) = n - R(\boldsymbol{A})$, i.e.,

$$R(\boldsymbol{A}^{\mathrm{T}}\boldsymbol{A}) = R(\boldsymbol{A})$$ ■

思 考 题

1. $\boldsymbol{Ax}=\boldsymbol{0}$ 的基础解系是否唯一?

2. $\boldsymbol{Ax}=\boldsymbol{0}$ 的两个不同的基础解系之间有什么关系?

3. $\boldsymbol{Ax}=\boldsymbol{0}$ 的自由未知量的选择是否唯一?

4. $\boldsymbol{Ax}=\boldsymbol{0}$ 的自由未知量的个数与 $R(\boldsymbol{A})$ 有何关系?

Questions

1. Is the system of fundamental solutions of $\boldsymbol{Ax}=\boldsymbol{0}$ unique?

2. What are the relations between two different systems of fundamental solutions of $\boldsymbol{Ax}=\boldsymbol{0}$?

3. Is the choice of free unknowns of $\boldsymbol{Ax}=\boldsymbol{0}$ unique?

4. What are the relations between the number of free unknowns of $\boldsymbol{Ax}=\boldsymbol{0}$ and $R(\boldsymbol{A})$?

习 题 4

Exercise 4

1. 已知向量组

1. Known that the vector set is

$$\boldsymbol{\alpha}_1=\begin{pmatrix}2\\5\\1\\3\end{pmatrix},\quad \boldsymbol{\alpha}_2=\begin{pmatrix}10\\1\\5\\10\end{pmatrix},\quad \boldsymbol{\alpha}_3=\begin{pmatrix}4\\1\\-1\\1\end{pmatrix}$$

且 $3(\boldsymbol{\alpha}_1-\boldsymbol{\alpha})+2(\boldsymbol{\alpha}_2+\boldsymbol{\alpha})=(\boldsymbol{\alpha}_3+\boldsymbol{\alpha})$. 求 $\boldsymbol{\alpha}$.

and that $3(\boldsymbol{\alpha}_1-\boldsymbol{\alpha})+2(\boldsymbol{\alpha}_2+\boldsymbol{\alpha})=(\boldsymbol{\alpha}_3+\boldsymbol{\alpha})$. Find $\boldsymbol{\alpha}$.

2. 已知向量组

2. Known that the vector set is

$$\boldsymbol{\beta}=\begin{pmatrix}2\\2\\-6\end{pmatrix},\quad \boldsymbol{\alpha}_1=\begin{pmatrix}1\\2\\3\end{pmatrix},\quad \boldsymbol{\alpha}_2=\begin{pmatrix}0\\2\\3\end{pmatrix},\quad \boldsymbol{\alpha}_3=\begin{pmatrix}0\\0\\3\end{pmatrix}$$

将 $\boldsymbol{\beta}$ 用 $\boldsymbol{\alpha}_1,\boldsymbol{\alpha}_2,\boldsymbol{\alpha}_3$ 线性表示.

Represent $\boldsymbol{\beta}$ linearly by $\boldsymbol{\alpha}_1,\boldsymbol{\alpha}_2,\boldsymbol{\alpha}_3$.

3. 已知向量组

3. Known that the vector set is

$$\boldsymbol{\alpha}_1=\begin{pmatrix}2\\1\\3\end{pmatrix},\quad \boldsymbol{\alpha}_2=\begin{pmatrix}x\\3\\2\end{pmatrix},\quad \boldsymbol{\alpha}_3=\begin{pmatrix}3\\2\\-1\end{pmatrix}$$

当 x 为何值时, 向量组 $\{\boldsymbol{\alpha}_1,\boldsymbol{\alpha}_2,\boldsymbol{\alpha}_3\}$ 线性无关?

What value does x take on such that the vector set $\{\boldsymbol{\alpha}_1,\boldsymbol{\alpha}_2,\boldsymbol{\alpha}_3\}$ is linearly independent?

4. 判别向量组

4. Decide that whether the vector set

$$\boldsymbol{\alpha}_1=\begin{pmatrix}1\\-2\\3\end{pmatrix},\quad \boldsymbol{\alpha}_2=\begin{pmatrix}0\\2\\-5\end{pmatrix},\quad \boldsymbol{\alpha}_3=\begin{pmatrix}-1\\0\\2\end{pmatrix}$$

是否线性相关.

is linearly dependent.

5. 设向量组 $\{\boldsymbol{\alpha}_1,\boldsymbol{\alpha}_2,\boldsymbol{\alpha}_3\}$ 线性无关, $\boldsymbol{\beta}_1=\boldsymbol{\alpha}_1,\boldsymbol{\beta}_2=\boldsymbol{\alpha}_1-\boldsymbol{\alpha}_2,\boldsymbol{\beta}_3=\boldsymbol{\alpha}_1-\boldsymbol{\alpha}_2-\boldsymbol{\alpha}_3$. 证明向量组 $\{\boldsymbol{\beta}_1,\boldsymbol{\beta}_2,\boldsymbol{\beta}_3\}$ 线性无关.

5. Suppose that the vector set $\{\boldsymbol{\alpha}_1,\boldsymbol{\alpha}_2,\boldsymbol{\alpha}_3\}$ is linearly independent and that $\boldsymbol{\beta}_1=\boldsymbol{\alpha}_1,\boldsymbol{\beta}_2=\boldsymbol{\alpha}_1-\boldsymbol{\alpha}_2,\boldsymbol{\beta}_3=\boldsymbol{\alpha}_1-\boldsymbol{\alpha}_2-\boldsymbol{\alpha}_3$. Prove that the vector set $\{\boldsymbol{\beta}_1,\boldsymbol{\beta}_2,\boldsymbol{\beta}_3\}$ is linearly independent.

6. 判断下列命题的正确性:

6. Decide the validity of the following propositions:

(1) 若向量组 $\{\boldsymbol{\alpha}_1,\boldsymbol{\alpha}_2,\cdots,\boldsymbol{\alpha}_m\}$ 线性相关, 则 $\boldsymbol{\alpha}_1$ 可由 $\{\boldsymbol{\alpha}_2,\cdots,\boldsymbol{\alpha}_m\}$ 线性表示;

(1) If the vector set $\{\boldsymbol{\alpha}_1,\boldsymbol{\alpha}_2,\cdots,\boldsymbol{\alpha}_m\}$ is linearly dependent, then $\boldsymbol{\alpha}_1$ can be linearly represented by $\{\boldsymbol{\alpha}_2,\cdots,\boldsymbol{\alpha}_m\}$;

(2) 若向量组 $\{\boldsymbol{\alpha}_1,\boldsymbol{\alpha}_2,\cdots,\boldsymbol{\alpha}_r\}$ 线性无关, 向量 $\boldsymbol{\alpha}_{r+1}$ 不能由 $\{\boldsymbol{\alpha}_1,\boldsymbol{\alpha}_2,\cdots,\boldsymbol{\alpha}_r\}$ 线性表示, 则向量组 $\{\boldsymbol{\alpha}_1,\boldsymbol{\alpha}_2,\cdots,\boldsymbol{\alpha}_r,\boldsymbol{\alpha}_{r+1}\}$ 线性无关;

(2) If the vector set $\{\boldsymbol{\alpha}_1,\boldsymbol{\alpha}_2,\cdots,\boldsymbol{\alpha}_r\}$ is linearly independent and if $\boldsymbol{\alpha}_{r+1}$ can not be linearly represented by $\{\boldsymbol{\alpha}_1,\boldsymbol{\alpha}_2,\cdots,\boldsymbol{\alpha}_r\}$, then the vector set $\{\boldsymbol{\alpha}_1,\boldsymbol{\alpha}_2,\cdots,\boldsymbol{\alpha}_r,\boldsymbol{\alpha}_{r+1}\}$ is linearly independent;

(3) 若向量组 $\{\boldsymbol{\alpha}_1,\boldsymbol{\alpha}_2,\cdots,\boldsymbol{\alpha}_m\}$ 线性相关, 向量组 $\{\boldsymbol{\beta}_1,\boldsymbol{\beta}_2,\cdots,\boldsymbol{\beta}_m\}$ 也线性相关, 则 $\{\boldsymbol{\alpha}_1+\boldsymbol{\beta}_1,\boldsymbol{\alpha}_2+\boldsymbol{\beta}_2,\cdots,\boldsymbol{\alpha}_m+\boldsymbol{\beta}_m\}$ 线性相关;

(3) If the vector set $\{\boldsymbol{\alpha}_1,\boldsymbol{\alpha}_2,\cdots,\boldsymbol{\alpha}_m\}$ is linearly dependent and if the vector set $\{\boldsymbol{\beta}_1,\boldsymbol{\beta}_2,\cdots,\boldsymbol{\beta}_m\}$ is also linearly dependent, then the vector set $\{\boldsymbol{\alpha}_1+\boldsymbol{\beta}_1,\boldsymbol{\alpha}_2+\boldsymbol{\beta}_2,\cdots,\boldsymbol{\alpha}_m+\boldsymbol{\beta}_m\}$ is linearly dependent;

(4) 若向量组 $\{\boldsymbol{\alpha}_1,\boldsymbol{\alpha}_2,\cdots,\boldsymbol{\alpha}_m\}$ 的部分组 $\{\boldsymbol{\alpha}_1,\boldsymbol{\alpha}_2,\cdots,\boldsymbol{\alpha}_r\}(r<m)$ 线性无关, 则 $\{\boldsymbol{\alpha}_1,\boldsymbol{\alpha}_2,\cdots,\boldsymbol{\alpha}_r\}$ 为向量组 $\{\boldsymbol{\alpha}_1,\boldsymbol{\alpha}_2,\cdots,\boldsymbol{\alpha}_m\}$ 的一个极大无关组;

(4) If the vector subset $\{\boldsymbol{\alpha}_1,\boldsymbol{\alpha}_2,\cdots,\boldsymbol{\alpha}_r\}$ $(r<m)$ of the vector set $\{\boldsymbol{\alpha}_1,\boldsymbol{\alpha}_2,\cdots,\boldsymbol{\alpha}_m\}$ is linearly independent, then it is a maximal linearly independent subset of $\{\boldsymbol{\alpha}_1,\boldsymbol{\alpha}_2,\cdots,\boldsymbol{\alpha}_m\}$;

(5) 等价向量组含有相同个数的向量.

(5) Equivalent vector sets have the same number of vectors.

7. 对于给定的 r 维向量组

7. For the given r dimensional vector set

$$\boldsymbol{\alpha}_i = \begin{pmatrix} a_{i1} \\ a_{i2} \\ \vdots \\ a_{ir} \end{pmatrix} \quad (i = 1, 2, \cdots, m)$$

及 $r+1$ 维向量组

and the $r+1$ dimensional vector set

$$\boldsymbol{\beta}_i = \begin{pmatrix} a_{i1} \\ a_{i2} \\ \vdots \\ a_{ir} \\ a_{i,r+1} \end{pmatrix} \quad (i = 1, 2, \cdots, m)$$

其中 $\boldsymbol{\beta}_i$ 是 $\boldsymbol{\alpha}_i$ 加上一个分量而得. 若 r 维向量组 $\{\boldsymbol{\alpha}_1, \boldsymbol{\alpha}_2, \cdots, \boldsymbol{\alpha}_m\}$ 线性无关, 证明 $r+1$ 维向量组 $\{\boldsymbol{\beta}_1, \boldsymbol{\beta}_2, \cdots, \boldsymbol{\beta}_m\}$ 也线性无关.

where $\boldsymbol{\beta}_i$ is obtained by adding a component on $\boldsymbol{\alpha}_i$. If the r dimensional vector set $\{\boldsymbol{\alpha}_1, \boldsymbol{\alpha}_2, \cdots, \boldsymbol{\alpha}_m\}$ is linearly independent, prove that the $r+1$ dimensional vector set $\{\boldsymbol{\beta}_1, \boldsymbol{\beta}_2, \cdots, \boldsymbol{\beta}_m\}$ is also linearly independent.

8. 设向量组 $\{\boldsymbol{\alpha}_1, \boldsymbol{\alpha}_2, \boldsymbol{\alpha}_3\}$ 满足 $k_1\boldsymbol{\alpha}_1 + k_2\boldsymbol{\alpha}_2 + k_3\boldsymbol{\alpha}_3 = \mathbf{0}$, 其中 $k_1k_3 \neq 0$, 试证两个向量组 $\{\boldsymbol{\alpha}_1, \boldsymbol{\alpha}_2\}$ 与 $\{\boldsymbol{\alpha}_2, \boldsymbol{\alpha}_3\}$ 等价.

8. Suppose that the vector set $\{\boldsymbol{\alpha}_1, \boldsymbol{\alpha}_2, \boldsymbol{\alpha}_3\}$ satisfies $k_1\boldsymbol{\alpha}_1 + k_2\boldsymbol{\alpha}_2 + k_3\boldsymbol{\alpha}_3 = \mathbf{0}$, where $k_1k_3 \neq 0$, prove that the two vector subsets $\{\boldsymbol{\alpha}_1, \boldsymbol{\alpha}_2\}$ and $\{\boldsymbol{\alpha}_2, \boldsymbol{\alpha}_3\}$ are equivalent.

9. 求下列向量组的秩, 并分别求其一个极大无关组.

9. Find the ranks of the following vector sets and find one of the maximal linearly independent subsets, respectively.

$$(1)\ \boldsymbol{\alpha}_1 = \begin{pmatrix} 1 \\ 1 \\ 0 \end{pmatrix}, \boldsymbol{\alpha}_2 = \begin{pmatrix} 0 \\ 2 \\ 0 \end{pmatrix}, \boldsymbol{\alpha}_3 = \begin{pmatrix} 0 \\ 1 \\ 3 \end{pmatrix};$$

(2) $\boldsymbol{\alpha}_1=\begin{pmatrix}1\\2\\-1\\4\end{pmatrix},\boldsymbol{\alpha}_2=\begin{pmatrix}0\\100\\10\\4\end{pmatrix},\boldsymbol{\alpha}_3=\begin{pmatrix}-2\\-4\\2\\-8\end{pmatrix}$;

(3) $\boldsymbol{\alpha}_1=\begin{pmatrix}1\\2\\1\\0\end{pmatrix},\boldsymbol{\alpha}_2=\begin{pmatrix}4\\1\\0\\2\end{pmatrix},\boldsymbol{\alpha}_3=\begin{pmatrix}1\\-1\\-3\\-6\end{pmatrix},\boldsymbol{\alpha}_4=\begin{pmatrix}0\\-3\\-1\\3\end{pmatrix}$.

10. 确定下面向量组的线性相关性.

10. Decide the linear dependence of the following vector set.

$$\boldsymbol{\alpha}_1=\begin{pmatrix}1\\0\\1\\1\end{pmatrix},\quad \boldsymbol{\alpha}_2=\begin{pmatrix}2\\1\\2\\1\end{pmatrix},\quad \boldsymbol{\alpha}_3=\begin{pmatrix}3\\b+4\\3\\1\end{pmatrix},\quad \boldsymbol{\alpha}_4=\begin{pmatrix}4\\5\\a-2\\-1\end{pmatrix}$$

11. 确定下列向量组的线性相关性. 若相关, 求其一个极大无关组, 并将其余向量用该极大无关组线性表示.

11. Decide the linear dependence of the following vector set. If it is linearly dependent, find a maximal linearly independent subset and represent the other vectors by this maximal linearly independent subset.

$$\boldsymbol{\alpha}_1=\begin{pmatrix}1\\0\\2\\1\end{pmatrix},\quad \boldsymbol{\alpha}_2=\begin{pmatrix}1\\2\\0\\1\end{pmatrix},\quad \boldsymbol{\alpha}_3=\begin{pmatrix}2\\1\\3\\0\end{pmatrix},\quad \boldsymbol{\alpha}_4=\begin{pmatrix}2\\5\\-1\\4\end{pmatrix}$$

12. 对于两个给定的向量组

12. For the two given vector sets

$$\boldsymbol{\alpha}_1=\begin{pmatrix}1\\2\\-3\end{pmatrix},\quad \boldsymbol{\alpha}_2=\begin{pmatrix}3\\0\\1\end{pmatrix},\quad \boldsymbol{\alpha}_3=\begin{pmatrix}9\\6\\-7\end{pmatrix}$$

和

and

$$\boldsymbol{\beta}_1=\begin{pmatrix}0\\1\\-1\end{pmatrix},\quad \boldsymbol{\beta}_2=\begin{pmatrix}a\\2\\1\end{pmatrix},\quad \boldsymbol{\beta}_3=\begin{pmatrix}b\\1\\0\end{pmatrix}$$

若它们有相同的秩, 且 $\boldsymbol{\beta}_3$ 可由 $\{\boldsymbol{\alpha}_1, \boldsymbol{\alpha}_2, \boldsymbol{\alpha}_3\}$ 线性表示, 求 a, b 的值.

if they have the same rank and if $\boldsymbol{\beta}_3$ can be linearly represented by $\{\boldsymbol{\alpha}_1, \boldsymbol{\alpha}_2, \boldsymbol{\alpha}_3\}$, find the values of a, b.

13. 求下列齐次线性方程组的基础解系和通解:

13. Find the systems of fundamental solutions and the general solutions of the following homogeneous linear systems:

(1) $x_1 + 2x_2 + 3x_3 + 4x_4 + 5x_5 = 0$; (2) $\begin{cases} x_1 + x_2 - x_3 - x_4 = 0, \\ 2x_1 - 5x_2 + 3x_3 + 2x_4 = 0, \\ 7x_1 - 7x_2 + 3x_3 + x_4 = 0; \end{cases}$

(3) $\begin{cases} x_1 - 8x_2 + 10x_3 + 2x_4 = 0, \\ 2x_1 + 4x_2 + 5x_3 - x_4 = 0, \\ 3x_1 + 8x_2 + 6x_3 - 2x_4 = 0; \end{cases}$ (4) $\begin{cases} x_1 + x_2 + x_5 = 0, \\ x_1 + x_2 - x_3 = 0, \\ x_3 + x_4 + x_5 = 0. \end{cases}$

14. 求解下列非齐次线性方程组的通解:

14. Find the general solutions of the following non-homogeneous linear system:

(1) $x_1 + 2x_2 + 3x_3 = 1$; (2) $\begin{cases} x_1 + x_2 - x_3 + 2x_4 = 3, \\ 2x_1 + x_2 \qquad - 3x_4 = 1, \\ -2x_1 - 2x_3 \qquad + 10x_4 = 4; \end{cases}$

(3) $\begin{cases} 2x_1 + 3x_2 + x_3 = 4, \\ x_1 - 2x_2 + 4x_3 = -5, \\ 3x_1 + 8x_2 - 2x_3 = 13, \\ 4x_1 - x_2 + 9x_3 = -6; \end{cases}$ (4) $\begin{cases} x_1 + x_2 + 2x_3 + x_4 + 2x_5 = 7, \\ x_1 + 2x_2 + 3x_3 + 4x_4 + 5x_5 = 15, \\ 2x_1 + 3x_2 + 5x_3 + 5x_4 + 7x_5 = 22. \end{cases}$

15. 对于一个四元非齐次线性方程组, 设对应的系数矩阵的秩为 3, $\boldsymbol{x}_1, \boldsymbol{x}_2, \boldsymbol{x}_3$ 是其三个解向量, 并且满足

15. For a non-homogeneous linear system in four unknowns, suppose that the rank of the corresponding coefficient matrix is 3 and that $\boldsymbol{x}_1, \boldsymbol{x}_2, \boldsymbol{x}_3$ are three solution vectors satisfying

$$\boldsymbol{x}_1 + \boldsymbol{x}_2 = \begin{pmatrix} 1 \\ 1 \\ 0 \\ 2 \end{pmatrix}, \quad \boldsymbol{x}_2 + \boldsymbol{x}_3 = \begin{pmatrix} 1 \\ 0 \\ 1 \\ 3 \end{pmatrix}$$

求该方程组的通解.

Find the general solution of the system.

16. 对于如下的线性方程组

16. For the following linear system

$$\begin{cases} ax_1 + x_2 + x_3 = 4 \\ x_1 + bx_2 + x_3 = 3 \\ x_1 + 3bx_2 + x_3 = 9 \end{cases}$$

当 a, b 取何值时, 方程组 (1) 有唯一解? (2) 无解? (3) 有无穷多组解? 在方程组有解的情况下, 求出唯一解或通解.

what values do a, b take on such that the system (1) has a unique solution? (2) has no solution? (3) has infinitely many solutions? In the case that the system has solutions, find the unique solution or the general solution.

17. 对于如下的线性方程组

17. For the following linear system

$$\begin{cases} x_1 + x_2 + 2x_3 + 3x_4 = 1 \\ x_1 + 3x_2 + 6x_3 + x_4 = 3 \\ 3x_1 - x_2 - k_1x_3 + 15x_4 = 3 \\ x_1 - 5x_2 - 10x_3 + 12x_4 = k_2 \end{cases}$$

当 k_1, k_2 取何值时, 方程组 (1) 有唯一解? (2) 无解? (3) 有无穷多组解? 在方程组有解的情况下, 求出唯一解或通解.

what values do k_1, k_2 take on such that the system (1) has a unique solution? (2) has no solution? (3) has infinitely many solutions? In the case that this system has solutions, find the unique solution or the general solution.

18. 设

18. Let

$$\boldsymbol{A} = \begin{pmatrix} 1 & 1 & 2 \\ 2 & 2 & 4 \\ 3 & 3 & 6 \end{pmatrix}$$

求一个秩为 2 的 3 阶矩阵 $\boldsymbol{B}$, 使 $\boldsymbol{AB} = \boldsymbol{O}$.

Find a third order square matrix $\boldsymbol{B}$ of rank 2 such that $\boldsymbol{AB} = \boldsymbol{O}$.

19. 设 $\boldsymbol{A}, \boldsymbol{B}$ 均为 n 阶矩阵, $\boldsymbol{AB} = \boldsymbol{O}$. 证明

19. Let $\boldsymbol{A}, \boldsymbol{B}$ be square matrices of order n, $\boldsymbol{AB} = \boldsymbol{O}$. Prove that

$$R(\boldsymbol{A}) + R(\boldsymbol{B}) \leqslant n$$

20. 设 $\boldsymbol{A}^2=\boldsymbol{A}$, 且 $\boldsymbol{A}$ 为 n 阶矩阵, 证明

20. If $\boldsymbol{A}^2=\boldsymbol{A}$ and if $\boldsymbol{A}$ is a square matrix of order n, prove that

$$R(\boldsymbol{A})+R(\boldsymbol{E}-\boldsymbol{A})=n$$

21. 设 $\boldsymbol{\eta}_1,\boldsymbol{\eta}_2,\cdots,\boldsymbol{\eta}_s$ 是非齐次线性方程组 $\boldsymbol{Ax}=\boldsymbol{b}$ 的 s 个解, $k_1,k_2,\cdots,k_s$ 为任意实数, 满足 $k_1+k_2+\cdots+k_s=1$, 证明 $\boldsymbol{x}=k_1\boldsymbol{\eta}_1+k_2\boldsymbol{\eta}_2+\cdots+k_s\boldsymbol{\eta}_s$ 也是 $\boldsymbol{Ax}=\boldsymbol{b}$ 的解.

21. Suppose that $\boldsymbol{\eta}_1,\boldsymbol{\eta}_2,\cdots,\boldsymbol{\eta}_s$ are s solutions of the non-homogeneous linear system $\boldsymbol{Ax}=\boldsymbol{b}$ and that $k_1, k_2,\cdots,k_s$ are arbitrary real numbers satisfying $k_1+k_2+\cdots+k_s=1$, prove that $\boldsymbol{x}=k_1\boldsymbol{\eta}_1+k_2\boldsymbol{\eta}_2+\cdots+k_s\boldsymbol{\eta}_s$ is also a solution of $\boldsymbol{Ax}=\boldsymbol{b}$.

22. 设 $\boldsymbol{\eta}^*$ 是 $\boldsymbol{Ax}=\boldsymbol{b}$ 的一个解, 向量组 $\{\boldsymbol{\xi}_1,\boldsymbol{\xi}_2,\cdots,\boldsymbol{\xi}_{n-r}\}$ 是对应齐次线性方程组 $\boldsymbol{Ax}=\boldsymbol{0}$ 的一个基础解系. 证明

22. Suppose that $\boldsymbol{\eta}^*$ is a solution of $\boldsymbol{Ax}=\boldsymbol{b}$, the vector set $\{\boldsymbol{\xi}_1,\boldsymbol{\xi}_2,\cdots,\boldsymbol{\xi}_{n-r}\}$ is a fundamental solutions associated with the system of homogeneous linear system $\boldsymbol{Ax}=\boldsymbol{0}$. Prove that

(1) 解向量组 $\{\boldsymbol{\eta}^*,\boldsymbol{\xi}_1,\boldsymbol{\xi}_2,\cdots,\boldsymbol{\xi}_{n-r}\}$ 线性无关;

(1) The solution vector set $\{\boldsymbol{\eta}^*,\boldsymbol{\xi}_1,\boldsymbol{\xi}_2,\cdots,\boldsymbol{\xi}_{n-r}\}$ is linearly independent;

(2) 解向量组 $\{\boldsymbol{\eta}^*,\boldsymbol{\eta}^*+\boldsymbol{\xi}_1,\boldsymbol{\eta}^*+\boldsymbol{\xi}_2,\cdots,\boldsymbol{\eta}^*+\boldsymbol{\xi}_{n-r}\}$ 线性无关.

(2) The solution vector set $\{\boldsymbol{\eta}^*,\boldsymbol{\eta}^*+\boldsymbol{\xi}_1,\boldsymbol{\eta}^*+\boldsymbol{\xi}_2,\cdots,\boldsymbol{\eta}^*+\boldsymbol{\xi}_{n-r}\}$ is linearly independent.

23. 设 $\boldsymbol{A}$ 是 n 阶矩阵 $(n\geqslant 2)$, $\boldsymbol{A}^*$ 是 $\boldsymbol{A}$ 的伴随矩阵. 证明

23. Let $\boldsymbol{A}$ be a square matrix of order $n\,(n\geqslant 2)$, and let $\boldsymbol{A}^*$ be an adjoint matrix of $\boldsymbol{A}$. Prove that

$$R(\boldsymbol{A}^*)=\begin{cases} n, & R(\boldsymbol{A})=n \\ 1, & R(\boldsymbol{A})=n-1 \\ 0, & R(\boldsymbol{A})<n-1 \end{cases}$$

补充题 4

Supplement Exercise 4

1. 设有向量组

1. Suppose that the vector set is given by

$$\boldsymbol{\alpha}_1=\begin{pmatrix}\lambda\\1\\1\end{pmatrix},\quad \boldsymbol{\alpha}_2=\begin{pmatrix}1\\\lambda\\1\end{pmatrix},\quad \boldsymbol{\alpha}_3=\begin{pmatrix}1\\1\\\lambda\end{pmatrix},\quad \boldsymbol{\beta}=\begin{pmatrix}\lambda-3\\-2\\-2\end{pmatrix}$$

已知 $\boldsymbol{\beta}$ 不能由向量组 $\{\boldsymbol{\alpha}_1,\boldsymbol{\alpha}_2,\boldsymbol{\alpha}_3\}$ 线性表示, 求 λ 的值.

It is known that $\boldsymbol{\beta}$ can not be linearly represented by $\{\boldsymbol{\alpha}_1,\boldsymbol{\alpha}_2,\boldsymbol{\alpha}_3\}$, find the value of λ.

2. 设有向量组

2. Suppose that the vector set is given by

$$\boldsymbol{\alpha}_1=\begin{pmatrix}1\\0\\2\\3\end{pmatrix},\quad \boldsymbol{\alpha}_2=\begin{pmatrix}1\\1\\3\\5\end{pmatrix},\quad \boldsymbol{\alpha}_3=\begin{pmatrix}1\\-1\\a+2\\1\end{pmatrix},$$

$$\boldsymbol{\alpha}_4=\begin{pmatrix}1\\2\\4\\a+8\end{pmatrix},\quad \boldsymbol{\beta}=\begin{pmatrix}1\\1\\b+3\\5\end{pmatrix}$$

(1) a,b 为何值时, $\boldsymbol{\beta}$ 不能由 $\{\boldsymbol{\alpha}_1,\boldsymbol{\alpha}_2,\boldsymbol{\alpha}_3,\boldsymbol{\alpha}_4\}$ 线性表示?

(1) What values do a,b take on such that $\boldsymbol{\beta}$ can be linearly represented by $\{\boldsymbol{\alpha}_1,\boldsymbol{\alpha}_2,\boldsymbol{\alpha}_3,\boldsymbol{\alpha}_4\}$?

(2) a,b 为何值时, $\boldsymbol{\beta}$ 能由 $\{\boldsymbol{\alpha}_1,\boldsymbol{\alpha}_2,\boldsymbol{\alpha}_3,\boldsymbol{\alpha}_4\}$ 唯一地线性表示?

(2) What values do a,b take on such that $\boldsymbol{\beta}$ can be linearly represented by $\{\boldsymbol{\alpha}_1,\boldsymbol{\alpha}_2,\boldsymbol{\alpha}_3,\boldsymbol{\alpha}_4\}$ uniquely?

(3) a,b 为何值时, $\boldsymbol{\beta}$ 能由 $\{\boldsymbol{\alpha}_1,\boldsymbol{\alpha}_2,\boldsymbol{\alpha}_3,\boldsymbol{\alpha}_4\}$ 线性表示, 但表示式不唯一?

(3) What values do a,b take on such that $\boldsymbol{\beta}$ can be linearly represented by $\{\boldsymbol{\alpha}_1,\boldsymbol{\alpha}_2,\boldsymbol{\alpha}_3,\boldsymbol{\alpha}_4\}$, but the expression is not unique?

3. 设 $\boldsymbol{\beta}$ 可由 $\{\boldsymbol{\alpha}_1,\boldsymbol{\alpha}_2,\cdots,\boldsymbol{\alpha}_{m-1},\boldsymbol{\alpha}_m\}$ 线性表示, 但不能由 $\{\boldsymbol{\alpha}_1,\boldsymbol{\alpha}_2,\cdots,\boldsymbol{\alpha}_{m-1}\}$ 线性表示, 判断

3. Suppose that $\boldsymbol{\beta}$ can be linearly represented by $\{\boldsymbol{\alpha}_1,\boldsymbol{\alpha}_2,\cdots,\boldsymbol{\alpha}_{m-1},\boldsymbol{\alpha}_m\}$, but can not be linearly represented by $\{\boldsymbol{\alpha}_1,\boldsymbol{\alpha}_2,\cdots,\boldsymbol{\alpha}_{m-1}\}$. Decide that

(1) $\boldsymbol{\alpha}_m$ 可否由 $\{\boldsymbol{\alpha}_1,\boldsymbol{\alpha}_2,\cdots,\boldsymbol{\alpha}_{m-1},\boldsymbol{\beta}\}$ 线性表示, 为什么?

(1) Whether $\boldsymbol{\alpha}_m$ can be linearly represented by $\{\boldsymbol{\alpha}_1,\boldsymbol{\alpha}_2,\cdots,\boldsymbol{\alpha}_{m-1},\boldsymbol{\beta}\}$, why?

(2) $\boldsymbol{\alpha}_m$ 可否由 $\{\boldsymbol{\alpha}_1,\boldsymbol{\alpha}_2,\cdots,\boldsymbol{\alpha}_{m-1}\}$ 线性表示, 为什么?

(2) Whether $\boldsymbol{\alpha}_m$ can be linearly represented by $\{\boldsymbol{\alpha}_1,\boldsymbol{\alpha}_2,\cdots,\boldsymbol{\alpha}_{m-1}\}$, why?

4. 设

$$A = \begin{pmatrix} 1 & 2 & -2 \\ 2 & 1 & 2 \\ 3 & 0 & 4 \end{pmatrix}, \quad \alpha = \begin{pmatrix} t \\ 1 \\ 1 \end{pmatrix}$$

若 $A\alpha$ 和 α 线性相关, 求 t 的值.

4. Suppose that A and α as above. If $A\alpha$ and α are linearly dependent, find the value of t.

5. 设向量组

$$\alpha_1 = \begin{pmatrix} 1 \\ 3 \\ 4 \\ -2 \end{pmatrix}, \quad \alpha_2 = \begin{pmatrix} 2 \\ 1 \\ 3 \\ t \end{pmatrix}, \quad \alpha_3 = \begin{pmatrix} 3 \\ 1 \\ 4 \\ 0 \end{pmatrix}$$

线性相关. 求 t 的值.

5. Suppose that the vector set above is linearly dependent. Find the value of t.

6. 设 $\boldsymbol{b}_1 = \boldsymbol{a}_1 + \boldsymbol{a}_2, \boldsymbol{b}_2 = \boldsymbol{a}_2 + \boldsymbol{a}_3, \boldsymbol{b}_3 = \boldsymbol{a}_3 + \boldsymbol{a}_4, \boldsymbol{b}_4 = \boldsymbol{a}_4 + \boldsymbol{a}_1$. 证明: 向量组 $\{\boldsymbol{b}_1, \boldsymbol{b}_2, \boldsymbol{b}_3, \boldsymbol{b}_4\}$ 线性相关.

6. Suppose that $\boldsymbol{b}_1 = \boldsymbol{a}_1 + \boldsymbol{a}_2, \boldsymbol{b}_2 = \boldsymbol{a}_2 + \boldsymbol{a}_3, \boldsymbol{b}_3 = \boldsymbol{a}_3 + \boldsymbol{a}_4, \boldsymbol{b}_4 = \boldsymbol{a}_4 + \boldsymbol{a}_1$. Prove that the vector set $\{\boldsymbol{b}_1, \boldsymbol{b}_2, \boldsymbol{b}_3, \boldsymbol{b}_4\}$ is linearly dependent.

7. 设 $\boldsymbol{b}_1 = \boldsymbol{a}_1, \boldsymbol{b}_2 = \boldsymbol{a}_1 + \boldsymbol{a}_2, \cdots, \boldsymbol{b}_r = \boldsymbol{a}_1 + \boldsymbol{a}_2 + \cdots + \boldsymbol{a}_r$ 且向量组 $\boldsymbol{a}_1, \boldsymbol{a}_2, \cdots, \boldsymbol{a}_r$ 线性无关, 证明: 向量组 $\{\boldsymbol{b}_1, \boldsymbol{b}_2, \cdots, \boldsymbol{b}_r\}$ 也线性无关.

7. Suppose that $\boldsymbol{b}_1 = \boldsymbol{a}_1, \boldsymbol{b}_2 = \boldsymbol{a}_1 + \boldsymbol{a}_2, \cdots, \boldsymbol{b}_r = \boldsymbol{a}_1 + \boldsymbol{a}_2 + \cdots + \boldsymbol{a}_r$ and that the vector set $\boldsymbol{a}_1, \boldsymbol{a}_2, \cdots, \boldsymbol{a}_r$ is linearly independent. Prove that the vector set $\{\boldsymbol{b}_1, \boldsymbol{b}_2, \cdots, \boldsymbol{b}_r\}$ is also linearly independent.

8. 设 $\boldsymbol{b}_1 = 3\boldsymbol{a}_1 + 2\boldsymbol{a}_2, \boldsymbol{b}_2 = \boldsymbol{a}_2 - \boldsymbol{a}_3, \boldsymbol{b}_3 = 4\boldsymbol{a}_3 - 5\boldsymbol{a}_1$, 且向量组 $\{\boldsymbol{a}_1, \boldsymbol{a}_2, \boldsymbol{a}_3\}$ 线性无关. 证明: 向量组 $\{\boldsymbol{b}_1, \boldsymbol{b}_2, \boldsymbol{b}_3\}$ 也线性无关.

8. Suppose that $\boldsymbol{b}_1 = 3\boldsymbol{a}_1 + 2\boldsymbol{a}_2, \boldsymbol{b}_2 = \boldsymbol{a}_2 - \boldsymbol{a}_3, \boldsymbol{b}_3 = 4\boldsymbol{a}_3 - 5\boldsymbol{a}_1$ and that the vector set $\{\boldsymbol{a}_1, \boldsymbol{a}_2, \boldsymbol{a}_3\}$ is linearly independent. Prove that the vector set $\{\boldsymbol{b}_1, \boldsymbol{b}_2, \boldsymbol{b}_3\}$ is also linearly independent.

9. 设

9. Suppose that

$$\alpha_1 = \begin{pmatrix} 6 \\ -1 \\ 1 \end{pmatrix}, \quad \alpha_2 = \begin{pmatrix} -7 \\ 4 \\ 2 \end{pmatrix}$$

是方程组

are two solutions of the system

$$\begin{cases} a_1x_1 + a_2x_2 + a_3x_3 = a_4 \\ x_1 + 3x_2 - 2x_3 = 1 \\ 2x_1 + 5x_2 + x_3 = 8 \end{cases}$$

的两个解. 求其通解.

Find its general solution.

10. 已知 4 阶方阵 $\boldsymbol{A} = (\boldsymbol{\alpha}_1, \boldsymbol{\alpha}_2, \boldsymbol{\alpha}_3, \boldsymbol{\alpha}_4)$, $\boldsymbol{\alpha}_1, \boldsymbol{\alpha}_2, \boldsymbol{\alpha}_3, \boldsymbol{\alpha}_4$ 为 4 维列向量, 其中 $\boldsymbol{\alpha}_2, \boldsymbol{\alpha}_3, \boldsymbol{\alpha}_4$ 线性无关, $\boldsymbol{\alpha}_1 = 2\boldsymbol{\alpha}_2 - \boldsymbol{\alpha}_3$. 如果 $\boldsymbol{\beta} = \boldsymbol{\alpha}_1 + \boldsymbol{\alpha}_2 + \boldsymbol{\alpha}_3 + \boldsymbol{\alpha}_4$, 求线性方程组 $\boldsymbol{A}\boldsymbol{x} = \boldsymbol{\beta}$ 的通解.

10. Let $\boldsymbol{A}$ be a square matrix of order 4, given by $\boldsymbol{A} = (\boldsymbol{\alpha}_1, \boldsymbol{\alpha}_2, \boldsymbol{\alpha}_3, \boldsymbol{\alpha}_4)$ and let $\boldsymbol{\alpha}_1, \boldsymbol{\alpha}_2, \boldsymbol{\alpha}_3, \boldsymbol{\alpha}_4$ be 4 dimensional column vectors, where $\boldsymbol{\alpha}_2, \boldsymbol{\alpha}_3, \boldsymbol{\alpha}_4$ are linearly independent and $\boldsymbol{\alpha}_1 = 2\boldsymbol{\alpha}_2 - \boldsymbol{\alpha}_3$. If $\boldsymbol{\beta} = \boldsymbol{\alpha}_1 + \boldsymbol{\alpha}_2 + \boldsymbol{\alpha}_3 + \boldsymbol{\alpha}_4$, find the general solution of the linear system $\boldsymbol{A}\boldsymbol{x} = \boldsymbol{\beta}$.

第 5 章
Chapter 5

方阵的特征值、相似与对角化
Eigenvalues, Similarity and Diagonalization of Square Matrices

工程技术领域中的振动问题和稳定性问题以及一些经济问题中的计算往往可归结为矩阵的特征值问题, 即对于方阵 $\boldsymbol{A}$, 求数 λ 和非零的向量 $\boldsymbol{\alpha}$, 使关系式 $\boldsymbol{A\alpha}=\lambda\boldsymbol{\alpha}$ 成立. 数 λ 称为 $\boldsymbol{A}$ 的特征值, $\boldsymbol{\alpha}$ 称为 $\boldsymbol{A}$ 的属于特征值 λ 的特征向量.

The vibration and stability problems in the field of engineering technology and the computation of some economic problems can be always reduced to the eigenvalue problems of matrices, namely, for a square matrix $\boldsymbol{A}$, find a number λ and a nonzero vector $\boldsymbol{\alpha}$ such that $\boldsymbol{A\alpha}=\lambda\boldsymbol{\alpha}$ is valid. The number λ is said to be an eigenvalue of $\boldsymbol{A}$, the vector $\boldsymbol{\alpha}$ is said to be an eigenvector of $\boldsymbol{A}$ associated with λ.

➢本章内容提要

Headline of this chapter

1. 方阵的特征值与特征向量的定义、性质、求法;

1. Definitions, properties and solutions of eigenvalues and eigenvectors of square matrices;

2. 相似矩阵的定义、性质, 方阵相似于对角矩阵的判定方法.

2. Definitions and properties of similar matrices, judgment methods of a square matrix similar to a diagonal matrix.

注 本章讨论的矩阵均为方阵.

Note The matrices discussed in this chapter are all square matrices.

5.1 方阵的特征值与特征向量

5.1 Eigenvalues and Eigenvectors of Square Matrices

5.1.1 特征值与特征向量的定义及计算

5.1.1 Definitions and Calculation of Eigenvalues and Eigenvectors

定义 1 设 $\boldsymbol{A}$ 是数域 P 上的 n 阶方阵. 如果在 P 中存在数 λ 和非零列向量 $\boldsymbol{\alpha}$, 使得 $\boldsymbol{A}\boldsymbol{\alpha}=\lambda\boldsymbol{\alpha}$ 成立, 则称 λ 为 $\boldsymbol{A}$ 的**特征值**, $\boldsymbol{\alpha}$ 称为 $\boldsymbol{A}$ 的属于 λ 的**特征向量**.

Definition 1 Let $\boldsymbol{A}$ be a square matrix of order n over a number field P. If there exist a number λ and a nonzero column vector $\boldsymbol{\alpha}$ over P such that $\boldsymbol{A}\boldsymbol{\alpha}=\lambda\boldsymbol{\alpha}$, then λ is called an **eigenvalue** of $\boldsymbol{A}$ and $\boldsymbol{\alpha}$ is called an **eigenvector** of $\boldsymbol{A}$ associated with λ.

根据上述定义, 容易验证

In accordance with the above definition, it is easy to verify that

(1) 若 $\boldsymbol{\alpha}$ 是 $\boldsymbol{A}$ 的属于特征值 λ 的特征向量, 则对任意非零数 k, $k\boldsymbol{\alpha}$ 也是 $\boldsymbol{A}$ 的属于 λ 的特征向量.

(1) If $\boldsymbol{\alpha}$ is an eigenvector of $\boldsymbol{A}$ associated with λ, so is $k\boldsymbol{\alpha}$ for any nonzero number k.

显然, $k\boldsymbol{\alpha}\neq\mathbf{0}$, 且有

Obviously, $k\boldsymbol{\alpha}\neq\mathbf{0}$ and

$$\boldsymbol{A}(k\boldsymbol{\alpha})=k(\boldsymbol{A}\boldsymbol{\alpha})=k(\lambda\boldsymbol{\alpha})=\lambda(k\boldsymbol{\alpha})$$

(2) 属于同一特征值的特征向量的任意非零线性组合仍是属于这个特征值的特征向量.

(2) Any nonzero linear combinations of eigenvectors associated with the same eigenvalue are also eigenvectors associated with the eigenvalue.

例 1 对于矩阵

Example 1 For the matrix

$$\boldsymbol{A}=\begin{pmatrix}1 & 2 & 2\\ 1 & -1 & 1\\ 4 & -12 & 1\end{pmatrix}$$

有

we have

$$\begin{pmatrix}1 & 2 & 2\\ 1 & -1 & 1\\ 4 & -12 & 1\end{pmatrix}\begin{pmatrix}3\\ 1\\ -1\end{pmatrix}=1\begin{pmatrix}3\\ 1\\ -1\end{pmatrix}$$

即 1 是 $\boldsymbol{A}$ 的特征值, $\begin{pmatrix} 3 \\ 1 \\ -1 \end{pmatrix}$ 是 $\boldsymbol{A}$ 的属于 1 的特征向量. 此外, 对任意非零数 k, $k\begin{pmatrix} 3 \\ 1 \\ -1 \end{pmatrix}$ 也是 $\boldsymbol{A}$ 的属于特征值 1 的特征向量. ■

that is to say, 1 is an eigenvalue of $\boldsymbol{A}$, and $\begin{pmatrix} 3 \\ 1 \\ -1 \end{pmatrix}$ is an eigenvector of $\boldsymbol{A}$ associated with 1. In addition, for any nonzero number k, $k\begin{pmatrix} 3 \\ 1 \\ -1 \end{pmatrix}$ is also an eigenvector of $\boldsymbol{A}$ associated with 1. ■

例 2 设

Example 2 Suppose that

$$\boldsymbol{A} = \begin{pmatrix} \lambda & & & \\ & \lambda & & \\ & & \ddots & \\ & & & \lambda \end{pmatrix}$$

求 $\boldsymbol{A}$ 的特征值与特征向量

Find the eigenvalues and eigenvectors of $\boldsymbol{A}$.

解 根据矩阵乘法, 对任意非零 n 维列向量 $\boldsymbol{\alpha}$ 有 $\boldsymbol{A\alpha} = (\lambda\boldsymbol{E})\boldsymbol{\alpha} = \lambda\boldsymbol{\alpha}$, 由定义 1 可知 λ 是 $\boldsymbol{A}$ 的特征值, 而 $\boldsymbol{\alpha}$ 是 $\boldsymbol{A}$ 的属于特征值 λ 的特征向量.

Solution In accordance with matrix multiplication, we have that $\boldsymbol{A\alpha} = (\lambda\boldsymbol{E})\boldsymbol{\alpha} = \lambda\boldsymbol{\alpha}$ is valid for any nonzero n dimensional column vector $\boldsymbol{\alpha}$. From Definition 1 we know that λ is an eigenvalue of $\boldsymbol{A}$ and that $\boldsymbol{\alpha}$ is an eigenvector of $\boldsymbol{A}$ associated with λ.

若矩阵 $\boldsymbol{A}$ 还有其他的特征值 $b(b \neq \lambda)$, 则应有 $\boldsymbol{\beta} \neq \boldsymbol{0}$, 使得 $\boldsymbol{A\beta} = b\boldsymbol{\beta}$. 由前面的分析知, 令 $\boldsymbol{\alpha} = \boldsymbol{\beta}$, 则有 $\boldsymbol{A\beta} = \lambda\boldsymbol{\beta}$. 所以, $\lambda\boldsymbol{\beta} = b\boldsymbol{\beta}$, 即 $(\lambda - b)\boldsymbol{\beta} = \boldsymbol{0}$. 由 $\boldsymbol{\beta} \neq \boldsymbol{0}$ 可得 $\lambda = b$, 与假设矛盾. 因此, λ 是 $\boldsymbol{A}$ 的唯一特征值. ■

If the matrix $\boldsymbol{A}$ has another eigenvalue b $(b \neq \lambda)$, then there exists a vector $\boldsymbol{\beta} \neq \boldsymbol{0}$ such that $\boldsymbol{A\beta} = b\boldsymbol{\beta}$. From the above analysis we know that $\boldsymbol{A\beta} = \lambda\boldsymbol{\beta}$ by letting $\boldsymbol{\alpha} = \boldsymbol{\beta}$. It leads to $\lambda\boldsymbol{\beta} = b\boldsymbol{\beta}$, i.e., $(\lambda - b)\boldsymbol{\beta} = \boldsymbol{0}$. From $\boldsymbol{\beta} \neq \boldsymbol{0}$ we have $\lambda = b$. It contradicts the assumption. Hence, λ is the unique eigenvalue of $\boldsymbol{A}$. ■

注 (1) 特征向量是非零列向量;

Note (1) Eigenvectors are nonzero column vectors;

(2) 特征向量总是相对于已知的特征值而言的;

(2) Eigenvectors are always associated with the known eigenvalues;

(3) 一个特征向量不能属于不同的特征值, 但一个特征值对应的特征向量不唯一.

(3) An eigenvector can not be associated with distinct eigenvalues, however, eigenvectors associated with an eigenvalue may not be unique.

现在存在两个问题亟待解决：任给数域 P 上的方阵 $\boldsymbol{A}$, 它是否一定有特征值? 当它有特征值时, 如何求出它的全部特征值和它对应的全部特征向量? 下面就来讨论这两个问题. 为了讨论方便, 我们先引入如下定义.

At present, there exist two problems to be solved, namely, for any square matrix $\boldsymbol{A}$ over a number field P, whether or not it must have eigenvalues? If so, how to find all the eigenvalues and all the corresponding eigenvectors? In the following part we discuss the above two problems. For convenience, we first introduce the following definition.

定义 2 设 $\boldsymbol{A}$ 是数域 P 上的 n 阶方阵, λ 是一个变量, 矩阵 $\lambda\boldsymbol{E}-\boldsymbol{A}$ 的行列式为

Definition 2 Let $\boldsymbol{A}$ be a square matrix of order n over a number field P and let λ be a variable, the determinant of the matrix $\lambda\boldsymbol{E}-\boldsymbol{A}$, given by

$$|\lambda\boldsymbol{E}-\boldsymbol{A}|=\begin{vmatrix}\lambda-a_{11} & -a_{12} & \cdots & -a_{1n}\\ -a_{21} & \lambda-a_{22} & \cdots & -a_{2n}\\ \vdots & \vdots & & \vdots\\ -a_{n1} & -a_{n2} & \cdots & \lambda-a_{nn}\end{vmatrix}$$

是 λ 的一个 n 次多项式, 称为 $\boldsymbol{A}$ 的**特征多项式**, 记为 $|\lambda\boldsymbol{E}-\boldsymbol{A}|=f(\lambda)$. 以 λ 为未知数的 n 次方程 $|\lambda\boldsymbol{E}-\boldsymbol{A}|=f(\lambda)=0$ 称为 $\boldsymbol{A}$ **的特征方程**.

is a polynomial of degree n with respect to λ, and it is called the **characteristic polynomial** of $\boldsymbol{A}$, written as $|\lambda\boldsymbol{E}-\boldsymbol{A}|=f(\lambda)$. The equation of degree n with respect to the unknown λ, given by $|\lambda\boldsymbol{E}-\boldsymbol{A}|=f(\lambda)=0$, is called the **characteristic equation** of $\boldsymbol{A}$.

注 n 阶方阵的特征方程在复数范围内恒有根, 其个数为方程的次数, 其中重根按重数计算. 因此 n 阶方阵在复数范围内有 n 个特征值 (根)$\lambda_1,\lambda_2,\cdots,\lambda_n$.

Note The characteristic equation of a square matrix of order n has roots in the complex field identically, and the root number is equal to the degree of the characteristic equation, where repeated roots count their multiplicity. Therefore, a square matrix of order n has exactly n eigenvalues $\lambda_1,\lambda_2,\cdots,\lambda_n$ in the complex field.

设 $\boldsymbol{A}$ 是数域 P 上的 n 阶方阵. 如果 λ 是 $\boldsymbol{A}$ 的特征值, $\boldsymbol{\alpha}$ 是 $\boldsymbol{A}$ 的属于 λ 的特征向量, 则

Let $\boldsymbol{A}$ be a square matrix of order n over a number field P. If λ is an eigenvalue of $\boldsymbol{A}$ and $\boldsymbol{\alpha}$ is the corresponding eigenvector, then

$$\boldsymbol{A}\boldsymbol{\alpha} = \lambda\boldsymbol{\alpha} \quad (\boldsymbol{\alpha} \neq \boldsymbol{0})$$

即

that is,

$$(\lambda\boldsymbol{E} - \boldsymbol{A})\boldsymbol{\alpha} = \boldsymbol{0}$$

这说明 $\boldsymbol{\alpha}$ 是齐次线性方程组 $(\lambda\boldsymbol{E} - \boldsymbol{A})\boldsymbol{x} = \boldsymbol{0}$ 的非零解, 从而有 $|\lambda\boldsymbol{E} - \boldsymbol{A}| = 0$, 即 λ 是特征方程 $|\lambda\boldsymbol{E} - \boldsymbol{A}| = 0$ 的根; 反之, 若 λ 是 $|\lambda\boldsymbol{E} - \boldsymbol{A}| = 0$ 的根, 则 $|\lambda\boldsymbol{E} - \boldsymbol{A}| = 0$, 从而 $(\lambda\boldsymbol{E} - \boldsymbol{A})\boldsymbol{x} = \boldsymbol{0}$ 有非零解 $\boldsymbol{\alpha}$, $(\lambda\boldsymbol{E} - \boldsymbol{A})\boldsymbol{\alpha} = \boldsymbol{0}$, 因此, $\boldsymbol{A}\boldsymbol{\alpha} = \lambda\boldsymbol{\alpha}$.

It shows that $\boldsymbol{\alpha}$ is a nonzero solution of the system of homogeneous linear equations, given by $(\lambda\boldsymbol{E} - \boldsymbol{A})\boldsymbol{x} = \boldsymbol{0}$, so we have $|\lambda\boldsymbol{E} - \boldsymbol{A}| = 0$, which implies that λ is a root of the characteristic equation $|\lambda\boldsymbol{E} - \boldsymbol{A}| = 0$. On the contrary, if λ is a root of $|\lambda\boldsymbol{E} - \boldsymbol{A}| = 0$, then $|\lambda\boldsymbol{E} - \boldsymbol{A}| = 0$, so $(\lambda\boldsymbol{E} - \boldsymbol{A})\boldsymbol{x} = \boldsymbol{0}$ has a nonzero solution $\boldsymbol{\alpha}$, namely, $(\lambda\boldsymbol{E} - \boldsymbol{A})\boldsymbol{\alpha} = \boldsymbol{0}$, then $\boldsymbol{A}\boldsymbol{\alpha} = \lambda\boldsymbol{\alpha}$.

由上面的讨论可得如下定理.

From the above discussion we obtain the following theorem.

定理 1 设 $\boldsymbol{A}$ 是数域 P 上的 n 阶方阵, 则 λ 是 $\boldsymbol{A}$ 的特征值, $\boldsymbol{\alpha}$ 是 $\boldsymbol{A}$ 的属于 λ 的特征向量的充要条件是, λ 是特征方程 $|\lambda\boldsymbol{E} - \boldsymbol{A}| = 0$ 在 P 中的根, 且 $\boldsymbol{\alpha}$ 是 $(\lambda\boldsymbol{E} - \boldsymbol{A})\boldsymbol{x} = \boldsymbol{0}$ 的非零解.

Theorem 1 Let $\boldsymbol{A}$ be a square matrix of order n over a number field P. Then λ is an eigenvalue of $\boldsymbol{A}$ and $\boldsymbol{\alpha}$ is an eigenvector of $\boldsymbol{A}$ associated with λ if and only if λ is a root of the characteristic equation $|\lambda\boldsymbol{E} - \boldsymbol{A}| = 0$ over P, and $\boldsymbol{\alpha}$ is a nonzero solution of $(\lambda\boldsymbol{E} - \boldsymbol{A})\boldsymbol{x} = \boldsymbol{0}$.

注 因为 $|\lambda\boldsymbol{E} - \boldsymbol{A}| = 0$ 与 $|\boldsymbol{A} - \lambda\boldsymbol{E}| = 0$ 有相同的根, 而且 $(\lambda\boldsymbol{E} - \boldsymbol{A})\boldsymbol{x} = \boldsymbol{0}$ 与 $(\boldsymbol{A} - \lambda\boldsymbol{E})\boldsymbol{x} = \boldsymbol{0}$ 有相同的解, 所以也可以用 $|\boldsymbol{A} - \lambda\boldsymbol{E}| = 0$ 求 $\boldsymbol{A}$ 的特征值, 用 $(\boldsymbol{A} - \lambda\boldsymbol{E})\boldsymbol{x} = \boldsymbol{0}$ 求 $\boldsymbol{A}$ 的特征向量.

Note Since the equations $|\lambda\boldsymbol{E} - \boldsymbol{A}| = 0$ and $|\boldsymbol{A} - \lambda\boldsymbol{E}| = 0$ have the same roots, moreover, $(\lambda\boldsymbol{E} - \boldsymbol{A})\boldsymbol{x} = \boldsymbol{0}$ and $(\boldsymbol{A} - \lambda\boldsymbol{E})\boldsymbol{x} = \boldsymbol{0}$ have the same solutions, we can also find the eigenvalues of $\boldsymbol{A}$ from $|\boldsymbol{A} - \lambda\boldsymbol{E}| = 0$ and find the corresponding eigenvectors from $(\boldsymbol{A} - \lambda\boldsymbol{E})\boldsymbol{x} = \boldsymbol{0}$.

例 3 设

Example 3 Suppose that

$$A=\begin{pmatrix}3 & 1\\5 & -1\end{pmatrix}$$

求 $\boldsymbol{A}$ 的特阵值与特征向量.

Find the eigenvalues and eigenvectors of $\boldsymbol{A}$.

解 $\boldsymbol{A}$ 的特征方程为

Solution The characteristic equation of $\boldsymbol{A}$ is

$$|\lambda \boldsymbol{E}-\boldsymbol{A}|=\begin{vmatrix}\lambda-3 & -1\\-5 & \lambda+1\end{vmatrix}=(\lambda-4)(\lambda+2)=0$$

所以, $\lambda_1=4,\lambda_2=-2$ 是 $\boldsymbol{A}$ 的两个不同的特征值.

Hence, $\lambda_1=4,\lambda_2=-2$ are two distinct eigenvalues of $\boldsymbol{A}$.

将 $\lambda_1=4$ 代入方程组 $(\lambda\boldsymbol{E}-\boldsymbol{A})\boldsymbol{x}=\boldsymbol{0}$, 得

Substituting $\lambda_1=4$ into the system $(\lambda\boldsymbol{E}-\boldsymbol{A})\boldsymbol{x}=\boldsymbol{0}$, we have

$$4\boldsymbol{E}-\boldsymbol{A}=\begin{pmatrix}1 & -1\\-5 & 5\end{pmatrix}\to\begin{pmatrix}1 & -1\\0 & 0\end{pmatrix}$$

基础解系为

The system of fundamental solutions is given by

$$\boldsymbol{\alpha}_1=\begin{pmatrix}1\\1\end{pmatrix}$$

所以 $\boldsymbol{A}$ 的属于 $\lambda_1=4$ 的全部特征向量是 $k\boldsymbol{\alpha}_1$(k 是任意的非零数).

Therefore, the total eigenvectors of $\boldsymbol{A}$ associated with $\lambda_1=4$ have the form $k\boldsymbol{\alpha}_1$, where k is an arbitrary nonzero number.

将 $\lambda_2=-2$ 代入方程组 $(\lambda\boldsymbol{E}-\boldsymbol{A})\boldsymbol{x}=\boldsymbol{0}$ 得

Substituting $\lambda_2=-2$ into the system $(\lambda\boldsymbol{E}-\boldsymbol{A})\boldsymbol{x}=\boldsymbol{0}$, we have

$$-2\boldsymbol{E}-\boldsymbol{A}=\begin{pmatrix}-5 & -1\\-5 & -1\end{pmatrix}\to\begin{pmatrix}5 & 1\\0 & 0\end{pmatrix}$$

基础解系为 $\boldsymbol{\alpha}_2=\begin{pmatrix}1\\-5\end{pmatrix}$. 所以 $\boldsymbol{A}$ 的属于 $\lambda_1=-2$ 的全部特征向量是 $k\boldsymbol{\alpha}_2$(k 是任意的非零数). ■

The system of fundamental solutions is given by $\boldsymbol{\alpha}_2=\begin{pmatrix}1\\-5\end{pmatrix}$. Therefore, the total eigenvectors of $\boldsymbol{A}$ associated with $\lambda_1=-2$ have the form $k\boldsymbol{\alpha}_2$, where k is an arbitrary nonzero number. ■

例 4 设

Example 4 Suppose that

$$A=\begin{pmatrix}-1&1&0\\-4&3&0\\1&0&2\end{pmatrix}$$

求 $\boldsymbol{A}$ 的特征值和特征向量.

Find the eigenvalues and the eigenvectors of $\boldsymbol{A}$.

解 $\boldsymbol{A}$ 的特征方程为

Solution The characteristic equation of $\boldsymbol{A}$ is

$$|\lambda \boldsymbol{E}-\boldsymbol{A}|=\begin{vmatrix}\lambda+1&-1&0\\4&\lambda-3&0\\-1&0&\lambda-2\end{vmatrix}=(\lambda-2)(\lambda-1)^2=0$$

所以, $\lambda_1=2,\lambda_2=\lambda_3=1$ 是 $\boldsymbol{A}$ 的特征值.

Hence, $\lambda_1=2,\lambda_2=\lambda_3=1$ are eigenvalues of $\boldsymbol{A}$.

将 $\lambda_1=2$ 代入方程组 $(\lambda\boldsymbol{E}-\boldsymbol{A})\boldsymbol{x}=\boldsymbol{0}$, 得

Substituting $\lambda_1=2$ into the system $(\lambda\boldsymbol{E}-\boldsymbol{A})\boldsymbol{x}=\boldsymbol{0}$, we have

$$2\boldsymbol{E}-\boldsymbol{A}=\begin{pmatrix}3&-1&0\\4&-1&0\\-1&0&0\end{pmatrix}\to\begin{pmatrix}1&0&0\\0&1&0\\0&0&0\end{pmatrix}$$

基础解系为

The system of fundamental solutions is given by

$$\boldsymbol{\alpha}_1=\begin{pmatrix}0\\0\\1\end{pmatrix}$$

所以 $\boldsymbol{A}$ 的属于 $\lambda_1=2$ 的全部特征向量是 $k\boldsymbol{\alpha}_1$(k 是任意的非零数).

Therefore, the total eigenvectors of $\boldsymbol{A}$ associated with $\lambda_1=2$ have the form $k\boldsymbol{\alpha}_1$, where k is an arbitrary nonzero number.

将 $\lambda_2=\lambda_3=1$ 代入方程组 $(\lambda\boldsymbol{E}-\boldsymbol{A})\boldsymbol{x}=\boldsymbol{0}$, 得

Substituting $\lambda_2=\lambda_3=1$ into the system $(\lambda\boldsymbol{E}-\boldsymbol{A})\boldsymbol{x}=\boldsymbol{0}$, we have

$$\boldsymbol{E}-\boldsymbol{A}=\begin{pmatrix}2&-1&0\\4&-2&0\\-1&0&-1\end{pmatrix}\to\begin{pmatrix}1&0&1\\0&1&2\\0&0&0\end{pmatrix}$$

基础解系为

The system of fundamental solutions is given by

$$\boldsymbol{\alpha}_2=\begin{pmatrix}1\\2\\-1\end{pmatrix}$$

所以 $\boldsymbol{A}$ 的属于 $\lambda_2=\lambda_3=1$ 的全部特征向量是 $k\boldsymbol{\alpha}_2$(k 是任意的非零数). ■

Therefore, the total eigenvectors of $\boldsymbol{A}$ associated with $\lambda_2=\lambda_3=1$ have the form $k\boldsymbol{\alpha}_2$, where k is an arbitrary nonzero number. ■

例 5 设

Example 5 Suppose that

$$\boldsymbol{A}=\begin{pmatrix}4&6&0\\-3&-5&0\\-3&-6&1\end{pmatrix}$$

求 $\boldsymbol{A}$ 的特征值和特征向量.

Find the eigenvalues and eigenvectors of $\boldsymbol{A}$.

解 $\boldsymbol{A}$ 的特征多方程为

Solution The characteristic equation of $\boldsymbol{A}$ is

$$|\lambda\boldsymbol{E}-\boldsymbol{A}|=\begin{vmatrix}\lambda-4&-6&0\\3&\lambda+5&0\\3&6&\lambda-1\end{vmatrix}=(\lambda+2)(\lambda-1)^2=0$$

所以, $\lambda_1=-2,\lambda_2=\lambda_3=1$ 是 $\boldsymbol{A}$ 的特征值.

Hence, the eigenvalues of $\boldsymbol{A}$ are $\lambda_1=-2,\lambda_2=\lambda_3=1$.

将 $\lambda_1=-2$ 代入方程组 $(\lambda\boldsymbol{E}-\boldsymbol{A})\boldsymbol{x}=\boldsymbol{0}$, 得

Substituting $\lambda_1=-2$ into the system $(\lambda\boldsymbol{E}-\boldsymbol{A})\boldsymbol{x}=\boldsymbol{0}$, we have

$$-2\boldsymbol{E}-\boldsymbol{A}=\begin{pmatrix}-6&-6&0\\3&3&0\\3&6&-3\end{pmatrix}\to\begin{pmatrix}1&0&1\\0&1&-1\\0&0&0\end{pmatrix}$$

基础解系为

The system of fundamental solutions is given by

$$\boldsymbol{\alpha}_1=\begin{pmatrix}-1\\1\\1\end{pmatrix}$$

所以, $\boldsymbol{A}$ 的属于 $\lambda_1=-2$ 的全部特征向量是 $k\boldsymbol{\alpha}_1$(k 是任意的非零数).

Therefore, the total eigenvectors of $\boldsymbol{A}$ associated with $\lambda_1=-2$ have the form $k\boldsymbol{\alpha}_1$, where k is an arbitrary nonzero number.

将 $\lambda_2=\lambda_3=1$ 代入方程组 $(\lambda\boldsymbol{E}-\boldsymbol{A})\boldsymbol{x}=\boldsymbol{0}$, 得

Substituting $\lambda_2=\lambda_3=1$ into the system $(\lambda\boldsymbol{E}-\boldsymbol{A})\boldsymbol{x}=\boldsymbol{0}$, we have

$$\boldsymbol{E}-\boldsymbol{A}=\begin{pmatrix}-3&-6&0\\3&6&0\\3&6&0\end{pmatrix}\to\begin{pmatrix}1&2&0\\0&0&0\\0&0&0\end{pmatrix}$$

基础解系为

The system of fundamental solutions is given by

$$\boldsymbol{\alpha}_2=\begin{pmatrix}-2\\1\\0\end{pmatrix},\boldsymbol{\alpha}_3=\begin{pmatrix}0\\0\\1\end{pmatrix}$$

所以, $\boldsymbol{A}$ 的属于 $\lambda_2=\lambda_3=1$ 的全部特征向量是 $k_2\boldsymbol{\alpha}_2+k_3\boldsymbol{\alpha}_3$ (k_2,k_3 不全为零). ■

Therefore, the total eigenvectors of $\boldsymbol{A}$ associated with $\lambda_2=\lambda_3=1$ have the form $k_2\boldsymbol{\alpha}_2+k_3\boldsymbol{\alpha}_3$, where the numbers k_2 and k_3 are not all zeros. ■

从上述例子可以归纳出具体计算方阵的特征值、特征向量的步骤:

From the above examples, we can summarize the practical process for finding eigenvalues and eigenvectors of square matrices:

第一步 计算特征多项式 $|\lambda\boldsymbol{E}-\boldsymbol{A}|$;

Step 1 Calculate the characteristic polynomial $|\lambda\boldsymbol{E}-\boldsymbol{A}|$;

第二步 求出特征方程 $|\lambda\boldsymbol{E}-\boldsymbol{A}|=0$ 的全部根, 它们就是 $\boldsymbol{A}$ 的全部特征值;

Step 2 Find all the roots of the characteristic equation $|\lambda\boldsymbol{E}-\boldsymbol{A}|=0$, they are the total eigenvalues of $\boldsymbol{A}$;

第三步 对于 $\boldsymbol{A}$ 的每一个特征值 λ_i, 求 $(\lambda_i\boldsymbol{E}-\boldsymbol{A})\boldsymbol{x}=\boldsymbol{0}$ 的一个基础解系

Step 3 For each eigenvalue λ_i of $\boldsymbol{A}$, find a system of fundamental solutions of $(\lambda_i\boldsymbol{E}-\boldsymbol{A})\boldsymbol{x}=\boldsymbol{0}$, written as

$$\boldsymbol{\alpha}_1,\boldsymbol{\alpha}_2,\cdots,\boldsymbol{\alpha}_r$$

则对应于 λ_i 的全部特征向量为

Then the eigenvectors associated with λ_i have the form

$$k_1\boldsymbol{\alpha}_1+k_2\boldsymbol{\alpha}_2+\cdots+k_r\boldsymbol{\alpha}_r$$

其中 $k_1, k_2, \cdots, k_r$ 为不全为零的数.

where the numbers $k_1, k_2, \cdots, k_r$ are not all zeros.

5.1.2 特征值与特征向量的基本性质

5.1.2 Basic Properties of Eigenvalues and Eigenvectors

定理 2 方阵 $\boldsymbol{A}$ 与它的转置矩阵 $\boldsymbol{A}^{\mathrm{T}}$ 有相同的特征值.

Theorem 2 A square matrix $\boldsymbol{A}$ and its transpose matrix $\boldsymbol{A}^{\mathrm{T}}$ have the same eigenvalues.

证 根据行列式的性质 1, 有

Proof In accordance with Property 1 of determinants, we have

$$|\lambda \boldsymbol{E} - \boldsymbol{A}| = \left|(\lambda \boldsymbol{E} - \boldsymbol{A})^{\mathrm{T}}\right| = \left|\lambda \boldsymbol{E} - \boldsymbol{A}^{\mathrm{T}}\right|$$

这说明矩阵 $\boldsymbol{A}$ 与 $\boldsymbol{A}^{\mathrm{T}}$ 有相同的特征多项式, 因而有相同的特征值. ▲

This shows that the matrices $\boldsymbol{A}$ and $\boldsymbol{A}^{\mathrm{T}}$ have the same characteristic polynomial, and so they have the same eigenvalues. ▲

定理 3 设 $\boldsymbol{A}$ 为 n 阶方阵, $\lambda_1, \lambda_2, \cdots, \lambda_n$ 是 $\boldsymbol{A}$ 的 n 个特征值, 则

Theorem 3 Let $\boldsymbol{A}$ be a square matrix of order n and let $\lambda_1, \lambda_2, \cdots, \lambda_n$ be n eigenvalues of $\boldsymbol{A}$, then

(1) $\lambda_1 + \lambda_2 + \cdots + \lambda_n = a_{11} + a_{22} + \cdots + a_{nn} = \mathrm{tr}(\boldsymbol{A})$ (称为矩阵 $\boldsymbol{A}$ 的**迹**).

(1) $\lambda_1 + \lambda_2 + \cdots + \lambda_n = a_{11} + a_{22} + \cdots + a_{nn} = \mathrm{tr}(\boldsymbol{A})$, which is called the **trace** of $\boldsymbol{A}$.

(2) $\lambda_1 \lambda_2 \cdots \lambda_n = |\boldsymbol{A}|$.

推论 1 $\boldsymbol{A}$ 可逆 $\Leftrightarrow \lambda_1 \lambda_2 \cdots \lambda_n = |\boldsymbol{A}| \neq 0 \Leftrightarrow$ 特征值 $\lambda_1, \lambda_2, \cdots, \lambda_n$ 都不为零.

Corollary 1 $\boldsymbol{A}$ is invertible $\Leftrightarrow \lambda_1 \lambda_2 \cdots \lambda_n = |\boldsymbol{A}| \neq 0 \Leftrightarrow$the eigenvalues $\lambda_1, \lambda_2, \cdots, \lambda_n$ are all nonzero.

定理 4 令 λ 是 $\boldsymbol{A}$ 的一个特征值. 当 $\boldsymbol{A}$ 可逆时, $\dfrac{1}{\lambda}$ 是 $\boldsymbol{A}^{-1}$ 的特征值.

Theorem 4 Let λ be an eigenvalue of $\boldsymbol{A}$. Then $\dfrac{1}{\lambda}$ is an eigenvalue of $\boldsymbol{A}^{-1}$ as $\boldsymbol{A}$ is invertible.

证 由已知条件可得 $\boldsymbol{A}\boldsymbol{\alpha} = \lambda\boldsymbol{\alpha}$. 因为 $\boldsymbol{A}$ 可逆, 所以 λ 不为零, 因此有

Proof From the known condition we have $\boldsymbol{A}\boldsymbol{\alpha} = \lambda\boldsymbol{\alpha}$. Since $\boldsymbol{A}$ is invertible, λ is not equal to zero. Hence,

$$\boldsymbol{A}^{-1}\boldsymbol{\alpha} = \frac{1}{\lambda}\boldsymbol{\alpha}$$

▲

定理 5 设 λ 是 $\boldsymbol{A}$ 的特征值, 则 λ^2 是 $\boldsymbol{A}^2$ 的特征值.

Theorem 5 Let λ be an eigenvalue of $\boldsymbol{A}$. Then λ^2 is an eigenvalue of $\boldsymbol{A}^2$.

证 由已知条件可得 $\boldsymbol{A\alpha} = \lambda\boldsymbol{\alpha}$. 易证

Proof From the known condition we have $\boldsymbol{A\alpha} = \lambda\boldsymbol{\alpha}$. It is easy to show that

$$\boldsymbol{A}^2\boldsymbol{\alpha} = \boldsymbol{A}(\boldsymbol{A\alpha}) = \lambda\boldsymbol{A\alpha} = \lambda^2\boldsymbol{\alpha}$$

这说明 λ^2 是 $\boldsymbol{A}^2$ 的特征值. ▲

It means that λ^2 is an eigenvalue of $\boldsymbol{A}^2$. ▲

同理, 不难证明下面的结论成立:

Similarly, it is not difficult to show that the following results are valid:

(1) 若 λ 是 $\boldsymbol{A}$ 的特征值, 则对任意的正整数 k, λ^k 是 $\boldsymbol{A}^k$ 的特征值.

(1) If λ is an eigenvalue of $\boldsymbol{A}$, then λ^k is an eigenvalue of $\boldsymbol{A}^k$ for any positive integer k.

(2) 设有 $f(\lambda) = a_0 + a_1\lambda + \cdots + a_m\lambda^m$, $f(\boldsymbol{A}) = a_0\boldsymbol{E} + a_1\boldsymbol{A} + \cdots + a_m\boldsymbol{A}^m$. 若 λ 是 $\boldsymbol{A}$ 的特征值, 则 $f(\lambda)$ 是 $f(\boldsymbol{A})$ 的特征值.

(2) Suppose that $f(\lambda) = a_0 + a_1\lambda + \cdots + a_m\lambda^m$, $f(\boldsymbol{A}) = a_0\boldsymbol{E} + a_1\boldsymbol{A} + \cdots + a_m\boldsymbol{A}^m$. If λ is an eigenvalue of $\boldsymbol{A}$, then $f(\lambda)$ is an eigenvalue of $f(\boldsymbol{A})$.

(3) 若 λ 是可逆方阵 $\boldsymbol{A}$ 的特征值, 则 $\dfrac{|\boldsymbol{A}|}{\lambda}$ 是 $\boldsymbol{A}^* = |\boldsymbol{A}|\,\boldsymbol{A}^{-1}$ 的特征值.

(3) If λ is an eigenvalue of the invertible square matrix $\boldsymbol{A}$, then $\dfrac{|\boldsymbol{A}|}{\lambda}$ is an eigenvalue of $\boldsymbol{A}^* = |\boldsymbol{A}|\,\boldsymbol{A}^{-1}$.

定理 6 设 $\boldsymbol{A}$ 为 n 阶方阵. 若 $\boldsymbol{\alpha}_1, \boldsymbol{\alpha}_2, \cdots, \boldsymbol{\alpha}_m$ 是 $\boldsymbol{A}$ 的属于互不相同特征值 $\lambda_1, \lambda_2, \cdots, \lambda_m$ 的特征向量, 则 $\{\boldsymbol{\alpha}_1, \boldsymbol{\alpha}_2, \cdots, \boldsymbol{\alpha}_m\}$ 必线性无关.

Theorem 6 Let $\boldsymbol{A}$ be a square matrix of order n. If $\boldsymbol{\alpha}_1, \boldsymbol{\alpha}_2, \cdots, \boldsymbol{\alpha}_m$ are eigenvectors of $\boldsymbol{A}$ associated with distinct eigenvalues $\lambda_1, \lambda_2, \cdots, \lambda_m$, then $\{\boldsymbol{\alpha}_1, \boldsymbol{\alpha}_2, \cdots, \boldsymbol{\alpha}_m\}$ must be linearly independent.

证法 1 反证法

Proof 1 Proof by contradiction.

设向量组 $\{\boldsymbol{\alpha}_1, \boldsymbol{\alpha}_2, \cdots, \boldsymbol{\alpha}_m\}$ 线性相关, 其极大无关组为 $\{\boldsymbol{\alpha}_1, \boldsymbol{\alpha}_2, \cdots, \boldsymbol{\alpha}_r\}(1 \leqslant r \leqslant m-1)$, 下面证明向量组 $\{\boldsymbol{\alpha}_1, \boldsymbol{\alpha}_2, \cdots, \boldsymbol{\alpha}_r, \boldsymbol{\alpha}_{r+1}\}$ 线性无关, 从而与 $\{\boldsymbol{\alpha}_1, \boldsymbol{\alpha}_2, \cdots, \boldsymbol{\alpha}_r\}$ 为极大无关组矛盾. 设

Assume that the vector set $\{\boldsymbol{\alpha}_1, \boldsymbol{\alpha}_2, \cdots, \boldsymbol{\alpha}_m\}$ is linearly dependent and that its maximal linearly independent subset is $\{\boldsymbol{\alpha}_1, \boldsymbol{\alpha}_2, \cdots, \boldsymbol{\alpha}_r\}(1 \leqslant r \leqslant m-1)$. In the following part we prove that the vector set $\{\boldsymbol{\alpha}_1, \boldsymbol{\alpha}_2, \cdots, \boldsymbol{\alpha}_r, \boldsymbol{\alpha}_{r+1}\}$ is linearly independent, which contradicts the fact that the set $\{\boldsymbol{\alpha}_1, \boldsymbol{\alpha}_2, \cdots, \boldsymbol{\alpha}_r\}$ is a maximal linearly independent subset. Let

$$k_1\boldsymbol{\alpha}_1 + k_2\boldsymbol{\alpha}_2 + \cdots + k_r\boldsymbol{\alpha}_r + k_{r+1}\boldsymbol{\alpha}_{r+1} = \mathbf{0} \tag{5.1}$$

式 (5.1) 两边左乘以 $\boldsymbol{A}$ 得

Premultiplying both sides of Equation (5.1) by $\boldsymbol{A}$, we have

$$k_1\lambda_1\boldsymbol{\alpha}_1 + k_2\lambda_2\boldsymbol{\alpha}_2 + \cdots + k_r\lambda_r\boldsymbol{\alpha}_r + k_{r+1}\lambda_{r+1}\boldsymbol{\alpha}_{r+1} = \mathbf{0} \tag{5.2}$$

式 (5.1) 两边同乘以 λ_{r+1} 得

Multiplying both sides of Equation (5.1) by λ_{r+1}, we get

$$k_1\lambda_{r+1}\boldsymbol{\alpha}_1 + k_2\lambda_{r+1}\boldsymbol{\alpha}_2 + \cdots + k_r\lambda_{r+1}\boldsymbol{\alpha}_r + k_{r+1}\lambda_{r+1}\boldsymbol{\alpha}_{r+1} = \mathbf{0} \tag{5.3}$$

式 (5.3) 减式 (5.2), 得

Subtracting Equation (5.3) from Equation.(5.2), we obtain

$$k_1(\lambda_{r+1} - \lambda_1)\boldsymbol{\alpha}_1 + \cdots + k_r(\lambda_{r+1} - \lambda_r)\boldsymbol{\alpha}_r = \mathbf{0}$$

由于向量组 $\{\boldsymbol{\alpha}_1, \boldsymbol{\alpha}_2, \cdots, \boldsymbol{\alpha}_r\}$ 线性无关, 且 $\lambda_i \neq \lambda_j (i \neq j)$, 因此必有 $k_1 = \cdots = k_r = 0$. 由式 (5.1) 可知, $k_{r+1}\boldsymbol{\alpha}_{r+1} = \mathbf{0}$, 但 $\boldsymbol{\alpha}_{r+1} \neq \mathbf{0}$, 从而 $k_{r+1} = 0$, 因此, 向量组 $\{\boldsymbol{\alpha}_1, \boldsymbol{\alpha}_2, \cdots, \boldsymbol{\alpha}_r, \boldsymbol{\alpha}_{r+1}\}$ 线性无关. 从而与向量组 $\{\boldsymbol{\alpha}_1, \boldsymbol{\alpha}_2, \cdots, \boldsymbol{\alpha}_r\}$ 为极大无关组矛盾.

Since the vector set $\{\boldsymbol{\alpha}_1, \boldsymbol{\alpha}_2, \cdots, \boldsymbol{\alpha}_m\}$ is linearly independent and $\lambda_i \neq \lambda_j (i \neq j)$, then there must have $k_1 = \cdots = k_r = 0$. From Equation (5.1) we have $k_{r+1}\boldsymbol{\alpha}_{r+1} = \mathbf{0}$. Since $\boldsymbol{\alpha}_{r+1} \neq \mathbf{0}$, it follows that $k_{r+1} = 0$. It means that the vector set $\{\boldsymbol{\alpha}_1, \boldsymbol{\alpha}_2, \cdots, \boldsymbol{\alpha}_r, \boldsymbol{\alpha}_{r+1}\}$ is linearly independent, which contradicts the fact that the vector set $\{\boldsymbol{\alpha}_1, \boldsymbol{\alpha}_2, \cdots, \boldsymbol{\alpha}_r\}$ is a maximal linearly independent subset.

证法 2 设存在数 $x_1, x_2, \cdots, x_m$, 使

Proof 2 Assume that there exist numbers $x_1, x_2, \cdots, x_m$ such that

$$x_1\boldsymbol{\alpha}_1 + x_2\boldsymbol{\alpha}_2 + \cdots + x_m\boldsymbol{\alpha}_m = \mathbf{0} \tag{5.4}$$

则

then

$$\boldsymbol{A}(x_1\boldsymbol{\alpha}_1 + x_2\boldsymbol{\alpha}_2 + \cdots + x_m\boldsymbol{\alpha}_m) = \mathbf{0}$$

即

namely,

$$\lambda_1 x_1\boldsymbol{\alpha}_1 + \lambda_2 x_2\boldsymbol{\alpha}_2 + \cdots + \lambda_m x_m\boldsymbol{\alpha}_m = \mathbf{0}$$

用 $\boldsymbol{A}$ 连续左乘式 (5.4), 类推有

Premultiplying Equation (5.4) by $\boldsymbol{A}$ continuously, we have

$$\lambda_1^k x_1\boldsymbol{\alpha}_1+\lambda_2^k x_2\boldsymbol{\alpha}_2+\cdots+\lambda_m^k x_m\boldsymbol{\alpha}_m=\mathbf{0}\quad(k=0,1,2,\cdots,m-1)$$

上列各式合写成矩阵形式, 得

Writing the above equations as a matrix form, we obtain

$$(x_1\boldsymbol{\alpha}_1,x_2\boldsymbol{\alpha}_2,\cdots,x_m\boldsymbol{\alpha}_m)\begin{pmatrix}1&\lambda_1&\cdots&\lambda_1^{m-1}\\1&\lambda_2&\cdots&\lambda_2^{m-1}\\\vdots&\vdots&&\vdots\\1&\lambda_m&\cdots&\lambda_m^{m-1}\end{pmatrix}=\mathbf{0}$$

上式中, 易见第二个矩阵的行列式等于范德蒙德行列式的转置, 当 $\lambda_i\neq\lambda_j(i\neq j)$ 时, 该矩阵可逆. 于是必有

In the above equation, it is easy to see that the determinant of the second matrix is equal to the transpose of the Vandermonde determinant, this implies that the matrix is invertible as $\lambda_i\neq\lambda_j(i\neq j)$. Consequently, there must be

$$(x_1\boldsymbol{\alpha}_1,x_2\boldsymbol{\alpha}_2,\cdots,x_m\boldsymbol{\alpha}_m)=\mathbf{0}$$

即 $x_i\boldsymbol{\alpha}_i=\mathbf{0}$. 由于 $\boldsymbol{\alpha}_i\neq\mathbf{0}$, 故 $x_i=0(i=1,2,\cdots,m)$, 所以向量组 $\{\boldsymbol{\alpha}_1,\boldsymbol{\alpha}_2,\cdots,\boldsymbol{\alpha}_m\}$ 线性无关. ▲

that is, $x_i\boldsymbol{\alpha}_i=\mathbf{0}$. Since $\boldsymbol{\alpha}_i\neq\mathbf{0}$, we have $x_i=0(i=1,2,\cdots,m)$. Hence, the vector set $\{\boldsymbol{\alpha}_1,\boldsymbol{\alpha}_2,\cdots,\boldsymbol{\alpha}_m\}$ is linearly independent.

▲

定理 6 可以推广为如下定理.

Theorem 6 can be generalized to the following theorem.

定理 7 令 $\boldsymbol{A}$ 是 n 阶方阵, 设 $\lambda_1,\lambda_2,\cdots,\lambda_m$ 是 $\boldsymbol{A}$ 的不同的特征值, $\boldsymbol{\alpha}_{i1},\boldsymbol{\alpha}_{i2},\cdots,\boldsymbol{\alpha}_{is_i}$ 是 $\boldsymbol{A}$ 的属于 λ_i 的线性无关的特征向量, 则分别对应于 $\lambda_1,\lambda_2,\cdots,\lambda_m$ 的特征向量组

Theorem 7 Let $\boldsymbol{A}$ be a square matrix of order n. Suppose that $\lambda_1,\lambda_2,\cdots,\lambda_m$ are distinct eigenvalues of $\boldsymbol{A}$ and that $\boldsymbol{\alpha}_{i1},\boldsymbol{\alpha}_{i2},\cdots,\boldsymbol{\alpha}_{is_i}$ are linearly independent eigenvectors of $\boldsymbol{A}$ associated with λ_i. Then the set consisted of the following eigenvectors, which are associated with $\lambda_1,\lambda_2,\cdots,\lambda_m$, respectively, given by

$$\{\boldsymbol{\alpha}_{11},\boldsymbol{\alpha}_{12},\cdots,\boldsymbol{\alpha}_{1s_1},\boldsymbol{\alpha}_{21},\boldsymbol{\alpha}_{22},\cdots,\boldsymbol{\alpha}_{2s_2},\cdots,\boldsymbol{\alpha}_{m1},\boldsymbol{\alpha}_{m2},\cdots,\boldsymbol{\alpha}_{ms_m}\}$$

也是线性无关的.

is also linearly independent.

注 对于一般的向量组, 即使各个部分组都线性无关, 则合并起来不一定线性无关, 定理 7 反映的是特征向量所独有的性质.

Note For a general vector set, even though each subset is linearly independent, their union may not be linearly independent. The property shown in Theorem 7 holds only for eigenvectors.

例 6 设 $\boldsymbol{A}$ 为 3 阶方阵, 其特征值为 $1,-1,2$. 求 $|\boldsymbol{A}^*+3\boldsymbol{A}-2\boldsymbol{E}|$.

Example 6 Let $\boldsymbol{A}$ be a square matrix of order 3 and let its eigenvalues be given by $1,-1,2$. Find $|\boldsymbol{A}^*+3\boldsymbol{A}-2\boldsymbol{E}|$.

解 由 $|\boldsymbol{A}|=\lambda_1\lambda_2\lambda_3=-2$ 知, $\boldsymbol{A}$ 可逆, 且 $\boldsymbol{A}^*=|\boldsymbol{A}|\boldsymbol{A}^{-1}$ 的特征值为 $-\dfrac{2}{\lambda}$(λ 是矩阵 $\boldsymbol{A}$ 的特征值). 令

Solution Since $|\boldsymbol{A}|=\lambda_1\lambda_2\lambda_3=-2$, we know that $\boldsymbol{A}$ is invertible, and that $-\dfrac{2}{\lambda}$ is an eigenvalue of $\boldsymbol{A}^*=|\boldsymbol{A}|\boldsymbol{A}^{-1}$, where λ is an eigenvalue of $\boldsymbol{A}$. Let

$$f(\boldsymbol{A})=\boldsymbol{A}^*+3\boldsymbol{A}-2\boldsymbol{E}=-2\boldsymbol{A}^{-1}+3\boldsymbol{A}-2\boldsymbol{E}$$

则 $f(\lambda)=-\dfrac{2}{\lambda}+3\lambda-2$ 为 $f(\boldsymbol{A})$ 的特征值, 即

Then $f(\lambda)=-\dfrac{2}{\lambda}+3\lambda-2$ is an eigenvalue of $f(\boldsymbol{A})$, that is, the eigenvalues of $f(\boldsymbol{A})$ are

$$f(2)=3,\quad f(1)=-1,\quad f(-1)=-3$$

于是

Hence,

$$|\boldsymbol{A}^*+3\boldsymbol{A}-2\boldsymbol{E}|=(-1)\cdot(-3)\cdot 3=9$$

■

例 7 设 λ_1,λ_2 是 $\boldsymbol{A}$ 的两个不同的特征值, 对应的特征向量分别为 $\boldsymbol{\alpha}_1$, $\boldsymbol{\alpha}_2$. 证明 $\boldsymbol{\alpha}_1+\boldsymbol{\alpha}_2$ 不是 $\boldsymbol{A}$ 的特征向量.

Example 7 Let λ_1, λ_2 be distinct eigenvalues of $\boldsymbol{A}$ and let $\boldsymbol{\alpha}_1$, $\boldsymbol{\alpha}_2$ be the corresponding eigenvectors, respectively. Prove that $\boldsymbol{\alpha}_1+\boldsymbol{\alpha}_2$ is not an eigenvector of $\boldsymbol{A}$.

证 按题设, 有 $\boldsymbol{A}\boldsymbol{\alpha}_1=\lambda_1\boldsymbol{\alpha}_1, \boldsymbol{A}\boldsymbol{\alpha}_2=\lambda_2\boldsymbol{\alpha}_2$, 故 $\boldsymbol{A}(\boldsymbol{\alpha}_1+\boldsymbol{\alpha}_2)=\lambda_1\boldsymbol{\alpha}_1+\lambda_2\boldsymbol{\alpha}_2$. 若 $\boldsymbol{\alpha}_1+\boldsymbol{\alpha}_2$ 是 $\boldsymbol{A}$ 的特征向量, 则存在数 λ, 使

Proof In accordance with the assumption we have $\boldsymbol{A}\boldsymbol{\alpha}_1=\lambda_1\boldsymbol{\alpha}_1, \boldsymbol{A}\boldsymbol{\alpha}_2=\lambda_2\boldsymbol{\alpha}_2$, then $\boldsymbol{A}(\boldsymbol{\alpha}_1+\boldsymbol{\alpha}_2)=\lambda_1\boldsymbol{\alpha}_1+\lambda_2\boldsymbol{\alpha}_2$. If $\boldsymbol{\alpha}_1+\boldsymbol{\alpha}_2$ is an eigenvector of $\boldsymbol{A}$, then there exists a number λ such that

$$\boldsymbol{A}(\boldsymbol{\alpha}_1+\boldsymbol{\alpha}_2)=\lambda(\boldsymbol{\alpha}_1+\boldsymbol{\alpha}_2)$$

于是

Thus,

$$\lambda(\boldsymbol{\alpha}_1+\boldsymbol{\alpha}_2)=\lambda_1\boldsymbol{\alpha}_1+\lambda_2\boldsymbol{\alpha}_2$$

即

that is,

$$(\lambda_1-\lambda)\boldsymbol{\alpha}_1+(\lambda_2-\lambda)\boldsymbol{\alpha}_2=\mathbf{0}$$

由定理 6 知 $\boldsymbol{\alpha}_1,\boldsymbol{\alpha}_2$ 线性无关, 故有

From Theorem 6 we know that $\boldsymbol{\alpha}_1,\boldsymbol{\alpha}_2$ are linearly independent. So we have

$$\lambda_1-\lambda=\lambda_2-\lambda=0$$

即 $\lambda_1=\lambda_2$, 与已知条件 $\lambda_1\neq\lambda_2$ 矛盾. 因此 $\boldsymbol{\alpha}_1+\boldsymbol{\alpha}_2$ 不是 $\boldsymbol{A}$ 的特征向量. ■

namely, $\lambda_1=\lambda_2$, which contradicts the known condition $\lambda_1\neq\lambda_2$. Hence $\boldsymbol{\alpha}_1+\boldsymbol{\alpha}_2$ is not an eigenvector of $\boldsymbol{A}$. ■

例 8 设 $\boldsymbol{A}$ 为 n 阶实方阵, $\boldsymbol{A}\boldsymbol{A}^{\mathrm{T}}=\boldsymbol{E},|\boldsymbol{A}|<0$. 求 $(\boldsymbol{A}^{-1})^*$ 的一个特征值.

Example 8 Let $\boldsymbol{A}$ be a real square matrix of order n, $\boldsymbol{A}\boldsymbol{A}^{\mathrm{T}}=\boldsymbol{E},|\boldsymbol{A}|<0$. Find an eigenvalue of $(\boldsymbol{A}^{-1})^*$.

解 由于 $(\boldsymbol{A}^{-1})^*=(\boldsymbol{A}^*)^{-1}$, 故可先算 $\boldsymbol{A}^*$ 的特征值, 再算 $(\boldsymbol{A}^*)^{-1}$ 的特征值.

Solution Since $(\boldsymbol{A}^{-1})^*=(\boldsymbol{A}^*)^{-1}$, we can first find the eigenvalues of $\boldsymbol{A}^*$, and then find the eigenvalues of $(\boldsymbol{A}^*)^{-1}$.

因为 $\boldsymbol{A}\boldsymbol{A}^{\mathrm{T}}=\boldsymbol{E}$, 有 $|\boldsymbol{A}|^2=1$, 即 $|\boldsymbol{A}|=\pm1$. 由条件 $|\boldsymbol{A}|<0$, 可得 $|\boldsymbol{A}|=-1$. 进而

Since $\boldsymbol{A}\boldsymbol{A}^{\mathrm{T}}=\boldsymbol{E}$, we have $|\boldsymbol{A}|^2=1$, that is, $|\boldsymbol{A}|=\pm1$. Using the condition $|\boldsymbol{A}|<0$, we have $|\boldsymbol{A}|=-1$. Furthermore,

$$|\boldsymbol{A}+\boldsymbol{E}|=\left|\boldsymbol{A}+\boldsymbol{A}\boldsymbol{A}^{\mathrm{T}}\right|=\left|\boldsymbol{A}(\boldsymbol{E}+\boldsymbol{A}^{\mathrm{T}})\right|=|\boldsymbol{A}|\left|(\boldsymbol{A}+\boldsymbol{E})^{\mathrm{T}}\right|=-|\boldsymbol{A}+\boldsymbol{E}|$$

故 $|\boldsymbol{A}+\boldsymbol{E}|=0$, 即 $\lambda=-1$ 是 $\boldsymbol{A}$ 的一个特征值. 因此, $\dfrac{|\boldsymbol{A}|}{\lambda}=\dfrac{-1}{-1}=1$, 即 1 是 $\boldsymbol{A}^*$ 的一个特征值. 从而, 1 是 $(\boldsymbol{A}^{-1})^*=(\boldsymbol{A}^*)^{-1}$ 的一个特征值. ■

This shows that $|\boldsymbol{A}+\boldsymbol{E}|=0$, that is to say, $\lambda=-1$ is an eigenvalue of $\boldsymbol{A}$. Therefore, $\dfrac{|\boldsymbol{A}|}{\lambda}=\dfrac{-1}{-1}=1$, i.e. 1 is an eigenvalue of $\boldsymbol{A}^*$, consequently, 1 is an eigenvalue of $(\boldsymbol{A}^{-1})^*=(\boldsymbol{A}^*)^{-1}$. ■

例 9 设 $\boldsymbol{A}$ 为 n 阶方阵, $(\boldsymbol{A}+\boldsymbol{E})^m=\boldsymbol{O}$, m 为正整数, 证明 $\boldsymbol{A}$ 可逆.

Example 9 Let $\boldsymbol{A}$ be a square matrix of order n, $(\boldsymbol{A}+\boldsymbol{E})^m=\boldsymbol{O}$, where m is a positive integer. Prove that $\boldsymbol{A}$ is invertible.

证 设 λ 为 $\boldsymbol{A}$ 的任意特征值, $\boldsymbol{\xi}$ 为对应的特征向量, 则

Proof Let λ be an arbitrary eigenvalue of $\boldsymbol{A}$ and let $\boldsymbol{\xi}$ be the corresponding eigenvector. Then

$$(\boldsymbol{A}+\boldsymbol{E})\boldsymbol{\xi}=(\lambda+1)\boldsymbol{\xi}$$

用 $\boldsymbol{A}+\boldsymbol{E}$ 连续左乘等式两边得到

Premultiplying both sides of the above equation by $\boldsymbol{A}+\boldsymbol{E}$ repeatedly, we have

$$(\boldsymbol{A}+\boldsymbol{E})^m\boldsymbol{\xi}=(\lambda+1)^m\boldsymbol{\xi}$$

因为 $(\boldsymbol{A}+\boldsymbol{E})^m=\boldsymbol{O}$, 且 $\boldsymbol{\xi}\neq\boldsymbol{0}$, 所以有 $(\lambda+1)^m=0$, 得 $\lambda=-1$. 故 -1 是 $\boldsymbol{A}$ 的唯一的特征值. 由定理 3 知, $|\boldsymbol{A}|=(-1)^n\neq 0$, 即可知 $\boldsymbol{A}$ 可逆. ■

Since $(\boldsymbol{A}+\boldsymbol{E})^m=\boldsymbol{O}$ and $\boldsymbol{\xi}\neq\boldsymbol{0}$, we have $(\lambda+1)^m=0$. which implies that $\lambda=-1$. Hence, -1 is the unique eigenvalue of $\boldsymbol{A}$. From Theorem 3 we have $|\boldsymbol{A}|=(-1)^n\neq 0$, that is to say, $\boldsymbol{A}$ is invertible. ■

思 考 题

Questions

1. $\boldsymbol{A}$ 的两个不同特征值能否对应同一个特征向量?

1. Can two distinct eigenvalues of $\boldsymbol{A}$ correspond to the same eigenvector?

2. 设 $\boldsymbol{A}$ 和 $\boldsymbol{B}$ 都是 n 阶方阵, $2n$ 阶分块对角阵 $\boldsymbol{C}=\begin{pmatrix}\boldsymbol{A}&\boldsymbol{O}\\\boldsymbol{O}&\boldsymbol{B}\end{pmatrix}$ 的特征值与 $\boldsymbol{A}$ 和 $\boldsymbol{B}$ 的特征值有什么关系?

2. Let $\boldsymbol{A}$ and B be square matrices of order n. What are the relations between the eigenvalues of the block diagonal matrix of order $2n$, given by $\boldsymbol{C}=\begin{pmatrix}\boldsymbol{A}&\boldsymbol{O}\\\boldsymbol{O}&\boldsymbol{B}\end{pmatrix}$, and the eigenvalues of $\boldsymbol{A}$ and $\boldsymbol{B}$.

3. 若用初等变换将 $\boldsymbol{A}$ 化为 $\boldsymbol{B}$, 则 $\boldsymbol{A}$ 和 $\boldsymbol{B}$ 的特征值是否相同?

3. If $\boldsymbol{A}$ can be reduced to $\boldsymbol{B}$ by elementary operations, whether $\boldsymbol{A}$ and $\boldsymbol{B}$ have the same eigenvalues?

4. 令 λ,μ 分别为 $\boldsymbol{A}$ 和 $\boldsymbol{B}$ 的特征值. 则 $\lambda+\mu$ 是否为 $\boldsymbol{A}+\boldsymbol{B}$ 的特征值?

4. Let λ,μ be eigenvalues of $\boldsymbol{A}$ and $\boldsymbol{B}$, respectively. whether $\lambda+\mu$ is an eigenvalue of $\boldsymbol{A}+\boldsymbol{B}$?

5. 令 $\boldsymbol{p}_1$ 和 $\boldsymbol{p}_2$ 都是 $\boldsymbol{A}$ 的特征值 μ 对应的特征向量, 则对任意实数 k_1 和 k_2. $k_1\boldsymbol{p}_1+k_2\boldsymbol{p}_2$ 是否为 μ 对应的特征向量?

5. Let $\boldsymbol{p}_1$ and $\boldsymbol{p}_2$ be eigenvectors associated with the eigenvalue μ of $\boldsymbol{A}$. Is $k_1\boldsymbol{p}_1+k_2\boldsymbol{p}_2$ an eigenvector associated with μ for any real numbers k_1 and k_2?

5.2 方阵的相似矩阵及对角化

定义 1 设 $\boldsymbol{A}$ 与 $\boldsymbol{B}$ 是 n 阶方阵. 如果存在可逆矩阵 $\boldsymbol{P}$, 使 $\boldsymbol{B}=\boldsymbol{P}^{-1}\boldsymbol{A}\boldsymbol{P}$, 则称 $\boldsymbol{A}$ 与 $\boldsymbol{B}$**相似**, 记为 $\boldsymbol{A}\sim\boldsymbol{B}$.

根据相似矩阵的定义, 容易验证如下两个常用运算表达式成立.

(1)

$$\boldsymbol{P}^{-1}\boldsymbol{A}\boldsymbol{B}\boldsymbol{P}=(\boldsymbol{P}^{-1}\boldsymbol{A}\boldsymbol{P})(\boldsymbol{P}^{-1}\boldsymbol{B}\boldsymbol{P})$$

(2)

$$\boldsymbol{P}^{-1}(k\boldsymbol{A}+l\boldsymbol{B})\boldsymbol{P}=k\boldsymbol{P}^{-1}\boldsymbol{A}\boldsymbol{P}+l\boldsymbol{P}^{-1}\boldsymbol{B}\boldsymbol{P}$$

其中 k,l 为任意实数.

相似是同阶矩阵间的一种等价关系, 即这种关系具有以下性质:

(1) **反身性** 对任意方阵 $\boldsymbol{A}$, 有 $\boldsymbol{A}\sim\boldsymbol{A}$;

(2) **对称性** 若 $\boldsymbol{A}\sim\boldsymbol{B}$, 则 $\boldsymbol{B}\sim\boldsymbol{A}$;

(3) **传递性** 若 $\boldsymbol{A}\sim\boldsymbol{B}$, $\boldsymbol{B}\sim\boldsymbol{C}$, 则 $\boldsymbol{A}\sim\boldsymbol{C}$.

证 (1) 因为 $\boldsymbol{A}=\boldsymbol{E}^{-1}\boldsymbol{A}\boldsymbol{E}$, 所以 $\boldsymbol{A}\sim\boldsymbol{A}$.

5.2 Similar Matrices and Diagonalization of Square Matrices

Definition 1 Let $\boldsymbol{A}$ and $\boldsymbol{B}$ be square matrices of order n. If there exists an invertible matrix $\boldsymbol{P}$ such that $\boldsymbol{B}=\boldsymbol{P}^{-1}\boldsymbol{A}\boldsymbol{P}$, then $\boldsymbol{A}$ is said to be **similar** to $\boldsymbol{B}$, written as $\boldsymbol{A}\sim\boldsymbol{B}$.

In accordance with the definition of similar matrices, it is easy to show that the following two commonly used operation expressions are valid, namely,

(1)

(2)

where k,l are arbitrary real numbers.

Similarity is an equivalence relation among matrices with the same order, namely, it has the following properties.

(1) **Reflexivity** For any square matrix $\boldsymbol{A}$, we have $\boldsymbol{A}\sim\boldsymbol{A}$;

(2) **Symmetry** If $\boldsymbol{A}\sim\boldsymbol{B}$, then $\boldsymbol{B}\sim\boldsymbol{A}$;

(3) **Transitivity** If $\boldsymbol{A}\sim\boldsymbol{B}$ and $\boldsymbol{B}\sim\boldsymbol{C}$, then $\boldsymbol{A}\sim\boldsymbol{C}$.

Proof (1) Since $\boldsymbol{A}=\boldsymbol{E}^{-1}\boldsymbol{A}\boldsymbol{E}$, $\boldsymbol{A}\sim\boldsymbol{A}$ is valid.

(2) 若 $\boldsymbol{A}\sim\boldsymbol{B}$, 则存在可逆矩阵 $\boldsymbol{P}$, 使得 $\boldsymbol{P}^{-1}\boldsymbol{A}\boldsymbol{P}=\boldsymbol{B}$, 于是 $\boldsymbol{A}=(\boldsymbol{P}^{-1})^{-1}\boldsymbol{B}\boldsymbol{P}^{-1}$, 从而 $\boldsymbol{B}\sim\boldsymbol{A}$.

(2) If $\boldsymbol{A}\sim\boldsymbol{B}$, then there exists an invertible matrix $\boldsymbol{P}$ such that $\boldsymbol{P}^{-1}\boldsymbol{A}\boldsymbol{P}=\boldsymbol{B}$. So $\boldsymbol{A}=(\boldsymbol{P}^{-1})^{-1}\boldsymbol{B}\boldsymbol{P}^{-1}$, which implies $\boldsymbol{B}\sim\boldsymbol{A}$.

(3) 若 $\boldsymbol{A}\sim\boldsymbol{B}$, $\boldsymbol{B}\sim\boldsymbol{C}$, 则有可逆矩阵 $\boldsymbol{P}_1$, $\boldsymbol{P}_2$ 使得 $\boldsymbol{P}_1^{-1}\boldsymbol{A}\boldsymbol{P}_1=\boldsymbol{B}$, $\boldsymbol{P}_2^{-1}\boldsymbol{B}\boldsymbol{P}_2=\boldsymbol{C}$, 于是 $\boldsymbol{C}=\boldsymbol{P}_2^{-1}\boldsymbol{P}_1^{-1}\boldsymbol{A}\boldsymbol{P}_1\boldsymbol{P}_2=(\boldsymbol{P}_1\boldsymbol{P}_2)^{-1}\boldsymbol{A}(\boldsymbol{P}_1\boldsymbol{P}_2)$. 因此, $\boldsymbol{A}\sim\boldsymbol{C}$. ▲

(3) If $\boldsymbol{A}\sim\boldsymbol{B}$ and $\boldsymbol{B}\sim\boldsymbol{C}$, then there exist two invertible matrices $\boldsymbol{P}_1$ and $\boldsymbol{P}_2$ such that $\boldsymbol{P}_1^{-1}\boldsymbol{A}\boldsymbol{P}_1=\boldsymbol{B}$, $\boldsymbol{P}_2^{-1}\boldsymbol{B}\boldsymbol{P}_2=\boldsymbol{C}$. So $\boldsymbol{C}=\boldsymbol{P}_2^{-1}\boldsymbol{P}_1^{-1}\boldsymbol{A}\boldsymbol{P}_1\boldsymbol{P}_2=(\boldsymbol{P}_1\boldsymbol{P}_2)^{-1}\boldsymbol{A}(\boldsymbol{P}_1\boldsymbol{P}_2)$, that is, $\boldsymbol{A}\sim\boldsymbol{C}$. ▲

此外, 相似矩阵还有下列简单性质:

Moreover, similar matrices also have the following simple properties:

(1) 若 $\boldsymbol{A}\sim\boldsymbol{B}$, 则 $|\boldsymbol{A}|=|\boldsymbol{B}|$;

(1) If $\boldsymbol{A}\sim\boldsymbol{B}$, then $|\boldsymbol{A}|=|\boldsymbol{B}|$;

(2) 若 $\boldsymbol{A}\sim\boldsymbol{B}$, 则 $\boldsymbol{A}$ 与 $\boldsymbol{B}$ 的特征多项式相同, 从而 $\boldsymbol{A}$ 与 $\boldsymbol{B}$ 有相同的特征值;

(2) If $\boldsymbol{A}\sim\boldsymbol{B}$, then $\boldsymbol{A}$ and $\boldsymbol{B}$ have the same characteristic polynomial, and hence have the same eigenvalues;

(3) 若 $\boldsymbol{A}\sim\boldsymbol{B}$, 则 $\mathrm{tr}(\boldsymbol{A})=\mathrm{tr}(\boldsymbol{B})$;

(3) If $\boldsymbol{A}\sim\boldsymbol{B}$, then $\mathrm{tr}(\boldsymbol{A})=\mathrm{tr}(\boldsymbol{B})$;

(4) 若 $\boldsymbol{A}\sim\mathrm{diag}(\lambda_1,\lambda_2,\cdots,\lambda_n)$, 则 $\lambda_1,\lambda_2,\cdots,\lambda_n$ 是 $\boldsymbol{A}$ 的特征值.

(4) If $\boldsymbol{A}\sim\mathrm{diag}(\lambda_1,\lambda_2,\cdots,\lambda_n)$, then $\lambda_1,\lambda_2,\cdots,\lambda_n$ are eigenvalues of $\boldsymbol{A}$.

证 设 $\boldsymbol{A}\sim\boldsymbol{B}$, 则有可逆矩阵 $\boldsymbol{P}$, 使得 $\boldsymbol{P}^{-1}\boldsymbol{A}\boldsymbol{P}=\boldsymbol{B}$.

Proof If $\boldsymbol{A}\sim\boldsymbol{B}$, then there exists an invertible matrix $\boldsymbol{P}$ such that $\boldsymbol{P}^{-1}\boldsymbol{A}\boldsymbol{P}=\boldsymbol{B}$.

(1) $|\boldsymbol{B}|=\left|\boldsymbol{P}^{-1}\right||\boldsymbol{A}||\boldsymbol{P}|=|\boldsymbol{A}|\left|\boldsymbol{P}^{-1}\right||\boldsymbol{P}|=|\boldsymbol{A}||\boldsymbol{E}|=|\boldsymbol{A}|$.

(2) 由于 $|\lambda\boldsymbol{E}-\boldsymbol{B}|=\left|\lambda\boldsymbol{E}-\boldsymbol{P}^{-1}\boldsymbol{A}\boldsymbol{P}\right|=\left|\boldsymbol{P}^{-1}(\lambda\boldsymbol{E}-\boldsymbol{A})\boldsymbol{P}\right|=|\lambda\boldsymbol{E}-\boldsymbol{A}|$, 所以 $\boldsymbol{A}$ 与 $\boldsymbol{B}$ 有相同的特征值.

(2) Since $|\lambda\boldsymbol{E}-\boldsymbol{B}|=\left|\lambda\boldsymbol{E}-\boldsymbol{P}^{-1}\boldsymbol{A}\boldsymbol{P}\right|=\left|\boldsymbol{P}^{-1}(\lambda\boldsymbol{E}-\boldsymbol{A})\boldsymbol{P}\right|=|\lambda\boldsymbol{E}-\boldsymbol{A}|$, it means that $\boldsymbol{A}$ and $\boldsymbol{B}$ have the same eigenvalues.

(3) 利用矩阵迹和特征值的关系有 $\mathrm{tr}(\boldsymbol{A})=\mathrm{tr}(\boldsymbol{B})$.

(3) From the relation between the trace and the eigenvalues of a matrix, we have $\mathrm{tr}(\boldsymbol{A})=\mathrm{tr}(\boldsymbol{B})$.

(4) 因 $\lambda_1,\lambda_2,\cdots,\lambda_n$ 是 $\mathrm{diag}(\lambda_1,\lambda_2,\cdots,\lambda_n)$ 的 n 个特征值. 由性质 2 可知 $\lambda_1,\lambda_2,\cdots,\lambda_n$ 也就是 $\boldsymbol{A}$ 的特征值.

(4) Obviously, $\lambda_1,\lambda_2,\cdots,\lambda_n$ are n eigenvalues of $\mathrm{diag}(\lambda_1,\lambda_2,\cdots,\lambda_n)$. From Property 2 we know that $\lambda_1,\lambda_2,\cdots,\lambda_n$ are also eigenvalues of $\boldsymbol{A}$.

注 有相同特征多项式的两个矩阵不一定相似. 例如,

Note Two matrices with the same characteristic polynomial may not be similar, for example,

$$\boldsymbol{A}=\begin{pmatrix}1&0\\0&1\end{pmatrix},\quad \boldsymbol{B}=\begin{pmatrix}1&0\\1&1\end{pmatrix}$$

问题 什么条件下方阵相似于一个对角矩阵?

Question What are the conditions such that a square matrix is similar to a diagonal matrix?

若 $\boldsymbol{A}$ 相似于对角矩阵, 则称 $\boldsymbol{A}$ 为**可对角化矩阵**.

If $\boldsymbol{A}$ is similar to a diagonal matrix, then $\boldsymbol{A}$ is said to be a **diagonalizable matrix**.

定理 1 令 $\boldsymbol{A}$ 为 n 阶方阵. $\boldsymbol{A}$ 与对角矩阵 $D=\mathrm{diag}(\lambda_1,\lambda_2,\cdots,\lambda_n)$ 相似的充分必要条件是, $\boldsymbol{A}$ 有 n 个线性无关的特征向量.

Theorem 1 Let $\boldsymbol{A}$ be a square matrix of order n. $\boldsymbol{A}$ is similar to a diagonal matrix $D=\mathrm{diag}(\lambda_1,\lambda_2,\cdots,\lambda_n)$ if and only if $\boldsymbol{A}$ has n linearly independent eigenvectors.

证 **必要性** 若 $\boldsymbol{A}\sim\mathrm{diag}(\lambda_1,\lambda_2,\cdots,\lambda_n)$, 则存在可逆矩阵 $\boldsymbol{P}$, 使

Proof **Necessity** If $\boldsymbol{A}\sim\mathrm{diag}(\lambda_1,\lambda_2,\cdots,\lambda_n)$, then there exists an invertible matrix $\boldsymbol{P}$ such that

$$\boldsymbol{P}^{-1}\boldsymbol{A}\boldsymbol{P}=\mathrm{diag}(\lambda_1,\lambda_2,\cdots,\lambda_n)$$

将 $\boldsymbol{P}$ 按列分块 $\boldsymbol{P}=(\boldsymbol{\alpha}_1,\boldsymbol{\alpha}_2,\cdots,\boldsymbol{\alpha}_n)$, 于是有

Partitioning $\boldsymbol{P}$ by columns, we have $\boldsymbol{P}=(\boldsymbol{\alpha}_1,\boldsymbol{\alpha}_2,\cdots,\boldsymbol{\alpha}_n)$, then

$$\boldsymbol{A}(\boldsymbol{\alpha}_1,\boldsymbol{\alpha}_2,\cdots,\boldsymbol{\alpha}_n)=(\boldsymbol{\alpha}_1,\boldsymbol{\alpha}_2,\cdots,\boldsymbol{\alpha}_n)\begin{pmatrix}\lambda_1&&&\\&\lambda_2&&\\&&\ddots&\\&&&\lambda_n\end{pmatrix}$$

由分块矩阵的乘法

From multiplication of block matrices, we have

$$(\boldsymbol{A}\boldsymbol{\alpha}_1,\boldsymbol{A}\boldsymbol{\alpha}_2,\cdots,\boldsymbol{A}\boldsymbol{\alpha}_n)=(\lambda_1\boldsymbol{\alpha}_1,\lambda_2\boldsymbol{\alpha}_2,\cdots,\lambda_n\boldsymbol{\alpha}_n)$$

所以

Hence

$$\boldsymbol{A}\boldsymbol{\alpha}_i=\lambda_i\boldsymbol{\alpha}_i\quad(i=1,2,\cdots,n)$$

因为 $\boldsymbol{P}$ 可逆, 所以向量组 $\{\boldsymbol{\alpha}_1, \boldsymbol{\alpha}_2, \cdots, \boldsymbol{\alpha}_n\}$ 线性无关, 因此, $\boldsymbol{A}$ 有 n 个线性无关的特征向量.

Since $\boldsymbol{P}$ is invertible, the vector set $\{\boldsymbol{\alpha}_1, \boldsymbol{\alpha}_2, \cdots, \boldsymbol{\alpha}_n\}$ is linearly independent. Therefore, $\boldsymbol{A}$ has n linearly independent eigenvectors.

充分性 设 $\boldsymbol{\alpha}_1, \boldsymbol{\alpha}_2, \cdots, \boldsymbol{\alpha}_n$ 分别是 $\boldsymbol{A}$ 的属于特征值 $\lambda_1, \lambda_2, \cdots, \lambda_n$ 的特征向量, 则有 $\boldsymbol{A}\boldsymbol{\alpha}_i = \lambda_i\boldsymbol{\alpha}_i (i = 1, 2, \cdots, n)$. 因为 $\{\boldsymbol{\alpha}_1, \boldsymbol{\alpha}_2, \cdots, \boldsymbol{\alpha}_n\}$ 线性无关, 所以矩阵

Sufficiency Let $\boldsymbol{\alpha}_1, \boldsymbol{\alpha}_2, \cdots, \boldsymbol{\alpha}_n$ be eigenvectors associated with the eigenvalues $\lambda_1, \lambda_2, \cdots, \lambda_n$ of $\boldsymbol{A}$, respectively. Then $\boldsymbol{A}\boldsymbol{\alpha}_i = \lambda_i\boldsymbol{\alpha}_i (i = 1, 2, \cdots, n)$. Since $\{\boldsymbol{\alpha}_1, \boldsymbol{\alpha}_2, \cdots, \boldsymbol{\alpha}_n\}$ are linearly independent, the matrix

$$\boldsymbol{P} = (\boldsymbol{\alpha}_1, \boldsymbol{\alpha}_2, \cdots, \boldsymbol{\alpha}_n)$$

可逆. 进而得到

is invertible. It follows that

$$\begin{aligned}\boldsymbol{AP} &= \boldsymbol{A}(\boldsymbol{\alpha}_1, \boldsymbol{\alpha}_2, \cdots, \boldsymbol{\alpha}_n) = (\boldsymbol{A}\boldsymbol{\alpha}_1, \boldsymbol{A}\boldsymbol{\alpha}_2, \cdots, \boldsymbol{A}\boldsymbol{\alpha}_n) = (\lambda_1\boldsymbol{\alpha}_1, \lambda_2\boldsymbol{\alpha}_2, \cdots, \lambda_n\boldsymbol{\alpha}_n) \\ &= (\boldsymbol{\alpha}_1, \boldsymbol{\alpha}_2, \cdots, \boldsymbol{\alpha}_n)\begin{pmatrix} \lambda_1 & & & \\ & \lambda_2 & & \\ & & \ddots & \\ & & & \lambda_n \end{pmatrix} = \boldsymbol{P}\begin{pmatrix} \lambda_1 & & & \\ & \lambda_2 & & \\ & & \ddots & \\ & & & \lambda_n \end{pmatrix}\end{aligned}$$

因此,

As a result,

$$\boldsymbol{P}^{-1}\boldsymbol{AP} = \mathrm{diag}(\lambda_1, \lambda_2, \cdots, \lambda_n)$$ ▲

注意到属于不同特征值的特征向量是线性无关的, 所以 n 阶方阵如果没有重特征值, 即都是单特征值, 那么它一定有 n 个线性无关的特征向量, 因此它一定相似于对角矩阵.

Note that the eigenvectors associated with distinct eigenvalues are linearly independent. Therefore, if a square matrix of order n has no repeated eigenvalues, that is, they are single eigenvalues, then it must have n linearly independent eigenvectors. Hence it must be similar to a diagonal matrix.

推论 1 若 n 阶方阵有 n 个互不相同的特征值, 则它一定相似于对角矩阵.

Corollary 1 If a square matrix of order n has n distinct eigenvalues, then it must be similar to a diagonal matrix.

注 推论 1 是方阵可对角化的充分条件而不是必要条件.

Note Corollary 1 is not a necessary condition but a sufficient condition to determine whether a square matrix is diagonalizable.

定理 2 令 $\boldsymbol{A}$ 为 n 阶方阵. $\boldsymbol{A}$ 与对角矩阵相似的充分必要条件是对每一个 n_i 重特征值 $\lambda_i \in P$, 都有 $R(\lambda_i \boldsymbol{E} - \boldsymbol{A}) = n - n_i$, 即 $(\lambda_i \boldsymbol{E} - \boldsymbol{A})\boldsymbol{x} = \boldsymbol{0}$ 基础解系所含向量的个数等于特征值 λ_i 的重数.

Theorem 2 Let $\boldsymbol{A}$ be a square matrix of order n. $\boldsymbol{A}$ is similar to a diagonal matrix if and only if $R(\lambda_i \boldsymbol{E} - \boldsymbol{A}) = n - n_i$ for any eigenvalue $\lambda_i \in P$ of multiplicity n_i, that is, the number of the vectors contained in the system of fundamental solutions of $(\lambda_i \boldsymbol{E} - \boldsymbol{A})\boldsymbol{x} = \boldsymbol{0}$ is equal to the multiplicity of the eigenvalue λ_i.

证明可参考 5.1 节定理 7 和本节定理 1.

The proof can refer to Theorem 7 in Section 5.1 and Theorem 1 in this Section.

在 5.1 节的例 4 中, 矩阵 $\boldsymbol{A}$ 的特征值 $\lambda_1 = 2$, $\lambda_2 = \lambda_3 = 1$. 显然, 1 是 $\boldsymbol{A}$ 的二重特征值. 因为 $R(\boldsymbol{E} - \boldsymbol{A}) = 2$, $n - R(\boldsymbol{E} - \boldsymbol{A}) = 3 - 2 = 1$, 即 $(\boldsymbol{E} - \boldsymbol{A})\boldsymbol{x} = \boldsymbol{0}$ 基础解系仅含 1 个向量, 不满足定理 7 的条件, 所以 $\boldsymbol{A}$ 不与对角矩阵相似.

In Example 4 of Section 5.1, the eigenvalues of the matrix, given by $\boldsymbol{A}$, are $\lambda_1 = 2$, $\lambda_2 = \lambda_3 = 1$. Obviously, 1 is an eigenvalue of multiplicity 2 of $\boldsymbol{A}$. Since $R(\boldsymbol{E} - \boldsymbol{A}) = 2$, $n - R(\boldsymbol{E} - \boldsymbol{A}) = 3 - 2 = 1$, namely, the system of fundamental solutions of $(\boldsymbol{E} - \boldsymbol{A})\boldsymbol{x} = \boldsymbol{0}$ contains only one vector, which does not satisfy the condition in Theorem 7, so $\boldsymbol{A}$ is not similar to a diagonal matrix.

在 5.1 节的例 5 中, 矩阵 $\boldsymbol{A}$ 的特征值 $\lambda_1 = -2$, $\lambda_2 = \lambda_3 = 1$. 显然, 1 也是 $\boldsymbol{A}$ 的 2 重特征值. 因为 $R(\boldsymbol{E} - \boldsymbol{A}) = 1$, $n - R(\boldsymbol{E} - \boldsymbol{A}) = 3 - 1 = 2$, 即 $(\boldsymbol{E} - \boldsymbol{A})\boldsymbol{x} = \boldsymbol{0}$ 的基础解系含 2 个向量, 满足定理 7 的条件, 所以 $\boldsymbol{A}$ 与对角矩阵相似.

In Example 5 of Section 5.1, the eigenvalues of the matrix $\boldsymbol{A}$, are $\lambda_1 = -2$, $\lambda_2 = \lambda_3 = 1$. Obviously, 1 is also an eigenvalue of multiplicity 2. Since $R(\boldsymbol{E} - \boldsymbol{A}) = 1$, $n - R(\boldsymbol{E} - \boldsymbol{A}) = 3 - 1 = 2$, namely, the system of fundamental solutions of $(\boldsymbol{E} - \boldsymbol{A})\boldsymbol{x} = \boldsymbol{0}$ contains 2 vectors, which satisfies the condition in Theorem 7, so $\boldsymbol{A}$ is similar to a diagonal matrix.

因此, 将方阵 $\boldsymbol{A}$ 对角化的方法如下:

Consequently, the approach for diagonalizing a matrix $\boldsymbol{A}$ is given as follows:

第一步 求出 $\boldsymbol{A}$ 的所有的特征值 $\lambda_1, \lambda_2, \cdots, \lambda_s (s \leqslant n)$.

Step 1 Find all the eigenvalues $\lambda_1, \lambda_2, \cdots, \lambda_s (s \leqslant n)$ of $\boldsymbol{A}$.

第二步 求出 $(\lambda_i \boldsymbol{E} - \boldsymbol{A})\boldsymbol{x} = \boldsymbol{0}$ 的基础解系 $(i = 1, 2, \cdots, s)$, 即特征值 $\lambda_1, \lambda_2, \cdots, \lambda_s$ 所对应的线性无关的特征向量, 设为

Step 2 Find the system of fundamental solutions of $(\lambda_i \boldsymbol{E} - \boldsymbol{A})\boldsymbol{x} = \boldsymbol{0} (i = 1, 2, \cdots, s)$, which are the linearly independent eigenvectors associated with $\lambda_1, \lambda_2, \cdots, \lambda_s$, written as

$$\boldsymbol{\alpha}_1, \boldsymbol{\alpha}_2, \cdots, \boldsymbol{\alpha}_{t_1}, \boldsymbol{\beta}_1, \boldsymbol{\beta}_2, \cdots, \boldsymbol{\beta}_{t_2}, \cdots, \boldsymbol{\gamma}_1, \boldsymbol{\gamma}_2, \cdots, \boldsymbol{\gamma}_{t_s}$$

其中 $t_1 + t_2 + \cdots + t_s \leqslant n$. 若 $t_1 + t_2 + \cdots + t_s = n$, 根据定理 1 知, 矩阵 $\boldsymbol{A}$ 可对角化; 若 $t_1 + t_2 + \cdots + t_s < n$, 则 $\boldsymbol{A}$ 不能对角化.

where $t_1 + t_2 + \cdots + t_s \leqslant n$. If $t_1 + t_2 + \cdots + t_s = n$, from Theorem 1 we know that $\boldsymbol{A}$ is diagonalizable; while if $t_1 + t_2 + \cdots + t_s < n$, then $\boldsymbol{A}$ can not be diagonalizable.

第三步 令

Step 3 Let

$$\boldsymbol{P} = (\boldsymbol{\alpha}_1, \boldsymbol{\alpha}_2, \cdots, \boldsymbol{\alpha}_{t_1}, \boldsymbol{\beta}_1, \boldsymbol{\beta}_2, \cdots, \boldsymbol{\beta}_{t_2}, \cdots, \boldsymbol{\gamma}_1, \boldsymbol{\gamma}_2, \cdots, \boldsymbol{\gamma}_{t_s})$$

则有

then

$$\boldsymbol{P}^{-1}\boldsymbol{A}\boldsymbol{P} = \boldsymbol{\Lambda}$$

其中 $\boldsymbol{\Lambda}$ 是对角矩阵, 且 $\boldsymbol{\Lambda}$ 的对角线元素为特征值, 它们依次对应于如下的线性无关的特征向量

where $\boldsymbol{\Lambda}$ is a diagonal matrix, and the diagonal elements of $\boldsymbol{\Lambda}$ are eigenvalues, which in turn correspond to the following linearly independent eigenvectors

$$\boldsymbol{\alpha}_1, \boldsymbol{\alpha}_2, \cdots, \boldsymbol{\alpha}_{t_1}, \boldsymbol{\beta}_1, \boldsymbol{\beta}_2, \cdots, \boldsymbol{\beta}_{t_2}, \cdots, \boldsymbol{\gamma}_1, \boldsymbol{\gamma}_2, \cdots, \boldsymbol{\gamma}_{t_s}$$

例 1 设

Example 1 Suppose that

$$\boldsymbol{A} = \begin{pmatrix} 1 & -3 & 3 \\ 3 & -5 & 3 \\ 6 & -6 & 4 \end{pmatrix}$$

判断矩阵 $\boldsymbol{A}$ 是否可对角化? 若可对角化, 试求出可逆矩阵 $\boldsymbol{P}$, 使得 $\boldsymbol{P}^{-1}\boldsymbol{A}\boldsymbol{P}$ 为对角矩阵.

Determine whether $\boldsymbol{A}$ is diagonalizable. If so, find an invertible matrix $\boldsymbol{P}$ such that $\boldsymbol{P}^{-1}\boldsymbol{A}\boldsymbol{P}$ is diagonal.

解 矩阵 $\boldsymbol{A}$ 的特征方程为

Solution The characteristic equation of the matrix $\boldsymbol{A}$ is

$$|\lambda \boldsymbol{E} - \boldsymbol{A}| = \begin{vmatrix} \lambda - 1 & 3 & -3 \\ -3 & \lambda + 5 & -3 \\ -6 & 6 & \lambda - 4 \end{vmatrix} = \begin{vmatrix} \lambda - 1 & 3 & -3 \\ -\lambda - 2 & \lambda + 2 & 0 \\ -6 & 6 & \lambda - 4 \end{vmatrix}$$

$$=(\lambda+2)\begin{vmatrix}\lambda-1 & 3 & -3\\ -1 & 1 & 0\\ -6 & 6 & \lambda-4\end{vmatrix}=(\lambda+2)^2(\lambda-4)=0$$

易见, $\boldsymbol{A}$ 的特征值为 $\lambda_1=\lambda_2=-2,\lambda_3=4$.

Obviously, the eigenvalues of $\boldsymbol{A}$ are $\lambda_1=\lambda_2=-2,\lambda_3=4$.

对于 $\lambda_1=\lambda_2=-2$, 解如下的线性方程组

For $\lambda_1=\lambda_2=-2$, solve the following linear system

$$(\lambda_1\boldsymbol{E}-\boldsymbol{A})\boldsymbol{X}=(-2\boldsymbol{E}-\boldsymbol{A})\boldsymbol{X}=\begin{pmatrix}-3 & 3 & -3\\ -3 & 3 & -3\\ -6 & 6 & -6\end{pmatrix}\begin{pmatrix}x_1\\ x_2\\ x_3\end{pmatrix}=\begin{pmatrix}0\\ 0\\ 0\end{pmatrix}$$

由于

Since

$$\begin{pmatrix}-3 & 3 & -3\\ -3 & 3 & -3\\ -6 & 6 & -6\end{pmatrix}\to\begin{pmatrix}1 & -1 & 1\\ 0 & 0 & 0\\ 0 & 0 & 0\end{pmatrix}$$

求得基础解系为

we get the system of fundamental solutions, given by

$$\boldsymbol{\alpha}_1=\begin{pmatrix}1\\ 1\\ 0\end{pmatrix},\quad \boldsymbol{\alpha}_2=\begin{pmatrix}-1\\ 0\\ 1\end{pmatrix}$$

对于 $\lambda_3=4$, 解如下的线性方程组

For $\lambda_3=4$, solve the following linear system

$$(\lambda_3\boldsymbol{E}-\boldsymbol{A})\boldsymbol{X}=(4\boldsymbol{E}-\boldsymbol{A})\boldsymbol{X}=\begin{pmatrix}3 & 3 & -3\\ -3 & 9 & -3\\ -6 & 6 & 0\end{pmatrix}\begin{pmatrix}x_1\\ x_2\\ x_3\end{pmatrix}=\begin{pmatrix}0\\ 0\\ 0\end{pmatrix}$$

由于

Since

$$\begin{pmatrix}3 & 3 & -3\\ -3 & 9 & -3\\ -6 & 6 & 0\end{pmatrix}\to\begin{pmatrix}1 & 1 & -1\\ 0 & 1 & -\frac{1}{2}\\ 0 & 0 & 0\end{pmatrix}\to\begin{pmatrix}1 & 0 & -\frac{1}{2}\\ 0 & 1 & -\frac{1}{2}\\ 0 & 0 & 0\end{pmatrix}$$

求得基础解系为

we obtain the system of fundamental solutions, given by

$$\boldsymbol{\alpha}_3 = \begin{pmatrix} 1 \\ 1 \\ 2 \end{pmatrix}$$

因此, 3 阶矩阵 $\boldsymbol{A}$ 有 3 个线性无关的特征向量 $\boldsymbol{\alpha}_1, \boldsymbol{\alpha}_2, \boldsymbol{\alpha}_3$, 故 $\boldsymbol{A}$ 可对角化. 令

Therefore, the matrix $\boldsymbol{A}$ of order 3 has three linearly independent eigenvectors $\boldsymbol{\alpha}_1, \boldsymbol{\alpha}_2, \boldsymbol{\alpha}_3$, that is to say, $\boldsymbol{A}$ is diagonalizable. Let

$$\boldsymbol{P} = (\boldsymbol{\alpha}_1, \boldsymbol{\alpha}_2, \boldsymbol{\alpha}_3) = \begin{pmatrix} 1 & -1 & 1 \\ 1 & 0 & 1 \\ 0 & 1 & 2 \end{pmatrix}$$

则

then

$$\boldsymbol{P}^{-1}\boldsymbol{A}\boldsymbol{P} = \begin{pmatrix} -2 & 0 & 0 \\ 0 & -2 & 0 \\ 0 & 0 & 4 \end{pmatrix}$$

例 2 设 $\boldsymbol{\alpha}$ 是 $\boldsymbol{A}$ 的一个特征向量, 其中

Example 2 Let $\boldsymbol{\alpha}$ be an eigenvector of $\boldsymbol{A}$, where

$$\boldsymbol{\alpha} = \begin{pmatrix} 1 \\ 1 \\ -1 \end{pmatrix}, \quad \boldsymbol{A} = \begin{pmatrix} 2 & -1 & 2 \\ 5 & a & 3 \\ -1 & b & -2 \end{pmatrix}$$

(1) 求 a, b 的值及 $\boldsymbol{A}$ 的与特征向量 $\boldsymbol{\alpha}$ 对应的特征值;

(1) Find the values of a, b and the eigenvalue of $\boldsymbol{A}$ corresponding to the eigenvector $\boldsymbol{\alpha}$;

(2) $\boldsymbol{A}$ 与对角阵是否相似?

(2) Is $\boldsymbol{A}$ similar to a diagonal matrix?

解 (1) 设 λ 为 $\boldsymbol{A}$ 的与特征向量 $\boldsymbol{\alpha}$ 相对应的特征值, $(\boldsymbol{A} - \lambda\boldsymbol{E})\,\boldsymbol{\alpha} = \boldsymbol{0}$, 即

Solution (1) Suppose that λ is an eigenvalue of $\boldsymbol{A}$ corresponding to the eigenvector $\boldsymbol{\alpha}$, we have $(\boldsymbol{A} - \lambda\boldsymbol{E})\,\boldsymbol{\alpha} = \boldsymbol{0}$, that is,

$$\begin{pmatrix} 2-\lambda & -1 & 2 \\ 5 & a-\lambda & 3 \\ -1 & b & -2-\lambda \end{pmatrix} \begin{pmatrix} 1 \\ 1 \\ -1 \end{pmatrix} = \begin{pmatrix} 0 \\ 0 \\ 0 \end{pmatrix}$$

得到如下方程组

it leads to the following system

$$\begin{cases} -\lambda - 1 = 0 \\ -\lambda + a + 2 = 0 \\ \lambda + b + 1 = 0 \end{cases}$$

解得

The solution is given by

$$\begin{cases} \lambda = -1 \\ a = -3 \\ b = 0 \end{cases}$$

(2) 由于

(2) Since

$$|\boldsymbol{A} - \lambda \boldsymbol{E}| = \begin{vmatrix} 2-\lambda & -1 & 2 \\ 5 & -3-\lambda & 3 \\ -1 & 0 & -2-\lambda \end{vmatrix} = -(\lambda+1)^3 = 0$$

可知 $\boldsymbol{A}$ 有三重特征值

we know that $\boldsymbol{A}$ has an eigenvalue of multiplicity 3, given by

$$\lambda_1 = \lambda_2 = \lambda_3 = -1$$

由于

Since

$$\boldsymbol{A} + \boldsymbol{E} = \begin{pmatrix} 3 & -1 & 2 \\ 5 & -2 & 3 \\ -1 & 0 & -1 \end{pmatrix} \to \begin{pmatrix} 1 & 0 & 1 \\ 0 & 1 & 1 \\ 0 & 0 & 0 \end{pmatrix}$$

易见,

Obviously,

$$R(\boldsymbol{A} + \boldsymbol{E}) = 2, \quad n - R(\boldsymbol{A} + \boldsymbol{E}) = 3 - 2 = 1$$

故 $\boldsymbol{A}$ 的三重特征值 $\lambda = -1$ 对应的线性无关的特征向量仅有 1 个. 所以 $\boldsymbol{A}$ 不与对角阵相似. ■

So the eigenvalue $\lambda = -1$ of multiplicity 3 of $\boldsymbol{A}$ has only 1 linearly independent eigenvector. Hence $\boldsymbol{A}$ is not similar to a diagonal matrix. ■

例 3 设

Example 3 Let

$$\boldsymbol{A} = \begin{pmatrix} 0 & 0 & 1 \\ 1 & 1 & a \\ 1 & 0 & 0 \end{pmatrix}$$

问 a 为何值时, 矩阵 $\boldsymbol{A}$ 能对角化?

Find the value of a such that $\boldsymbol{A}$ is diagonalizable.

解 由于

Solution Since

$$|\lambda\boldsymbol{E}-\boldsymbol{A}|=\begin{vmatrix}\lambda & 0 & -1\\ -1 & \lambda-1 & -a\\ -1 & 0 & \lambda\end{vmatrix}=(\lambda-1)^2(\lambda+1)=0$$

可得 $\lambda_1=-1,\lambda_2=\lambda_3=1$.

we have $\lambda_1=-1$, $\lambda_2=\lambda_3=1$

对于单根 $\lambda_1=-1$, 仅对应于一个线性无关的特征向量. 对于二重根 $\lambda_2=\lambda_3=1$, 欲使 $\boldsymbol{A}$ 能够对角化, $\boldsymbol{A}$ 应有两个线性无关的特征向量, 也就是说, 方程组 $(\boldsymbol{E}-\boldsymbol{A})\boldsymbol{x}=\boldsymbol{0}$ 有两个线性无关的解, 即 $R(\boldsymbol{E}-\boldsymbol{A})=1$. 由于

For the single root $\lambda_1=-1$, it corresponds to only one linearly independent eigenvector. For the double root $\lambda_2=\lambda_3=1$, in order to make $\boldsymbol{A}$ be diagonalizable, $\boldsymbol{A}$ must have two linearly independent eigenvectors, that is, the system $(\boldsymbol{E}-\boldsymbol{A})\boldsymbol{x}=\boldsymbol{0}$ has two linear independent solutions, which implies $R(\boldsymbol{E}-\boldsymbol{A})=1$. Since

$$\boldsymbol{E}-\boldsymbol{A}=\begin{pmatrix}1 & 0 & -1\\ -1 & 0 & -a\\ -1 & 0 & 1\end{pmatrix}\rightarrow\begin{pmatrix}1 & 0 & -1\\ 0 & 0 & a+1\\ 0 & 0 & 0\end{pmatrix}$$

要使 $R(\boldsymbol{E}-\boldsymbol{A})=1$, 有 $a+1=0$, 即 $a=-1$. 因此, 当 $a=-1$ 时, 矩阵 $\boldsymbol{A}$ 可以对角化.■

To make $R(\boldsymbol{E}-\boldsymbol{A})=1$, we have $a+1=0$, namely, $a=-1$. So $\boldsymbol{A}$ is diagonalizable as $a=-1$. ■

例 4 设

Example 4 Suppose that

$$\boldsymbol{A}=\begin{pmatrix}2 & -1\\ -1 & 2\end{pmatrix}$$

求 $\boldsymbol{A}^n$.

Find $\boldsymbol{A}^n$.

解 由于

Solution Since

$$|\boldsymbol{A}-\lambda\boldsymbol{E}|=\begin{vmatrix}2-\lambda & -1\\ -1 & 2-\lambda\end{vmatrix}=\lambda^2-4\lambda+3=(\lambda-1)(\lambda-3)=0$$

$\boldsymbol{A}$ 的特征值为 $\lambda_1=1,\lambda_2=3$.

the eigenvalues of $\boldsymbol{A}$ are $\lambda_1=1,\lambda_2=3$.

根据定理 1 的推论 1 可知, 存在可逆矩阵 $\boldsymbol{P}$ 及对角阵 $\boldsymbol{\Lambda}$, 使 $\boldsymbol{P}^{-1}\boldsymbol{A}\boldsymbol{P}=\boldsymbol{\Lambda}$. 于是 $\boldsymbol{A}=\boldsymbol{P}\boldsymbol{\Lambda}\boldsymbol{P}^{-1}$, 进而 $\boldsymbol{A}^n=\boldsymbol{P}\boldsymbol{\Lambda}^n\boldsymbol{P}^{-1}$. 显然,

From Corollary 1 of Theorem 1 we know that there exists an invertible matrix $\boldsymbol{P}$ and a diagonal matrix $\boldsymbol{\Lambda}$ such that $\boldsymbol{P}^{-1}\boldsymbol{A}\boldsymbol{P}=\boldsymbol{\Lambda}$. So $\boldsymbol{A}=\boldsymbol{P}\boldsymbol{\Lambda}\boldsymbol{P}^{-1}$ and then $\boldsymbol{A}^n=\boldsymbol{P}\boldsymbol{\Lambda}^n\boldsymbol{P}^{-1}$. Obviously,

$$\boldsymbol{\Lambda}=\begin{pmatrix}1&0\\0&3\end{pmatrix},\quad \boldsymbol{\Lambda}^n=\begin{pmatrix}1&0\\0&3^n\end{pmatrix}$$

对于 $\lambda_1=1$, 由 $(\boldsymbol{A}-\boldsymbol{E})\boldsymbol{x}=\boldsymbol{0}$ 解得对应特征向量

For $\lambda_1=1$, using $(\boldsymbol{A}-\boldsymbol{E})\boldsymbol{x}=\boldsymbol{0}$, we get the corresponding eigenvector

$$\boldsymbol{\alpha}_1=\begin{pmatrix}1\\1\end{pmatrix}$$

对于 $\lambda_2=3$, 由 $(\boldsymbol{A}-3\boldsymbol{E})\boldsymbol{x}=\boldsymbol{0}$ 解得对应特征向量

For $\lambda_2=3$, using $(\boldsymbol{A}-3\boldsymbol{E})\boldsymbol{x}=\boldsymbol{0}$, we get the corresponding eigenvector

$$\boldsymbol{\alpha}_2=\begin{pmatrix}1\\-1\end{pmatrix}$$

令

Let

$$\boldsymbol{P}=(\boldsymbol{\alpha}_1,\boldsymbol{\alpha}_2)=\begin{pmatrix}1&1\\1&-1\end{pmatrix}$$

求得

We obtain

$$\boldsymbol{P}^{-1}=\frac{1}{2}\begin{pmatrix}1&1\\1&-1\end{pmatrix}$$

于是

Consequently

$$\boldsymbol{A}^n=\boldsymbol{P}\boldsymbol{\Lambda}^n\boldsymbol{P}^{-1}=\frac{1}{2}\begin{pmatrix}1&1\\1&-1\end{pmatrix}\begin{pmatrix}1&0\\0&3^n\end{pmatrix}\begin{pmatrix}1&1\\1&-1\end{pmatrix}=\frac{1}{2}\begin{pmatrix}1+3^n&1-3^n\\1-3^n&1+3^n\end{pmatrix}$$ ■

例 5 已知

Example 5 Known that

$$A=\begin{pmatrix}-2&0&0\\2&a&2\\3&1&1\end{pmatrix},\quad B=\begin{pmatrix}-1&0&0\\0&2&0\\0&0&b\end{pmatrix}$$

若 $\boldsymbol{A}$ 与 $\boldsymbol{B}$ 相似, 求 a, b 的值及矩阵 $\boldsymbol{P}$, 使得 $\boldsymbol{P}^{-1}\boldsymbol{AP}=\boldsymbol{B}$.

If $\boldsymbol{A}$ and $\boldsymbol{B}$ are similar, find the values of a, b and the matrix $\boldsymbol{P}$ such that $\boldsymbol{P}^{-1}\boldsymbol{AP}=\boldsymbol{B}$.

解 由相似矩阵的必要条件 $\mathrm{tr}(\boldsymbol{A})=\mathrm{tr}(\boldsymbol{B})$ 和 $|\boldsymbol{A}|=|\boldsymbol{B}|$ 只能得到方程 $a=b+2$. 必须寻找第二个方程来进一步确定 a, b 的值. 因为相似的矩阵具有相同的特征值, -1 是 $\boldsymbol{B}$ 的特征值, 所以 -1 也是 $\boldsymbol{A}$ 的特征值, 即

Solution The necessary conditions of similar matrices, $\mathrm{tr}(\boldsymbol{A})=\mathrm{tr}(\boldsymbol{B})$ and $|\boldsymbol{A}|=|\boldsymbol{B}|$, can only lead to the equaiton $a=b+2$. So we must find the second equation to further determine a and b. Obviously, -1 is an eigenvalue of $\boldsymbol{B}$. Since the similar matrices have the same eigenvalues, -1 is also an eigenvalue of $\boldsymbol{A}$, that is,

$$|\boldsymbol{A}-(-1)\boldsymbol{E}|=\begin{vmatrix}-1&0&0\\2&a+1&2\\3&1&2\end{vmatrix}=-2a=0$$

所以有

So we have

$$a=0,\quad b=-2$$

于是, $\boldsymbol{A}$ 的三个特征值为 $\lambda_1=-1, \lambda_2=2, \lambda_3=-2$, 解得对应的特征向量分别为

Therefore, the three eigenvalues of $\boldsymbol{A}$ are $\lambda_1=-1, \lambda_2=2, \lambda_3=-2$, and the corresponding eigenvectors are

$$\boldsymbol{\alpha}_1=\begin{pmatrix}0\\-2\\1\end{pmatrix},\quad \boldsymbol{\alpha}_2=\begin{pmatrix}0\\1\\1\end{pmatrix},\quad \boldsymbol{\alpha}_3=\begin{pmatrix}-1\\0\\1\end{pmatrix}$$

令

Let

$$\boldsymbol{P}=(\boldsymbol{\alpha}_1,\boldsymbol{\alpha}_2,\boldsymbol{\alpha}_3)=\begin{pmatrix}0&0&-1\\-2&1&0\\1&1&1\end{pmatrix}$$

则必有

Then

$$P^{-1}AP = B$$ ■

思 考 题

Questions

1. 设 A 为 n 阶方阵. 若存在可逆阵 P, 使得 $P^{-1}AP = \mathrm{diag}(\lambda_1, \lambda_2, \cdots, \lambda_n)$, 则 P 是否唯一? 为什么?

1. Let A be a square matrix of order n. If there exists an invertible matrix P such that $P^{-1}AP = \mathrm{diag}(\lambda_1, \lambda_2, \cdots, \lambda_n)$, is P unique? Why?

2. 举例说明特征值完全相同的两个矩阵不一定相似.

2. Illustrate by an example that two matrices with the same eigenvalues may not be similar.

3. 有人断定方阵 A 的秩等于其非零特征值的个数, 这个结论是否成立? 若不成立, 加上什么条件就一定成立?

3. Someone concludes that the rank of a square matrix A is equal to the number of the nonzero eigenvalues, whether the result is valid? If not, what conditions should be appended?

5.3 向量的内积

5.3 Inner Product of Vectors

在空间解析几何中, 向量的内积可以用于描述向量的度量性质, 如长度、夹角等. 由内积的定义

In Space Analytic Geometry, the inner product of vectors can be used to describe the metrical properties of vectors, such as length and angle, and so on. From the definition of an inner product

$$\boldsymbol{x} \cdot \boldsymbol{y} = |\boldsymbol{x}|\,|\boldsymbol{y}|\cos\theta$$

可得

we have

$$|\boldsymbol{x}| = \sqrt{\boldsymbol{x} \cdot \boldsymbol{x}}, \quad \cos\theta = \frac{\boldsymbol{x} \cdot \boldsymbol{y}}{|\boldsymbol{x}|\,|\boldsymbol{y}|}$$

在直角坐标系中, 有

Based on the system of rectangular coordinates, we have

$$(x_1, x_2, x_3) \cdot (y_1, y_2, y_3) = x_1y_1 + x_2y_2 + x_3y_3.$$

将上述三维向量的内积概念可以推广到 n 维向量上.

The above concept on an inner product for 3 dimensional vectors can be generalized to the case of n dimensional vectors.

定义 1 给定两个 n 维向量

Definition 1 For the two given n-dimensional vectors

$$\boldsymbol{x}=\begin{pmatrix}x_1\\x_2\\\vdots\\x_n\end{pmatrix},\quad \boldsymbol{y}=\begin{pmatrix}y_1\\y_2\\\vdots\\y_n\end{pmatrix}$$

表示式

the expression

$$[\boldsymbol{x},\boldsymbol{y}]=x_1y_1+x_2y_2+\cdots+x_ny_n$$

称为 $\boldsymbol{x}$ 与 $\boldsymbol{y}$ 的**内积**.

is called an **inner product** of $\boldsymbol{x}$ and $\boldsymbol{y}$.

内积是向量的一种运算, 用矩阵形式可表为 $[\boldsymbol{x},\boldsymbol{y}]=\boldsymbol{x}^{\mathrm{T}}\boldsymbol{y}$.

The inner product is an operation of vectors, in matrix form, it can be expressed by $[\boldsymbol{x},\boldsymbol{y}]=\boldsymbol{x}^{\mathrm{T}}\boldsymbol{y}$.

例 1 计算 $[\boldsymbol{x},\boldsymbol{y}]$, 其中 $\boldsymbol{x},\boldsymbol{y}$ 如下

Example 1 Calculate $[\boldsymbol{x},\boldsymbol{y}]$, where

$$(1)\ \boldsymbol{x}=\begin{pmatrix}0\\1\\5\\-2\end{pmatrix},\ \boldsymbol{y}=\begin{pmatrix}-2\\0\\-1\\3\end{pmatrix};\qquad (2)\ \boldsymbol{x}=\begin{pmatrix}-2\\1\\0\\3\end{pmatrix},\ \boldsymbol{y}=\begin{pmatrix}3\\-6\\8\\4\end{pmatrix}.$$

解

Solution

(1) $[\boldsymbol{x},\boldsymbol{y}]=0\cdot(-2)+1\cdot0+5\cdot(-1)+(-2)\cdot3=-11.$

(2) $[\boldsymbol{x},\boldsymbol{y}]=(-2)\cdot3+1\cdot(-6)+0\cdot8+3\cdot4=0.$ ■

若 $\boldsymbol{x},\boldsymbol{y},\boldsymbol{z}$ 为 n 维实向量, λ 为实数, 则从内积的定义可立刻推得下列性质:

If $\boldsymbol{x},\boldsymbol{y},\boldsymbol{z}$ are n dimensional real vectors, λ a is real number, then we can get the following properties from the definition of the inner product:

(1) $[\boldsymbol{x},\boldsymbol{y}]=[\boldsymbol{y},\boldsymbol{x}]$;

(2) $[\lambda\boldsymbol{x},\boldsymbol{y}]=\lambda[\boldsymbol{x},\boldsymbol{y}]$;

(3) $[\boldsymbol{x}+\boldsymbol{y},\boldsymbol{z}]=[\boldsymbol{x},\boldsymbol{z}]+[\boldsymbol{y},\boldsymbol{z}]$.

同三维向量空间一样, 可用内积的运算定义 n 维向量的长度和夹角.

Similar to the 3 dimensional vector space, we can define the length and the angle of n dimensional vectors by the operation of an inner product.

定义 2 称 $\|\boldsymbol{x}\|$ 为向量 $\boldsymbol{x}$ 的长度 (或范数), 其中

Definition 2 $\|\boldsymbol{x}\|$ is called the length (or norm) of $\boldsymbol{x}$,where

$$\|\boldsymbol{x}\| =\sqrt{[\boldsymbol{x},\boldsymbol{x}]} = \sqrt{x_1^2+x_2^2+\cdots+x_n^2}$$

当 $\|\boldsymbol{x}\|=1$ 时, 称 $\boldsymbol{x}$ 为**单位向量**.

The vector $\boldsymbol{x}$ satisfying $\|\boldsymbol{x}\|=1$ is called a **unit vector**.

从向量长度的定义可推得以下基本性质:

From the definition of the vector length we can get the following basic properties:

(1) **非负性** 当 $\boldsymbol{x}\neq\boldsymbol{0}$ 时, $\|\boldsymbol{x}\|>0$, 当 $\boldsymbol{x}=\boldsymbol{0}$ 时, $\|\boldsymbol{x}\|=0$;

(2) **齐次性** $\|\lambda\boldsymbol{x}\|=|\lambda|\ \|\boldsymbol{x}\|$ (λ 为实数);

(3) **三角不等式** $\|\boldsymbol{x}+\boldsymbol{y}\|\leqslant\|\boldsymbol{x}\|+\|\boldsymbol{y}\|$;

(4) **柯西 -施瓦茨不等式** $[\boldsymbol{x},\boldsymbol{y}]\leqslant\|\boldsymbol{x}\|\cdot\|\boldsymbol{y}\|$.

(1) **Nonnegativity** $\|\boldsymbol{x}\|>0$ for $\boldsymbol{x}\neq\boldsymbol{0}$, and $\|\boldsymbol{x}\|=0$ for $\boldsymbol{x}=\boldsymbol{0}$;

(2) **Homogeneity** $\|\lambda\boldsymbol{x}\|=|\lambda|\ \|\boldsymbol{x}\|$ for real λ;

(3) **Triangle inequality** $\|\boldsymbol{x}+\boldsymbol{y}\|\leqslant\|\boldsymbol{x}\|+\|\boldsymbol{y}\|$;

(4) **Cauchy-Schwarz inequality** $[\boldsymbol{x},\boldsymbol{y}]\leqslant\|\boldsymbol{x}\|\cdot\|\boldsymbol{y}\|$.

由柯西-施瓦茨不等式可得

From the Cauchy-Schwarz inequality, we have

$$\left|\frac{[\boldsymbol{x},\boldsymbol{y}]}{\|\boldsymbol{x}\|\cdot\|\boldsymbol{y}\|}\right|\leqslant 1 \quad (\|\boldsymbol{x}\|\cdot\|\boldsymbol{y}\|\neq 0)$$

定义 3 若 $\|\boldsymbol{x}\|\neq 0$, $\|\boldsymbol{y}\|\neq 0$, 称

Definition 3 If $\|\boldsymbol{x}\|\neq 0$ and $\|\boldsymbol{y}\|\neq 0$,

$$\boldsymbol{\theta}=\arccos\frac{[\boldsymbol{x},\boldsymbol{y}]}{\|\boldsymbol{x}\|\cdot\|\boldsymbol{y}\|}$$

为 $\boldsymbol{x}$ 与 $\boldsymbol{y}$ 的夹角. 当 $[\boldsymbol{x},\boldsymbol{y}]=0$ 时, 称 $\boldsymbol{x}$ 与 $\boldsymbol{y}$ 正交.

is called the included angle between $\boldsymbol{x}$ and $\boldsymbol{y}$. If $[\boldsymbol{x},\boldsymbol{y}]=0$, then $\boldsymbol{x}$ and $\boldsymbol{y}$ are said to be orthogonal.

显然, n 维零向量与任意 n 维向量正交.

Obviously, an n dimensional zero vector is orthogonal to any n dimensional vectors.

定义 4 若非零向量组中的向量彼此正交, 则称之为**正交向量组**.

Definition 4 If the vectors in the nonzero vector set are orthogonal to each other, the set is called an **orthogonal vector set**.

定理 1 若 n 维非零向量组 $\{\boldsymbol{\alpha}_1, \boldsymbol{\alpha}_2, \cdots, \boldsymbol{\alpha}_r\}$ 为正交向量组, 则它们一定是线性无关向量组.

Theorem 1 If the n dimensional nonzero vector set $\{\boldsymbol{\alpha}_1, \boldsymbol{\alpha}_2, \cdots, \boldsymbol{\alpha}_r\}$ is an orthogonal vector set, it must be a linearly independent set.

证 设有 $\lambda_1, \lambda_2, \cdots, \lambda_r$ 使 $\sum\limits_{i=1}^{r} \lambda_i \boldsymbol{\alpha}_i = \mathbf{0}$, 分别用 $\boldsymbol{\alpha}_k\,(k=1,2,\cdots,r)$ 与 $\sum\limits_{i=1}^{r} \lambda_i \boldsymbol{\alpha}_i = \mathbf{0}$ 两端作内积运算, 得

Proof Suppose that there exist numbers $\lambda_1, \lambda_2, \cdots, \lambda_r$ such that $\sum\limits_{i=1}^{r} \lambda_i \boldsymbol{\alpha}_i = \mathbf{0}$. Performing the operation of inner product on both sides of $\sum\limits_{i=1}^{r} \lambda_i \boldsymbol{\alpha}_i = \mathbf{0}$ by $\boldsymbol{\alpha}_k\,(k=1,2,\cdots,r)$, respectively, we have

$$\lambda_k\,[\boldsymbol{\alpha}_k, \boldsymbol{\alpha}_k] = [\boldsymbol{\alpha}_k, \mathbf{0}] = \mathbf{0}$$

由 $\boldsymbol{\alpha}_k \neq \mathbf{0}$, 故 $[\boldsymbol{\alpha}_k, \boldsymbol{\alpha}_k] = \|\boldsymbol{\alpha}_k\|^2 \neq 0$, 从而必有 $\lambda_k = 0\ (k=1,2,\cdots,r)$. 因此, 向量组 $\{\boldsymbol{\alpha}_1, \boldsymbol{\alpha}_2, \cdots, \boldsymbol{\alpha}_r\}$ 线性无关. ▲

Since $\boldsymbol{\alpha}_k \neq \mathbf{0}$, we have $[\boldsymbol{\alpha}_k, \boldsymbol{\alpha}_k] = \|\boldsymbol{\alpha}_k\|^2 \neq 0$, this means that there must have $\lambda_k = 0(k=1,2,\cdots,r)$. Therefore, the vector set $\{\boldsymbol{\alpha}_1, \boldsymbol{\alpha}_2, \cdots, \boldsymbol{\alpha}_r\}$is linearly independent. ▲

例 2 已知

Example 2 Known that

$$\boldsymbol{\alpha}_1 = \begin{pmatrix} 1 \\ 1 \\ 1 \end{pmatrix}, \boldsymbol{\alpha}_2 = \begin{pmatrix} 1 \\ -2 \\ 1 \end{pmatrix}$$

正交, 求一个非零向量 $\boldsymbol{\alpha}_3$, 使 $\boldsymbol{\alpha}_1, \boldsymbol{\alpha}_2, \boldsymbol{\alpha}_3$ 两两正交.

are orthogonal. Find a nonzero vector $\boldsymbol{\alpha}_3$ such that $\boldsymbol{\alpha}_1, \boldsymbol{\alpha}_2, \boldsymbol{\alpha}_3$ are orthogonal to each other.

解 令 $\boldsymbol{\alpha}_3 = (x_1, x_2, x_3)^{\mathrm{T}}$, 则问题等价于求解如下方程组

Solution Let $\boldsymbol{\alpha}_3 = (x_1, x_2, x_3)^{\mathrm{T}}$. The problem is equivalent to solving the following system

$$\begin{pmatrix} 1 & 1 & 1 \\ 1 & -2 & 1 \end{pmatrix}\begin{pmatrix} x_1 \\ x_2 \\ x_3 \end{pmatrix}=\begin{pmatrix} 0 \\ 0 \end{pmatrix}$$

不难求得基础解系为 $\begin{pmatrix} -1 \\ 0 \\ 1 \end{pmatrix}$, 取 $\boldsymbol{\alpha}_3 = \begin{pmatrix} -1 \\ 0 \\ 1 \end{pmatrix}$, 则 $\boldsymbol{\alpha}_3$ 即为所求. ■

It is not difficult to obtain that the system of fundamental solutions is $\begin{pmatrix} -1 \\ 0 \\ 1 \end{pmatrix}$. Let $\boldsymbol{\alpha}_3 = \begin{pmatrix} -1 \\ 0 \\ 1 \end{pmatrix}$, then $\boldsymbol{\alpha}_3$ is our desired. ■

下面介绍将向量组 $\{\boldsymbol{\alpha}_1, \boldsymbol{\alpha}_2, \cdots, \boldsymbol{\alpha}_r\}$ 转换为正交向量组的格拉姆-施密特正交化方法, 其具体步骤如下：

In the following part we introduce the Gram-Schmidt orthogonalization process which transforms the vector set $\{\boldsymbol{\alpha}_1, \boldsymbol{\alpha}_2, \cdots, \boldsymbol{\alpha}_r\}$ into an orthogonal vector set. The practical procedure is as follows:

令

Let

$$\begin{aligned}
&\boldsymbol{\beta}_1 = \boldsymbol{\alpha}_1 \\
&\boldsymbol{\beta}_2 = \boldsymbol{\alpha}_2 - \frac{[\boldsymbol{\beta}_1, \boldsymbol{\alpha}_2]}{[\boldsymbol{\beta}_1, \boldsymbol{\beta}_1]}\boldsymbol{\beta}_1 \\
&\qquad\cdots\cdots \\
&\boldsymbol{\beta}_r = \boldsymbol{\alpha}_r - \frac{[\boldsymbol{\beta}_1, \boldsymbol{\alpha}_r]}{[\boldsymbol{\beta}_1, \boldsymbol{\beta}_1]}\boldsymbol{\beta}_1 - \frac{[\boldsymbol{\beta}_2, \boldsymbol{\alpha}_r]}{[\boldsymbol{\beta}_2, \boldsymbol{\beta}_2]}\boldsymbol{\beta}_2 - \cdots - \frac{[\boldsymbol{\beta}_{r-1}, \boldsymbol{\alpha}_r]}{[\boldsymbol{\beta}_{r-1}, \boldsymbol{\beta}_{r-1}]}\boldsymbol{\beta}_{r-1}
\end{aligned}$$

容易验证 $\boldsymbol{\beta}_1, \boldsymbol{\beta}_2, \cdots, \boldsymbol{\beta}_r$ 两两正交.

It is easy to verify that $\boldsymbol{\beta}_1, \boldsymbol{\beta}_2, \cdots, \boldsymbol{\beta}_r$ are orthogonal to each other.

还可以将它们单位化, 即令

Moreover, we can normalize them, namely,

$$\boldsymbol{e}_1 = \frac{\boldsymbol{\beta}_1}{\|\boldsymbol{\beta}_1\|}, \boldsymbol{e}_2 = \frac{\boldsymbol{\beta}_2}{\|\boldsymbol{\beta}_2\|}, \cdots, \boldsymbol{e}_r = \frac{\boldsymbol{\beta}_r}{\|\boldsymbol{\beta}_r\|}$$

称向量组 $\{\boldsymbol{e}_1, \boldsymbol{e}_2, \cdots, \boldsymbol{e}_r\}$ 为**标准正交向量组**.

The vector set $\{\boldsymbol{e}_1, \boldsymbol{e}_2, \cdots, \boldsymbol{e}_r\}$ is an **orthonormal vector set.**

例 3 设有向量组

Example 3 Suppose that the vector set is given by

$$\boldsymbol{\alpha}_1=\begin{pmatrix}1\\-1\\0\end{pmatrix},\quad \boldsymbol{\alpha}_2=\begin{pmatrix}1\\0\\1\end{pmatrix},\quad \boldsymbol{\alpha}_3=\begin{pmatrix}1\\-1\\1\end{pmatrix}$$

将其化为正交单位向量组.

Transform it into an orthonormal vector set.

解 对向量组 $\{\boldsymbol{\alpha}_1,\boldsymbol{\alpha}_2,\boldsymbol{\alpha}_3\}$ 进行如下运算, 即

Solution Performing the following operations on the vector set $\{\boldsymbol{\alpha}_1,\boldsymbol{\alpha}_2,\boldsymbol{\alpha}_3\}$, namely,

$$\boldsymbol{\beta}_1=\boldsymbol{\alpha}_1=\begin{pmatrix}1\\-1\\0\end{pmatrix}$$

$$\boldsymbol{\beta}_2=\boldsymbol{\alpha}_2-\frac{[\boldsymbol{\beta}_1,\boldsymbol{\alpha}_2]}{[\boldsymbol{\beta}_1,\boldsymbol{\beta}_1]}\boldsymbol{\beta}_1=\begin{pmatrix}1\\0\\1\end{pmatrix}-\frac{1}{2}\begin{pmatrix}1\\-1\\0\end{pmatrix}=\begin{pmatrix}\frac{1}{2}\\\frac{1}{2}\\1\end{pmatrix}$$

$$\begin{aligned}\boldsymbol{\beta}_3&=\boldsymbol{\alpha}_3-\frac{[\boldsymbol{\beta}_1,\boldsymbol{\alpha}_3]}{[\boldsymbol{\beta}_1,\boldsymbol{\beta}_1]}\boldsymbol{\beta}_1-\frac{[\boldsymbol{\beta}_2,\boldsymbol{\alpha}_3]}{[\boldsymbol{\beta}_2,\boldsymbol{\beta}_2]}\boldsymbol{\beta}_2\\&=\begin{pmatrix}1\\-1\\0\end{pmatrix}-\begin{pmatrix}1\\-1\\0\end{pmatrix}-\frac{2}{3}\begin{pmatrix}\frac{1}{2}\\\frac{1}{2}\\1\end{pmatrix}=\begin{pmatrix}-\frac{1}{3}\\-\frac{1}{3}\\\frac{1}{3}\end{pmatrix}\end{aligned}$$

再将 $\boldsymbol{\beta}_1,\boldsymbol{\beta}_2,\boldsymbol{\beta}_3$ 单位化, 得到正交单位向量组

and then normaling $\boldsymbol{\beta}_1,\boldsymbol{\beta}_2,\boldsymbol{\beta}_3$, we obtain the orthonormal vector set, as follows

$$\boldsymbol{e}_1=\frac{\boldsymbol{\beta}_1}{\|\boldsymbol{\beta}_1\|}=\begin{pmatrix}\frac{1}{\sqrt{2}}\\-\frac{1}{\sqrt{2}}\\0\end{pmatrix},\quad \boldsymbol{e}_2=\frac{\boldsymbol{\beta}_2}{\|\boldsymbol{\beta}_2\|}=\begin{pmatrix}\frac{1}{\sqrt{6}}\\\frac{1}{\sqrt{6}}\\\frac{2}{\sqrt{6}}\end{pmatrix},\quad \boldsymbol{e}_3=\frac{\boldsymbol{\beta}_3}{\|\boldsymbol{\beta}_3\|}=\begin{pmatrix}-\frac{1}{\sqrt{3}}\\-\frac{1}{\sqrt{3}}\\\frac{1}{\sqrt{3}}\end{pmatrix}$$

■

定义 5 如果方阵 $\boldsymbol{A}$ 满足 $\boldsymbol{A}^{\mathrm{T}}\boldsymbol{A}=\boldsymbol{E}$, 即 $\boldsymbol{A}^{-1}=\boldsymbol{A}^{\mathrm{T}}$, 则称 $\boldsymbol{A}$ 为**正交矩阵**.

Definition 5 If a square matrix $\boldsymbol{A}$ satisfies $\boldsymbol{A}^{\mathrm{T}}\boldsymbol{A}=\boldsymbol{E}$, that is, $\boldsymbol{A}^{-1}=\boldsymbol{A}^{\mathrm{T}}$, then $\boldsymbol{A}$ is said to be an **orthogonal matrix**.

$\boldsymbol{A}$ 用列向量表示, 则有

Representing $\boldsymbol{A}$ by column vectors, we get

$$\begin{pmatrix} \boldsymbol{\alpha}_1^{\mathrm{T}} \\ \boldsymbol{\alpha}_2^{\mathrm{T}} \\ \vdots \\ \boldsymbol{\alpha}_n^{\mathrm{T}} \end{pmatrix} (\boldsymbol{\alpha}_1, \boldsymbol{\alpha}_2, \cdots, \boldsymbol{\alpha}_n) = \boldsymbol{E}$$

亦即

that is,

$$(\boldsymbol{\alpha}_i^{\mathrm{T}} \boldsymbol{\alpha}_j) = (\delta_{ij})$$

由此得到 n^2 个关系式

Then we obtain the following n^2 equations

$$\boldsymbol{\alpha}_i^{\mathrm{T}} \boldsymbol{\alpha}_j = \delta_{ij} = \begin{cases} 1, & i = j \\ 0, & i \neq j \end{cases} \quad (i, j = 1, 2, \cdots, n)$$

这说明, 方阵 $\boldsymbol{A}$ 为正交矩阵的充分必要条件是: $\boldsymbol{A}$ 的列向量组是正交单位向量组.

It means that the square matrix $\boldsymbol{A}$ is an orthogonal matrix if and only if the column vector set of $\boldsymbol{A}$ is an orthonormal set.

注 (1) 因为 $\boldsymbol{A}^{\mathrm{T}}\boldsymbol{A} = \boldsymbol{E} = \boldsymbol{A}\boldsymbol{A}^{\mathrm{T}}$, 所以上述结论对 $\boldsymbol{A}$ 的行向量组也成立.

Note (1) Since $\boldsymbol{A}^{\mathrm{T}}\boldsymbol{A} = \boldsymbol{E} = \boldsymbol{A}\boldsymbol{A}^{\mathrm{T}}$, the above result are also valid for the row vector set of $\boldsymbol{A}$.

(2) 因为 $\boldsymbol{A}^{\mathrm{T}}\boldsymbol{A} = \boldsymbol{E}$, 两边取行列式得到 $\left|\boldsymbol{A}^{\mathrm{T}}\right| |\boldsymbol{A}| = |\boldsymbol{E}|$, 即 $|\boldsymbol{A}|^2 = 1$. 这表明: 若 $\boldsymbol{A}$ 为正交矩阵, 则 $|\boldsymbol{A}| = \pm 1$.

(2) Since $\boldsymbol{A}^{\mathrm{T}}\boldsymbol{A} = \boldsymbol{E}$, taking the determinant on its both sides yields $\left|\boldsymbol{A}^{\mathrm{T}}\right| |\boldsymbol{A}| = |\boldsymbol{E}|$, i.e., $|\boldsymbol{A}|^2 = 1$. This shows that if $\boldsymbol{A}$ is an orthogonal matrix, we have $|\boldsymbol{A}| = \pm 1$.

例 4 判别下列矩阵是否为正交阵.

Example 4 Determine whether the following matrices are orthogonal matrices.

$$(1) \begin{pmatrix} 1 & -1/2 & 1/3 \\ -1/2 & 1 & 1/2 \\ 1/3 & 1/2 & -1 \end{pmatrix}; \quad (2) \begin{pmatrix} 1/9 & -8/9 & -4/9 \\ -8/9 & 1/9 & -4/9 \\ -4/9 & -4/9 & 7/9 \end{pmatrix}.$$

解 (1) 考察矩阵的第一列和第二列, 因为

Solution (1) Examine the first and the second columns of the matrix. Since

$$1 \times \left(-\frac{1}{2}\right) + \left(-\frac{1}{2}\right) \times 1 + \frac{1}{3} \times \frac{1}{2} \neq 0$$

所以它不是正交矩阵.

the given matrix is not an orthogonal matrix.

(2) 因为

(2) Since

$$\begin{pmatrix} 1/9 & -8/9 & -4/9 \\ -8/9 & 1/9 & -4/9 \\ -4/9 & -4/9 & 7/9 \end{pmatrix}\begin{pmatrix} 1/9 & -8/9 & -4/9 \\ -8/9 & 1/9 & -4/9 \\ -4/9 & -4/9 & 7/9 \end{pmatrix}^{\mathrm{T}} = \begin{pmatrix} 1 & 0 & 0 \\ 0 & 1 & 0 \\ 0 & 0 & 1 \end{pmatrix}$$

由正交矩阵的定义知它是正交矩阵. ■

from the definition of orthogonal matrices we know that it is an orthogonal matrix. ■

思 考 题

Questions

1. *若方阵 $\boldsymbol{A}$ 的列向量组是正交向量组, 则 $\boldsymbol{A}$ 是否为可逆矩阵? 是否为正交阵?*

1. If the column vector set of a square matrix $\boldsymbol{A}$ is an orthogonal vector set, whether $\boldsymbol{A}$ is an invertible matrix and is an orthogonal matrix?

2. *若 $\boldsymbol{A}, \boldsymbol{B}$ 分别为 m 阶和 n 阶正交阵, 则 $\begin{pmatrix} \boldsymbol{A} & \boldsymbol{O} \\ \boldsymbol{O} & \boldsymbol{B} \end{pmatrix}$ 是否为正交阵?*

2. Let $\boldsymbol{A}$ and $\boldsymbol{B}$ be orthogonal matrices of orders m and n, respectively, whether $\begin{pmatrix} \boldsymbol{A} & \boldsymbol{O} \\ \boldsymbol{O} & \boldsymbol{B} \end{pmatrix}$ is an orthogonal matrix?

5.4 实对称矩阵的对角化

5.4 Diagonalization of Real Symmetric Matrices

由 5.2 节中的讨论获知, 并不是数域 P 上的所有方阵都可以对角化. 本节讨论一类可以对角化的矩阵类 —— 实对称矩阵.

From the discussion in Section 5.2 we know that not all square matrices over a number field P can be diagonalizable. In this section, we discuss a class of diagonalizable matrices, that is, real symmetric matrices.

实对称矩阵不仅可对角化, 而且还可要求可逆矩阵 $\boldsymbol{P}$ 是正交矩阵, 即对于任意的实对称矩阵一定存在同阶正交矩阵 $\boldsymbol{P}$, 使得 $\boldsymbol{P}^{-1}\boldsymbol{A}\boldsymbol{P} = \boldsymbol{P}^{\mathrm{T}}\boldsymbol{A}\boldsymbol{P}$ 成为对角矩阵. 由于矩阵的对角化问题与特征值和特征向量密切相关, 因此, 首先来讨论实对称矩阵特征值与特征向量的一些特殊性质.

For real symmetric matrices, not only they are diagonalizable, but also they can restrict the invertible matrix $\boldsymbol{P}$ to be an orthogonal matrix, that is to say, there must exist an orthogonal matrix $\boldsymbol{P}$ with the same order such that $\boldsymbol{P}^{-1}\boldsymbol{A}\boldsymbol{P} = \boldsymbol{P}^{\mathrm{T}}\boldsymbol{A}\boldsymbol{P}$ is a diagonal matrix. Since the diagonalization problem of matrices is frequently related to the eigenvalues and eigenvectors, we first discuss some special properties of the eigenvalues and eigenvectors for real symmetric matrices.

定义 1 设 $\boldsymbol{A}=(a_{ij})_{m\times n}$ 为复矩阵, $\overline{\boldsymbol{A}}=(\bar{a}_{ij})_{m\times n}$ 称为 $\boldsymbol{A}$ 的共轭矩阵.

Definition 1 Let $\boldsymbol{A}=(a_{ij})_{m\times n}$ be a complex matrix, $\overline{\boldsymbol{A}}=(\bar{a}_{ij})_{m\times n}$ is said to be the conjugate matrix of $\boldsymbol{A}$.

定理 1 设 $\boldsymbol{A}$ 是 n 阶实对称矩阵, 则 $\boldsymbol{A}$ 的特征值是实数.

Theorem 1 Let $\boldsymbol{A}$ be a real symmetric matrix of order n. The eigenvalues of $\boldsymbol{A}$ are real numbers.

证 令 λ 是 $\boldsymbol{A}$ 的任意特征值, $\boldsymbol{\alpha}$ 是 $\boldsymbol{A}$ 的属于 λ 的特征向量. 则

Proof Let λ be an arbitrary eigenvalue of $\boldsymbol{A}$ and $\boldsymbol{\alpha}$ be an eigenvector associated with λ. Then

$$\boldsymbol{A\alpha}=\lambda\boldsymbol{\alpha} \tag{5.5}$$

在式 (5.5) 的两边进行转置并共轭的运算, 得

Performing the operations of transpose and conjugate on both sides of Equation (5.5), we get

$$\bar{\boldsymbol{\alpha}}^{\mathrm{T}}\boldsymbol{A}=\bar{\lambda}\bar{\boldsymbol{\alpha}}^{\mathrm{T}} \tag{5.6}$$

在式 (5.6) 两边右乘以 $\boldsymbol{\alpha}$ 得

Postmultiplying both sides of Equation (5.6) by $\boldsymbol{\alpha}$, we obtain

$$\bar{\boldsymbol{\alpha}}^{\mathrm{T}}\boldsymbol{A\alpha}=\bar{\lambda}\bar{\boldsymbol{\alpha}}^{\mathrm{T}}\boldsymbol{\alpha} \tag{5.7}$$

式 (5.5) 两边左乘以 $\bar{\boldsymbol{\alpha}}^{\mathrm{T}}$ 得

Meanwhile, premultiplying both sides of (5.5) by $\bar{\boldsymbol{\alpha}}^{\mathrm{T}}$, we get

$$\bar{\boldsymbol{\alpha}}^{\mathrm{T}}\boldsymbol{A\alpha}=\bar{\boldsymbol{\alpha}}^{\mathrm{T}}\lambda\boldsymbol{\alpha}=\lambda\bar{\boldsymbol{\alpha}}^{\mathrm{T}}\boldsymbol{\alpha} \tag{5.8}$$

由式 (5.7), (5.8) 可知 $(\lambda-\bar{\lambda})\bar{\boldsymbol{\alpha}}^{\mathrm{T}}\boldsymbol{\alpha}=\boldsymbol{0}$, 而 $\bar{\boldsymbol{\alpha}}^{\mathrm{T}}\boldsymbol{\alpha}\neq\boldsymbol{0}$ 所以 $\lambda=\bar{\lambda}$, 从而 λ 是实数. ▲

Comparing Equations (5.7) and (5.8), we see that $(\lambda-\bar{\lambda})\bar{\boldsymbol{\alpha}}^{\mathrm{T}}\boldsymbol{\alpha}=\boldsymbol{0}$. Since $\bar{\boldsymbol{\alpha}}^{\mathrm{T}}\boldsymbol{\alpha}\neq\boldsymbol{0}$, we have $\lambda=\bar{\lambda}$, that is, λ is a real number. ▲

定理 2 实对称矩阵 $\boldsymbol{A}$ 的属于不同特征值的特征向量彼此正交.

Theorem 2 The eigenvectors of a real symmetric matrix $\boldsymbol{A}$ associated with distinct eigenvalues are orthogonal to each other.

证 令 λ_1, λ_2 是 $\boldsymbol{A}$ 的不同特征值, $\boldsymbol{\alpha}_1, \boldsymbol{\alpha}_2$ 是对应的特征向量. 容易计算如下等式成立, 即

Proof Let λ_1, λ_2 be distinct eigenvalues of $\boldsymbol{A}$, and $\boldsymbol{\alpha}_1, \boldsymbol{\alpha}_2$ be the corresponding eigenvectors. It is easy to calculate that the following equations are valid, namely,

$$\begin{aligned}\lambda_1[\boldsymbol{\alpha}_1, \boldsymbol{\alpha}_2] =&[\lambda_1\boldsymbol{\alpha}_1, \boldsymbol{\alpha}_2] = [\boldsymbol{A}\boldsymbol{\alpha}_1, \boldsymbol{\alpha}_2] = (\boldsymbol{A}\boldsymbol{\alpha}_1)^{\mathrm{T}}\boldsymbol{\alpha} \\ =&\boldsymbol{\alpha}_1^{\mathrm{T}}\boldsymbol{A}\boldsymbol{\alpha}_2 = \boldsymbol{\alpha}_1^{\mathrm{T}}\lambda_2\boldsymbol{\alpha}_2 = \lambda_2[\boldsymbol{\alpha}_1, \boldsymbol{\alpha}_2]\end{aligned}$$

于是, $(\lambda_1 - \lambda_2)[\boldsymbol{\alpha}_1, \boldsymbol{\alpha}_2] = 0$. 由于 $\lambda_1 \neq \lambda_2$, 因此有 $[\boldsymbol{\alpha}_1, \boldsymbol{\alpha}_2] = 0$. ▲

Consequently, $(\lambda_1 - \lambda_2)[\boldsymbol{\alpha}_1, \boldsymbol{\alpha}_2] = 0$. Since $\lambda_1 \neq \lambda_2$, we have $[\boldsymbol{\alpha}_1, \boldsymbol{\alpha}_2] = 0$. ▲

定理 3 设 $\boldsymbol{A}$ 为实对称矩阵, 则必存在正交矩阵 $\boldsymbol{P}$, 使

Theorem 3 Suppose that $\boldsymbol{A}$ is a real symmetric matrix. Then there must exist an orthogonal matrix $\boldsymbol{P}$ such that

$$\boldsymbol{P}^{-1}\boldsymbol{A}\boldsymbol{P} = \boldsymbol{P}^{\mathrm{T}}\boldsymbol{A}\boldsymbol{P} = \boldsymbol{\Lambda} = \begin{pmatrix} \lambda_1 & & & \\ & \lambda_2 & & \\ & & \ddots & \\ & & & \lambda_n \end{pmatrix}$$

其中 $\lambda_1, \lambda_2, \cdots, \lambda_n$ 是 $\boldsymbol{A}$ 的特征值.

where $\lambda_1, \lambda_2, \cdots, \lambda_n$ are eigenvalues of $\boldsymbol{A}$.

由于 $\boldsymbol{P}$ 是正交矩阵, 所以 $\boldsymbol{P}$ 的列向量组是规范正交的向量组. 如前所述, $\boldsymbol{P}$ 的列向量组是由 $\boldsymbol{A}$ 的 n 个线性无关的特征向量组成. 因此对 $\boldsymbol{P}$ 的列向量组有三条要求, 即

Since $\boldsymbol{P}$ is an orthogonal matrix, the column vector set of $\boldsymbol{P}$ is an orthonormal set. Meanwhile, as mentioned above, the column vector set of $\boldsymbol{P}$ is consisted of n linearly independent eigenvectors of $\boldsymbol{A}$. Hence, there are three requirements for the column vector set of $\boldsymbol{P}$, as follows.

(1) 每个列向量是特征向量;

(1) Each column vector is an eigenvector;

(2) 任意两个列向量正交;

(2) Any two column vectors are orthogonal to each other;

(3) 每个列向量是单位向量.

(3) Each column vector is a unit vector.

于是求正交矩阵 $\boldsymbol{P}$ 使 $\boldsymbol{P}^{-1}\boldsymbol{A}\boldsymbol{P}$ 为对角矩阵的具体步骤如下:

As a result, for finding an orthogonal matrix $\boldsymbol{P}$ such that $\boldsymbol{P}^{-1}\boldsymbol{A}\boldsymbol{P}$ is a diagonal matrix, the practical procedure is given as follows:

第一步 求出 $\boldsymbol{A}$ 的所有的特征值 $\lambda_1, \lambda_2, \cdots, \lambda_n$(可能有重根).

Step 1 Find all eigenvalues λ_1, $\lambda_2, \cdots, \lambda_n$ (they may be multiple roots) of $\boldsymbol{A}$.

第二步 求出 $\boldsymbol{A}$ 对应于每个特征值 λ_i 的一组线性无关的特征向量, 即求出 $(\lambda_i \boldsymbol{E} - \boldsymbol{A})\boldsymbol{x} = \mathbf{0}$ 的一个基础解系, 并将此组基础解系规范正交化. 注意, 将此组基础解系规范正交化以后得到的向量, 它还是特征值 λ_i 的一组线性无关的特征向量.

Step 2 Find a set of linearly independent eigenvectors associated with each eigenvalue λ_i of $\boldsymbol{A}$, that is, find a system of fundamental solutions of $(\lambda_i \boldsymbol{E} - \boldsymbol{A})\boldsymbol{x} = \mathbf{0}$, and then orthonormalize them. Note that the vectors, obtained by orthonormalizing the system of fundamental solutions, are still a set of linearly independent eigenvectors associated with the eigenvalue λ_i.

第三步 将所有特征值 $\lambda_1, \lambda_2, \cdots, \lambda_n$ 对应的 n 个规范正交的特征向量作为列向量所得的 n 阶方阵, 即为所求的正交矩阵 $\boldsymbol{P}$. 以相应的特征值作为主对角线元素得到对角矩阵 $\boldsymbol{\Lambda}$, 即

Step 3 The matrix of order n, obtained by taking n orthonormal eigenvectors associated with all the eigenvalues $\lambda_1, \lambda_2, \cdots, \lambda_n$ to be columns, is just the wanted orthogonal matrix $\boldsymbol{P}$. The diagonal matrix $\boldsymbol{\Lambda}$ can be obtained by taking the corresponding eigenvalues to be the diagonal elements, that is,

$$\boldsymbol{P}^{-1}\boldsymbol{A}\boldsymbol{P} = \boldsymbol{P}^{\mathrm{T}}\boldsymbol{A}\boldsymbol{P} = \boldsymbol{\Lambda} = \begin{pmatrix} \lambda_1 & & & \\ & \lambda_2 & & \\ & & \ddots & \\ & & & \lambda_n \end{pmatrix}$$

例 1 设

Example 1 Let

$$\boldsymbol{A} = \begin{pmatrix} 1 & 2 & 2 \\ 2 & 1 & 2 \\ 2 & 2 & 1 \end{pmatrix}$$

求正交矩阵 $\boldsymbol{P}$, 使 $\boldsymbol{P}^{-1}\boldsymbol{A}\boldsymbol{P}$ 为对角矩阵.

Find an orthogonal matrix $\boldsymbol{P}$ such that $\boldsymbol{P}^{-1}\boldsymbol{A}\boldsymbol{P}$ is a diagonal matrix.

解 **第一步** 先求 $\boldsymbol{A}$ 的特征值, 由

Solution **Step 1** Find the eigenvalues of $\boldsymbol{A}$. From

$$|\boldsymbol{A}-\lambda\boldsymbol{E}|=\begin{vmatrix}1-\lambda & 2 & 2\\ 2 & 1-\lambda & 2\\ 2 & 2 & 1-\lambda\end{vmatrix}=\begin{vmatrix}5-\lambda & 2 & 2\\ 5-\lambda & 1-\lambda & 2\\ 5-\lambda & 2 & 1-\lambda\end{vmatrix}$$
$$=(5-\lambda)\begin{vmatrix}1 & 2 & 2\\ 0 & -(\lambda+1) & 0\\ 0 & 0 & -(\lambda+1)\end{vmatrix}$$
$$=(5-\lambda)(\lambda+1)^2=0$$

求得 $\boldsymbol{A}$ 的特征值为 $\lambda_1=\lambda_2-1$(二重), $\lambda_3=5$.

we obtain that the eigenvalues of $\boldsymbol{A}$ are $\lambda_1=\lambda_2-1$ (of multiplicity 2) and $\lambda_3=5$.

第二步 对于 $\lambda_1=-1$, 求解 $(\boldsymbol{A}+\boldsymbol{E})\boldsymbol{x}=\boldsymbol{0}$. 由

Step 2 For $\lambda_1=-1$, solve $(\boldsymbol{A}+\boldsymbol{E})\boldsymbol{x}=\boldsymbol{0}$. Since

$$\boldsymbol{A}+\boldsymbol{E}=\begin{pmatrix}2&2&2\\2&2&2\\2&2&2\end{pmatrix}\to\begin{pmatrix}1&1&1\\0&0&0\\0&0&0\end{pmatrix}$$

求得一基础解系为

we get a system of fundamental solutions, given by

$$\boldsymbol{\alpha}_1=\begin{pmatrix}-1\\1\\0\end{pmatrix},\quad \boldsymbol{\alpha}_2=\begin{pmatrix}-1\\0\\1\end{pmatrix}$$

显然, 它们不是正交的, 将其正交化, 令

Obviously, they are not orthogonal and need to be orthogonalized. Let

$$\boldsymbol{\beta}_1=\boldsymbol{\alpha}_1=\begin{pmatrix}-1\\1\\0\end{pmatrix}$$
$$\boldsymbol{\beta}_2=\boldsymbol{\alpha}_2-\frac{[\boldsymbol{\beta}_1,\boldsymbol{\alpha}_2]}{[\boldsymbol{\beta}_1,\boldsymbol{\beta}_1]}\boldsymbol{\beta}_1=\begin{pmatrix}-1\\0\\1\end{pmatrix}-\frac{1}{2}\begin{pmatrix}-1\\1\\0\end{pmatrix}=\begin{pmatrix}-\frac{1}{2}\\-\frac{1}{2}\\1\end{pmatrix}$$

将其单位化, 得

Normalizing them, we get

$$\boldsymbol{\eta}_1=\frac{\boldsymbol{\beta}_1}{\|\boldsymbol{\beta}_1\|}=\begin{pmatrix}-\frac{1}{\sqrt{2}}\\\frac{1}{\sqrt{2}}\\0\end{pmatrix},\quad \boldsymbol{\eta}_2=\frac{\boldsymbol{\beta}_2}{\|\boldsymbol{\beta}_2\|}=\begin{pmatrix}-\frac{\sqrt{6}}{6}\\-\frac{\sqrt{6}}{6}\\\frac{\sqrt{6}}{3}\end{pmatrix}$$

对于 $\lambda_3=5$, 求解 $(\boldsymbol{A}-5\boldsymbol{E})\boldsymbol{x}=\boldsymbol{0}$, 由

For $\lambda_3=5$, solve $(\boldsymbol{A}-5\boldsymbol{E})\boldsymbol{x}=\boldsymbol{0}$. Since

$$\boldsymbol{A}-5\boldsymbol{E}=\begin{pmatrix}-4&2&2\\2&-4&2\\2&2&-4\end{pmatrix}\to\begin{pmatrix}1&0&-1\\0&1&-1\\0&0&0\end{pmatrix}$$

求得它的一基础解系为

we get a system of fundamental solutions, given by

$$\boldsymbol{\alpha}_3=\begin{pmatrix}1\\1\\1\end{pmatrix}$$

这里只有一个向量, 将其单位化, 得

It contains only one vector. Normalizing it, we get

$$\boldsymbol{\eta}_3=\frac{\boldsymbol{\alpha}_3}{\|\boldsymbol{\alpha}_3\|}=\begin{pmatrix}\frac{1}{\sqrt{3}}\\\frac{1}{\sqrt{3}}\\\frac{1}{\sqrt{3}}\end{pmatrix}$$

第三步 以正交单位向量组 $\boldsymbol{\eta}_1,\boldsymbol{\eta}_2,\boldsymbol{\eta}_3$ 为列向量的矩阵即为所求的正交矩阵 $\boldsymbol{P}$, 即

Step 3 Take the orthogonal vector set $\boldsymbol{\eta}_1,\boldsymbol{\eta}_2,\boldsymbol{\eta}_3$ to be the column vectors of a matrix, which is the wanted orthogonal matrix $\boldsymbol{P}$, as follows,

$$\boldsymbol{P}=(\boldsymbol{\eta}_1,\boldsymbol{\eta}_2,\boldsymbol{\eta}_3)=\begin{pmatrix}-\frac{1}{\sqrt{2}}&-\frac{1}{\sqrt{6}}&\frac{1}{\sqrt{3}}\\\frac{1}{\sqrt{2}}&-\frac{1}{\sqrt{6}}&\frac{1}{\sqrt{3}}\\0&\frac{\sqrt{6}}{3}&\frac{1}{\sqrt{3}}\end{pmatrix}$$

从而有

As a result, we have

$$\boldsymbol{P}^{-1}\boldsymbol{A}\boldsymbol{P}=\begin{pmatrix}-1&0&0\\0&-1&0\\0&0&5\end{pmatrix}$$ ■

例 2 已知

Example 2 Known that

$$\boldsymbol{A}=\begin{pmatrix}0&1&1&-1\\1&0&-1&1\\1&-1&0&1\\-1&1&1&0\end{pmatrix}$$

求一个正交矩阵 P, 使 $P^{-1}AP$ 为对角矩阵.

Find an orthogonal matrix P such that $P^{-1}AP$ is a diagonal matrix.

解 **第一步** 求 A 的特征值, 由于

Solution **Step 1** Find the eigenvalues of A. Since

$$|\lambda E - A| = \begin{vmatrix} \lambda & -1 & -1 & 1 \\ -1 & \lambda & 1 & -1 \\ -1 & 1 & \lambda & -1 \\ 1 & -1 & -1 & \lambda \end{vmatrix} = \begin{vmatrix} 0 & \lambda-1 & \lambda-1 & 1-\lambda^2 \\ 0 & \lambda-1 & 0 & \lambda-1 \\ 0 & 0 & \lambda-1 & \lambda-1 \\ 1 & -1 & -1 & \lambda \end{vmatrix}$$

$$= -(\lambda-1)^3 \begin{vmatrix} 1 & 1 & -1-\lambda \\ 1 & 0 & 1 \\ 0 & 1 & 1 \end{vmatrix} = (\lambda-1)^3(\lambda+3) = 0$$

A 的特征值为 $\lambda_1 = \lambda_2 = \lambda_3 = 1$(三重), $\lambda_4 = -3$.

the eigenvalues of A are $\lambda_1 = \lambda_2 = \lambda_3 = 1$ (of multiplicity 3) and $\lambda_4 = -3$.

第二步 求属于 $\lambda_1 = \lambda_2 = \lambda_3 = 1$ 的特征向量, 将 $\lambda_1 = 1$ 代入 $(\lambda E - A)x = 0$, 得

Step 2 Find the eigenvectors associated with $\lambda_1 = \lambda_2 = \lambda_3 = 1$. Substituting $\lambda_1 = 1$ into $(\lambda E - A)x = 0$, yields

$$E - A = \begin{pmatrix} 1 & -1 & -1 & 1 \\ -1 & 1 & 1 & -1 \\ -1 & 1 & 1 & -1 \\ 1 & -1 & -1 & 1 \end{pmatrix} \to \begin{pmatrix} 1 & -1 & -1 & 1 \\ 0 & 0 & 0 & 0 \\ 0 & 0 & 0 & 0 \\ 0 & 0 & 0 & 0 \end{pmatrix}$$

求得基础解系为

We get its system of fundamental solutions, given by

$$\alpha_1 = \begin{pmatrix} 1 \\ 1 \\ 0 \\ 0 \end{pmatrix}, \quad \alpha_2 = \begin{pmatrix} 0 \\ 0 \\ 1 \\ 1 \end{pmatrix}, \quad \alpha_3 = \begin{pmatrix} 1 \\ -1 \\ 1 \\ -1 \end{pmatrix}$$

显然, 向量组 $\{\alpha_1, \alpha_2, \alpha_3\}$ 是正交向量组 (注意: 自由未知量的取法可以既保证 $\alpha_1, \alpha_2, \alpha_3$ 线性无关, 又保证 $\alpha_1, \alpha_2, \alpha_3$ 是正交的, 这样可以避免正交化过程), 将它们单位化, 得到

Obviously, the vector set $\{\alpha_1, \alpha_2, \alpha_3\}$ is an orthogonal vector set (Note. The choice of free unknowns can ensure not only $\alpha_1, \alpha_2, \alpha_3$ are linearly independent, but also they are orthogonal, which can avoid the orthogonalization process). Normalizing them, we get

$$\boldsymbol{\eta}_1=\begin{pmatrix}\frac{1}{\sqrt{2}}\\ \frac{1}{\sqrt{2}}\\ 0\\ 0\end{pmatrix},\quad \boldsymbol{\eta}_2=\begin{pmatrix}0\\ 0\\ \frac{1}{\sqrt{2}}\\ \frac{1}{\sqrt{2}}\end{pmatrix},\quad \boldsymbol{\eta}_3=\begin{pmatrix}\frac{1}{2}\\ -\frac{1}{2}\\ \frac{1}{2}\\ -\frac{1}{2}\end{pmatrix}$$

再求属于 $\lambda_4=-3$ 的特征向量. 将 $\lambda_4=-3$ 代入 $(\lambda\boldsymbol{E}-\boldsymbol{A})\boldsymbol{x}=\boldsymbol{0}$, 得

Next, find the eigenvector associated with $\lambda_4=-3$. Substituting $\lambda_4=-3$ into $(\lambda\boldsymbol{E}-\boldsymbol{A})\boldsymbol{x}=\boldsymbol{0}$ leads to

$$-3\boldsymbol{E}-\boldsymbol{A}=\begin{pmatrix}-3&-1&-1&1\\-1&-3&1&-1\\-1&1&-3&-1\\1&-1&-1&-3\end{pmatrix}\to\begin{pmatrix}1&-1&-1&-3\\0&-4&0&-4\\0&0&-4&-4\\0&-4&-4&-8\end{pmatrix}$$

$$\to\begin{pmatrix}1&-1&-1&-3\\0&1&0&1\\0&0&1&1\\0&1&1&2\end{pmatrix}\to\begin{pmatrix}1&0&-1&-2\\0&1&0&1\\0&0&1&1\\0&0&1&1\end{pmatrix}\to\begin{pmatrix}1&0&0&-1\\0&1&0&1\\0&0&1&1\\0&0&0&0\end{pmatrix}$$

求得基础解系为

We get the system of fundamental solutions, given by

$$\boldsymbol{\alpha}_4=\begin{pmatrix}1\\-1\\-1\\1\end{pmatrix}$$

将它单位化, 得

Normalizing it, we get

$$\boldsymbol{\eta}_4=\begin{pmatrix}\frac{1}{2}\\ \frac{-1}{2}\\ \frac{-1}{2}\\ \frac{1}{2}\end{pmatrix}$$

第三步 以正交单位向量组 $\eta_1, \eta_2, \eta_3, \eta_4$ 为列向量的矩阵, 即为所求的正交矩阵 $\boldsymbol{P}$

Step 3 Take the orthonormal vector set $\eta_1, \eta_2, \eta_3, \eta_4$ to be the column vectors of a matrix, which is the wanted orthogonal matrix $\boldsymbol{P}$

$$\boldsymbol{P} = (\boldsymbol{\eta}_1, \boldsymbol{\eta}_2, \boldsymbol{\eta}_3, \boldsymbol{\eta}_4) = \begin{pmatrix} \frac{1}{\sqrt{2}} & 0 & \frac{1}{2} & \frac{1}{2} \\ \frac{1}{\sqrt{2}} & 0 & -\frac{1}{2} & -\frac{1}{2} \\ 0 & \frac{1}{\sqrt{2}} & \frac{1}{2} & -\frac{1}{2} \\ 0 & \frac{1}{\sqrt{2}} & -\frac{1}{2} & \frac{1}{2} \end{pmatrix}$$

从而有

As a result, we have

$$\boldsymbol{P}^{-1}\boldsymbol{A}\boldsymbol{P} = \boldsymbol{P}^{\mathrm{T}}\boldsymbol{A}\boldsymbol{P} = \begin{pmatrix} 1 & & & \\ & 1 & & \\ & & 1 & \\ & & & -3 \end{pmatrix} \qquad \blacksquare$$

例 3 已知 1 为矩阵 $\boldsymbol{A}$ 的一个特征值, 其中

Example 3 Known that 1 is an eigenvalue of the matrix $\boldsymbol{A}$, where

$$\boldsymbol{A} = \begin{pmatrix} 2 & 0 & 0 \\ 0 & a & 2 \\ 0 & 2 & a \end{pmatrix} (a > 0)$$

求正交矩阵 $\boldsymbol{P}$, 使得 $\boldsymbol{P}^{-1}\boldsymbol{A}\boldsymbol{P}$ 为对角矩阵.

Find an orthogonal matrix $\boldsymbol{P}$ such that $\boldsymbol{P}^{-1}\boldsymbol{A}\boldsymbol{P}$ is a diagonal matrix.

解 $\boldsymbol{A}$ 的特征方程为

Solution The characteristic equation of $\boldsymbol{A}$ is

$$|\lambda\boldsymbol{E} - \boldsymbol{A}| = \begin{vmatrix} \lambda - 2 & 0 & 0 \\ 0 & \lambda - a & -2 \\ 0 & -2 & \lambda - a \end{vmatrix} = (\lambda - 2)(\lambda - a + 2)(\lambda - a - 2) = 0$$

由于 $\boldsymbol{A}$ 有特征值 1, 故有两种情形:

Since 1 is an eigenvalue of $\boldsymbol{A}$, there may be the following two cases:

(1) 若 $a-2=1$, 则 $a=3$;

(2) 若 $a+2=1$, 则 $a=-1$.

由于 $a>0$, 所以 $a=3$. 从而, $\boldsymbol{A}$ 的特征值为 2, 1, 5.

对 $\lambda_1=2$, 由 $(2\boldsymbol{E}-\boldsymbol{A})\boldsymbol{x}=\boldsymbol{0}$ 得对应的基础解系为

(1) $a-2=1$, which implies $a=3$;

(2) $a+2=1$, which implies $a=-1$.

Since $a>0$, we get $a=3$. Therefore, the eigenvalues of $\boldsymbol{A}$ are given by 2, 1, 5.

For $\lambda_1=2$, using $(2\boldsymbol{E}-\boldsymbol{A})\boldsymbol{x}=\boldsymbol{0}$ we get the corresponding system of fundamental solutions, given by

$$\boldsymbol{\alpha}_1=\begin{pmatrix}1\\0\\0\end{pmatrix}$$

对 $\lambda_2=1$, 由 $(\boldsymbol{E}-\boldsymbol{A})\boldsymbol{x}=\boldsymbol{0}$ 得对应的基础解系为

For $\lambda_2=1$, using $(\boldsymbol{E}-\boldsymbol{A})\boldsymbol{x}=\boldsymbol{0}$ we get the corresponding system of fundamental solutions, given by

$$\boldsymbol{\alpha}_2=\begin{pmatrix}0\\1\\-1\end{pmatrix}$$

对 $\lambda_3=5$, 由 $(5\boldsymbol{E}-\boldsymbol{A})\boldsymbol{x}=\boldsymbol{0}$ 得对应的基础解系为

For $\lambda_3=5$, using $(5\boldsymbol{E}-\boldsymbol{A})\boldsymbol{x}=\boldsymbol{0}$ we get the corresponding system of fundamental solutions, given by

$$\boldsymbol{\alpha}_3=\begin{pmatrix}0\\1\\1\end{pmatrix}$$

因为实对称矩阵的属于不同特征值的特征向量必相互正交, 故特征向量 $\boldsymbol{\alpha}_1,\boldsymbol{\alpha}_2,\boldsymbol{\alpha}_3$ 已是正交向量, 将它们单位化, 得

Since the eigenvectors associated with distinct eigenvalues of symmetric matrices must be orthogonal to each other, $\boldsymbol{\alpha}_1,\boldsymbol{\alpha}_2,\boldsymbol{\alpha}_3$ are already orthogonal vectors. Normalizing them, we get

$$\boldsymbol{\eta}_1=\begin{pmatrix}1\\0\\0\end{pmatrix},\quad \boldsymbol{\eta}_2=\begin{pmatrix}0\\ \dfrac{1}{\sqrt{2}}\\ -\dfrac{1}{\sqrt{2}}\end{pmatrix},\quad \boldsymbol{\eta}_3=\begin{pmatrix}0\\ \dfrac{1}{\sqrt{2}}\\ \dfrac{1}{\sqrt{2}}\end{pmatrix}$$

令 | Let

$$
\boldsymbol{P}=(\boldsymbol{\eta}_1,\boldsymbol{\eta}_2,\boldsymbol{\eta}_3)=\begin{pmatrix} 1 & 0 & 0 \\ 0 & \dfrac{1}{\sqrt{2}} & \dfrac{1}{\sqrt{2}} \\ 0 & -\dfrac{1}{\sqrt{2}} & \dfrac{1}{\sqrt{2}} \end{pmatrix}
$$

从而有 | Then

$$
\boldsymbol{P}^{-1}\boldsymbol{A}\boldsymbol{P}=\boldsymbol{P}^{\mathrm{T}}\boldsymbol{A}\boldsymbol{P}=\begin{pmatrix} 2 & 0 & 0 \\ 0 & 1 & 0 \\ 0 & 0 & 5 \end{pmatrix}
$$

■

思 考 题

1. 实对称矩阵 $\boldsymbol{A}$ 的非零特征值的个数是否为 $R(\boldsymbol{A})$?

2. 已知实对称阵 $\boldsymbol{A}$ 的 k 重 $(k \geqslant 2)$ 特征值 λ_j 对应的 k 个的特征向量是线性无关的. 将它们正交化后, 得到的 k 个新向量是否还是 λ_j 对应的特征向量? 为什么?

Questions

1. Is the number of the nonzero eigenvalues of a real symmetric matrix $\boldsymbol{A}$ equal to $R(\boldsymbol{A})$?

2. Known that the k eigenvectors associated with the eigenvalue λ_j of multiplicity $k\,(k \geqslant 2)$ of a real symmetric matrix $\boldsymbol{A}$ are linearly independent. Whether the k new vectors obtained by orthogonalizing the k original eigenvectors are still the eigenvectors associated with λ_j? Why?

习 题 5

1. 求下列矩阵的特征值和特征向量:

Exercise 5

1. Find the eigenvalues and eigenvectors of the following matrices:

(1) $\begin{pmatrix} 2 & -1 & 2 \\ 5 & -3 & 3 \\ -1 & 0 & -2 \end{pmatrix}$; (2) $\begin{pmatrix} 1 & 2 & 3 \\ 2 & 1 & 3 \\ 3 & 3 & 6 \end{pmatrix}$;

(3) $\begin{pmatrix} 0 & 0 & 0 & 1 \\ 0 & 0 & 1 & 0 \\ 0 & 1 & 0 & 0 \\ 1 & 0 & 0 & 0 \end{pmatrix}$; (4) $\begin{pmatrix} 3 & -2 & -4 \\ -2 & 6 & -2 \\ -4 & -2 & 3 \end{pmatrix}$.

2. 设

2. Suppose that

$$A = \begin{pmatrix} 1 & -\sqrt{3} \\ \sqrt{3} & 1 \end{pmatrix}$$

分别在复数域上和实数域上求 A 的特征值和特征向量.

Find the eigenvalues and eigenvectors of A over the complex field and the real field, respectively.

3. 令 A 为 n 阶方阵. 若 A 满足条件 $A^2 = A$, 则 A 的特征值只可能是 0 或 1 .

3. Let A be a square matrix of order n. If A satisfies $A^2 = A$, prove that the eigenvalues of A can only be 0 or 1.

4. 证明: 一个向量 x 不可能是矩阵 A 的不同特征值的特征向量.

4. Prove that a vector x can not be eigenvectors associated with distinct eigenvalues.

5. 令 A 为 n 阶方阵. 若 A 的任一行中 n 个元素之和皆为 a, 证明 a 是 A 的特征值, 并且 n 维向量 $\alpha = (1, 1, \cdots, 1)^{\mathrm{T}}$ 是 A 的属于特征值 a 的特征向量.

5. Let A be a square matrix of order n. If the sum of n elements on each row of A is equal to a, prove that a is an eigenvalue of A, and the n dimensional vector $\alpha = (1, 1, \cdots, 1)^{\mathrm{T}}$ is an eigenvector of A associated with the eigenvalue a.

6. 设 A, B 均为 n 阶矩阵, A 可逆. 证明 $AB \sim BA$.

6. Suppose that A and B are all square matrices of order n and that A is invertible. Prove that $AB \sim BA$.

7. 判断 1 题中的矩阵是否可以对角化.

7. Determine whether the matrices in Exercise 1 can be diagonalizable.

8. 设

8. Let

$$A = \begin{pmatrix} 3 & 1 & 1 \\ 1 & 2 & 0 \\ 1 & 0 & 2 \end{pmatrix}$$

(1) 求 A 的特征值与特征向量;

(1) Find the eigenvalues and eigenvectors of A;

(2) 判断 A 能否对角化. 若能对角化, 则求出可逆矩阵 P, 将 A 对角化.

(2) Determine whether or not A is diagonalizable. If so, find an invertible matrix P to diagonalize A.

9. 令 $\boldsymbol{A}$ 为三阶方阵. 已知 $\boldsymbol{A}$ 的特征值分别为 1, -1, 2, $\boldsymbol{B}=\boldsymbol{A}^3-5\boldsymbol{A}^2$.

9. Let $\boldsymbol{A}$ be a square matrix of order 3. Known that the eigenvalues of $\boldsymbol{A}$ are given by 1, -1 and 2, respectively, and that $\boldsymbol{B}=\boldsymbol{A}^3-5\boldsymbol{A}^2$.

(1) 求 $\boldsymbol{B}$ 的特征值及与其相似的对角阵;

(1) Find the eigenvalues of $\boldsymbol{B}$ and find the diagonal matrix similar to $\boldsymbol{B}$;

(2) 求行列式 $|\boldsymbol{B}|$ 和 $|\boldsymbol{A}-5\boldsymbol{E}|$.

(2) Find the determinants $|\boldsymbol{B}|$ and $|\boldsymbol{A}-5\boldsymbol{E}|$.

10. 设

10. Suppose that

$$\boldsymbol{A}=\begin{pmatrix}1 & 1 & -1\\ 0 & 0 & 1\\ 0 & -2 & 3\end{pmatrix}$$

求 $\boldsymbol{A}^{100}$.

Find $\boldsymbol{A}^{100}$.

11. 利用施密特正交化方法将下面的向量组正交化:

11. Orthogonalize the following vector set by the Gram-Schmidt orthogonalization process:

$$\boldsymbol{\alpha}_1=\begin{pmatrix}1\\0\\-1\\1\end{pmatrix},\quad \boldsymbol{\alpha}_2=\begin{pmatrix}1\\-1\\0\\1\end{pmatrix},\quad \boldsymbol{\alpha}_3=\begin{pmatrix}-1\\1\\1\\0\end{pmatrix}$$

12. 求下列向量的长度:

12. Find the length of the following vectors:

$$\boldsymbol{\alpha}_1=\begin{pmatrix}1\\0\\2\\1\end{pmatrix},\quad \boldsymbol{\alpha}_2=\begin{pmatrix}2\\1\\2\\3\end{pmatrix}$$

13. 判断下列矩阵是否为正交矩阵:

13. Decide whether or not the following matrices are orthogonal matrices:

(1) $\begin{pmatrix} \frac{\sqrt{3}}{2} & -\frac{1}{2} \\ \frac{1}{2} & \frac{\sqrt{3}}{2} \end{pmatrix}$; (2) $\begin{pmatrix} \frac{\sqrt{2}}{2} & \frac{\sqrt{6}}{6} & \frac{\sqrt{3}}{6} & \frac{1}{2} \\ \frac{\sqrt{2}}{2} & -\frac{\sqrt{6}}{6} & -\frac{\sqrt{3}}{6} & \frac{1}{2} \\ 0 & \frac{\sqrt{6}}{3} & \frac{-\sqrt{3}}{6} & \frac{1}{2} \\ 0 & 0 & \frac{\sqrt{3}}{2} & \frac{1}{2} \end{pmatrix}$.

14. 若 $\boldsymbol{A}_1, \boldsymbol{A}_2, \cdots, \boldsymbol{A}_m$ 都是 n 阶正交矩阵, 证明乘积 $\boldsymbol{A}_1\boldsymbol{A}_2\cdots\boldsymbol{A}_m$ 也是正交矩阵.

14. If $\boldsymbol{A}_1, \boldsymbol{A}_2, \cdots, \boldsymbol{A}_m$ are all orthogonal matrices of order n, prove that the product $\boldsymbol{A}_1\boldsymbol{A}_2\cdots\boldsymbol{A}_m$ is also an orthogonal matrix.

15. 若 $\boldsymbol{A}$ 是正交矩阵, 证明 $\boldsymbol{A}^{\mathrm{T}}$ 也是正交矩阵.

15. If $\boldsymbol{A}$ is an orthogonal matrix, prove that $\boldsymbol{A}^{\mathrm{T}}$ is also an orthogonal matrix.

16. 令 $\boldsymbol{A}$ 是实对称矩阵, $\boldsymbol{P}$ 是正交矩阵. 证明 $\boldsymbol{P}^{-1}\boldsymbol{A}\boldsymbol{P}$ 也是实对称矩阵.

16. Let $\boldsymbol{A}$ be a real symmetric matrix and $\boldsymbol{P}$ be an orthogonal matrix. Prove that $\boldsymbol{P}^{-1}\boldsymbol{A}\boldsymbol{P}$ is also a real symmetric matrix.

17. 设 $\boldsymbol{x}$ 为 n 维列向量, $\boldsymbol{x}^{\mathrm{T}}\boldsymbol{x} = 1$, 令 $\boldsymbol{H} = \boldsymbol{E} - 2\boldsymbol{x}\boldsymbol{x}^{\mathrm{T}}$. 证明 $\boldsymbol{H}$ 是对称的正交矩阵.

17. Let $\boldsymbol{x}$ be an n dimensional column vector satisfying $\boldsymbol{x}^{\mathrm{T}}\boldsymbol{x} = 1$ and $\boldsymbol{H} = \boldsymbol{E} - 2\boldsymbol{x}\boldsymbol{x}^{\mathrm{T}}$. Prove that $\boldsymbol{H}$ is a symmetric orthogonal matrix.

18. 求正交矩阵, 分别将下面的对称矩阵化为对角矩阵.

18. Find orthogonal matrices to reduce following symmetric matrices to diagonal matrices, respectively.

(1) $\boldsymbol{A} = \begin{pmatrix} 0 & -2 & 2 \\ -2 & -3 & 4 \\ 2 & 4 & -3 \end{pmatrix}$; (2) $\boldsymbol{A} = \begin{pmatrix} 2 & -1 & -1 \\ -1 & 2 & -1 \\ -1 & -1 & 2 \end{pmatrix}$;

(3) $\boldsymbol{A} = \begin{pmatrix} 4 & 1 & 0 & -1 \\ 1 & 4 & -1 & 0 \\ 0 & -1 & 4 & 1 \\ -1 & 0 & 1 & 4 \end{pmatrix}$; (4) $\boldsymbol{A} = \begin{pmatrix} 3 & -2 & 0 \\ -2 & 2 & -2 \\ 0 & -2 & 1 \end{pmatrix}$.

补充题 5

Supplement Exercise 5

1. 已知 1 是矩阵 $\boldsymbol{A}$ 的一个特征值, 其中

1. Known that 1 is an eigenvalue of the matrix $\boldsymbol{A}$, where

$$\boldsymbol{A}=\begin{pmatrix}1&1&1\\1&3&a\\1&a&1\end{pmatrix}$$

求 a 的值.

Find the value of a.

2. 已知矩阵 $\boldsymbol{A}$ 只有一个线性无关的特征向量, 其中

2. Known that the matrix $\boldsymbol{A}$ has only a linearly independent eigenvector, where

$$\boldsymbol{A}=\begin{pmatrix}1&-1&a\\1&3&5\\0&0&2\end{pmatrix}$$

求 $\boldsymbol{A}$ 的特征值.

Find the eigenvalues of $\boldsymbol{A}$.

3. 已知

3. Known that

$$\boldsymbol{A}=\begin{pmatrix}3&2&2\\2&3&2\\2&2&3\end{pmatrix}$$

求 $\boldsymbol{A}$ 的伴随矩阵 $\boldsymbol{A}^*$ 的特征值.

Find the eigenvalues of the adjoint matrix $\boldsymbol{A}^*$ of $\boldsymbol{A}$.

4. 设方阵 $\boldsymbol{A}$ 满足 $\boldsymbol{A}^2-3\boldsymbol{A}+2\boldsymbol{E}=\boldsymbol{O}$, 其中 $\boldsymbol{E}$ 为 n 阶单位矩阵, 求 $\boldsymbol{A}$ 的特征值.

4. Suppose that the square matrix $\boldsymbol{A}$ satisfies $\boldsymbol{A}^2-3\boldsymbol{A}+2\boldsymbol{E}=\boldsymbol{O}$, where $\boldsymbol{E}$ is a unit matrix of order n. Find the eigenvalues of $\boldsymbol{A}$.

5. 设 3 阶奇异矩阵 $\boldsymbol{A}$ 的特征值分别为 1 和 2, $\boldsymbol{B}=\boldsymbol{A}^2-2\boldsymbol{A}+3\boldsymbol{E}$. 求 $|\boldsymbol{B}|$.

5. Suppose that the eigenvalues of a singular matrix $\boldsymbol{A}$ of order 3 are given by 1 and 2, respectively, and that $\boldsymbol{B}=\boldsymbol{A}^2-2\boldsymbol{A}+3\boldsymbol{E}$. Find $|\boldsymbol{B}|$.

6. 已知矩阵 $\boldsymbol{A}$ 只有 2 个线性无关的特征向量, 其中

6. Known that the matrix $\boldsymbol{A}$ has only two linearly independent eigenvectors, where

$$A = \begin{pmatrix} 2 & 0 & 3 \\ 3 & 3 & a \\ 0 & 0 & 3 \end{pmatrix}$$

求 a 的值.

Find the value of a.

7. 令 $\boldsymbol{A}$ 为 3 阶方阵. 设 $\boldsymbol{A}$ 的特征值分别为 $1, 0, -1$, 对应的特征向量依次为

7. Let $\boldsymbol{A}$ be a square matrix of order 3. Suppose that the eigenvalues of $\boldsymbol{A}$ are given by 1, 0 and -1, respectively, and that the corresponding eigenvectors are in turn as follows

$$\boldsymbol{p}_1 = \begin{pmatrix} 1 \\ 2 \\ 2 \end{pmatrix}, \quad \boldsymbol{p}_2 = \begin{pmatrix} 2 \\ -2 \\ 1 \end{pmatrix}, \quad \boldsymbol{p}_3 = \begin{pmatrix} -2 \\ -1 \\ 2 \end{pmatrix}$$

求 $\boldsymbol{A}$.

Find $\boldsymbol{A}$.

8. 设矩阵 $\boldsymbol{A}$ 有 3 个线性无关的特征向量, 其中

8. Suppose that the matrix $\boldsymbol{A}$ has three linearly independent eigenvectors, where

$$A = \begin{pmatrix} 1 & -1 & 1 \\ x & 4 & y \\ -3 & 3 & 5 \end{pmatrix}$$

且 $\lambda = 2$ 是 $\boldsymbol{A}$ 的二重特征值. 求可逆矩阵 $\boldsymbol{P}$ 使 $\boldsymbol{P}^{-1}\boldsymbol{AP}$ 成为对角矩阵.

and that $\lambda = 2$ is an eigenvalue of multiplicity 2 of $\boldsymbol{A}$. Find an invertible matrix $\boldsymbol{P}$ such that $\boldsymbol{P}^{-1}\boldsymbol{AP}$ is a diagonal matrix.

9. 设矩阵 $\boldsymbol{A}$ 与矩阵 $\boldsymbol{B}$ 相似, 其中

9. Suppose that the matrices $\boldsymbol{A}$ and $\boldsymbol{B}$ are similar, where

$$A = \begin{pmatrix} -2 & 0 & 0 \\ 2 & x & 2 \\ 3 & 1 & 1 \end{pmatrix}, \quad B = \begin{pmatrix} -1 & 0 & 0 \\ 0 & 2 & 0 \\ 0 & 0 & y \end{pmatrix}$$

(1) 求 x, y 的值; (2) 求可逆矩阵 $\boldsymbol{P}$, 使 $\boldsymbol{P}^{-1}\boldsymbol{AP} = \boldsymbol{B}$.

(1) Find the vuales of x, y; (2) Find an invertible matrix $\boldsymbol{P}$ such that $\boldsymbol{P}^{-1}\boldsymbol{AP} = \boldsymbol{B}$.

10. 设 $\boldsymbol{A}$ 和 $\boldsymbol{B}$ 是两个 n 阶方阵, $\boldsymbol{B}$ 的特征多项式为 $f(\lambda)$. 证明 $f(\boldsymbol{A})$ 可逆的充要条件是 $\boldsymbol{B}$ 的任一特征值都不是 $\boldsymbol{A}$ 的特征值.

10. Let $\boldsymbol{A}$ and $\boldsymbol{B}$ be two square matrices of order n and let $f(\lambda)$ be the characteristic polynomial of $\boldsymbol{B}$. Prove that $f(\boldsymbol{A})$ is invertible if and only if each eigenvalue of $\boldsymbol{B}$ is not an eigenvalue of $\boldsymbol{A}$.

第 6 章 Chapter 6

二次型 Quadratic Forms

二次型的研究起源于解析几何中化二次曲面的方程为标准形式的问题. 二次型不但在几何中出现, 而且在数学的其他分支以及物理、力学中也常常会遇到. 在本章里, 我们用矩阵知识来讨论二次型的一般理论, 如二次型的化简和正定二次型的判定以及一些基本性质.

The study about quadratic forms originated in how to transform the equation of a quadratic surface into a canonical form in Analytic Geometry. Quadratic forms not only arise in Geometry, but also run into other branches of Mathematics, Physics and Mechanics. In this chapter, we use the known knowledge about matrices to discuss the general theory of quadratic forms, such as simplifying the quadratic forms, determining the positive definite quadratic forms and some basic properties.

➢本章内容提要

Headline of this chapter

1. 二次型的矩阵表示、二次型的标准形的求法;

1. Matrix representations of quadratic forms, methods for finding canonical forms of quadratic forms;

2. 正定二次型的定义及其判定.

2. Definitions and determination of positive definite quadratic forms.

6.1 二次型的概念及其矩阵表示

6.1 Concepts and Matrix Representations of Quadratic Forms

在解析几何中, 坐标原点与中心重合时的有心二次曲线的一般方程是

In Analytic Geometry, the general equation of a quadratic curve whose origin coincides with its center has the following form

$$ax^2 + bxy + cy^2 = 1 \tag{6.1}$$

为了研究这个二次曲线的几何性质, 可以选择适当的角度 θ, 作旋转变换

In order to study the geometric properties of the quadratic curve, we can choose a suitable angle θ and use the following rotation transformation

$$\begin{cases} x = x'\cos\theta - y'\sin\theta \\ y = x'\sin\theta + y'\cos\theta \end{cases} \tag{6.2}$$

将方程 (6.1) 化成标准方程. 对二次曲面的研究也有类似的情况. 将上面的作法加以一般化, 便导致了对二次型的研究.

to transform Equation(6.1) into a standard equation. The study of quadratic surfaces also has the similar case. Generalizing the above methods leads to the study of quadratic forms.

定义 1 令 P 是一个数域. 含有 n 个变量 $x_1, x_2, \cdots, x_n$ 的二次齐次多项式

Definition 1 Let P be a number field. A quadratic homogeneous polynomial with respect to n variables $x_1, x_2, \cdots, x_n$, given by

$$\begin{aligned} f(x_1, x_2, \cdots, x_n) = a_{11}x_1^2 + 2a_{12}x_2 + \cdots + 2a_{1n}x_1x_n \\ + a_{22}x_2^2 + \cdots + 2a_{2n}x_2x_n \\ + \cdots + \cdots \\ + a_{nn}x_n^2 \end{aligned} \tag{6.3}$$

称为数域 P 上的一个 n **元二次型**.

is said to be a **quadratic form in** n **variables** over P.

系数全为实数的二次型称为实二次型. 简称**二次型**. 本书只讨论实二次型.

A quadratic form with real coefficients is called a real quadratic form, for short, a **quadratic form**. In this textbook, we only discuss real quadratic forms.

例如,

For example,

$$x_1^2+x_1x_2+3x_1x_3+2x_2^2+4x_2x_3+3x_3^2$$

是一个三元二次型. 不难验证这个二次型可写成下面的矩阵形式

is a quadratic form in three variables. It is not difficult to verify that the quadratic form can be written as the following matrix form

$$\begin{aligned}&x_1^2+x_1x_2+3x_1x_3+2x_2^2+4x_2x_3+3x_3^2\\ =&(x_1,x_2,x_3)\begin{pmatrix}1&\frac{1}{2}&\frac{3}{2}\\ \frac{1}{2}&2&2\\ \frac{3}{2}&2&3\end{pmatrix}\begin{pmatrix}x_1\\x_2\\x_3\end{pmatrix}\end{aligned}$$

其中

where

$$\boldsymbol{A}=\begin{pmatrix}1&\frac{1}{2}&\frac{3}{2}\\ \frac{1}{2}&2&2\\ \frac{3}{2}&2&3\end{pmatrix}$$

是一个对称矩阵.

is a symmetric matrix.

一般地, 在式 (6.3) 中令 $a_{ij}=a_{ji}(i<j)$, 则有

In general, let $a_{ij}=a_{ji}(i<j)$ in Equation(6.3), then we have

$$\begin{aligned}f(x_1,x_2,\cdots,x_n)=&a_{11}x_1^2+a_{12}x_1x_2+\cdots+a_{1n}x_1x_n\\ &+a_{21}x_2x_1+a_{22}x_2^2+\cdots+a_{2n}x_2x_n\\ &+\cdots\\ &+a_{n1}x_nx_1+a_{n2}x_nx_2+\cdots+a_{nn}x_n^2\end{aligned}\tag{6.4}$$

令

Let

$$\boldsymbol{A}=\begin{pmatrix}a_{11}&a_{12}&\cdots&a_{1n}\\ a_{21}&a_{22}&\cdots&a_{2n}\\ \vdots&\vdots&&\vdots\\ a_{n1}&a_{n2}&\cdots&a_{nn}\end{pmatrix}$$

则

then

$$\boldsymbol{A}=\boldsymbol{A}^{\mathrm{T}}$$

令

Let

$$\boldsymbol{x}=\begin{pmatrix} x_1 \\ x_2 \\ \vdots \\ x_n \end{pmatrix}$$

于是

it leads to

$$f=\boldsymbol{x}^{\mathrm{T}}\boldsymbol{A}\boldsymbol{x}=(x_1,x_2,\cdots,x_n)\begin{pmatrix} a_{11} & a_{12} & \cdots & a_{1n} \\ a_{21} & a_{22} & \cdots & a_{2n} \\ \vdots & \vdots & & \vdots \\ a_{n1} & a_{n2} & \cdots & a_{nn} \end{pmatrix}\begin{pmatrix} x_1 \\ x_2 \\ \vdots \\ x_n \end{pmatrix}$$

称 $\boldsymbol{x}^{\mathrm{T}}\boldsymbol{A}\boldsymbol{x}$ **为二次型的矩阵表示**.

$\boldsymbol{x}^{\mathrm{T}}\boldsymbol{A}\boldsymbol{x}$ is called the **matrix representation of a quadratic form.**

显然这种矩阵表示是唯一的, 即任给一个二次型就唯一确定一个对称矩阵. 反之, 任给一个对称矩阵也可唯一确定一个二次型. 换句话说, 二者之间存在一一对应关系. $\boldsymbol{A}$ 称为**二次型**$f=\boldsymbol{x}^{\mathrm{T}}\boldsymbol{A}\boldsymbol{x}$**的对称矩阵**, $\boldsymbol{A}$ 的秩称为 $f=\boldsymbol{x}^{\mathrm{T}}\boldsymbol{A}\boldsymbol{x}$ 的**秩**. 称 $f=\boldsymbol{x}^{\mathrm{T}}\boldsymbol{A}\boldsymbol{x}$ **为对称矩阵**A**的二次型**.

Obviously, the above matrix representation is unique, that is, for any given quadratic form, we can determine a symmetric matrix uniquely. Conversely, for any given symmetric matrix we can determine a quadratic form uniquely. In other words, there exists a one-to-one correspondence between them. $\boldsymbol{A}$ is called the **symmetric matrix associated with the quadratic form** $f=\boldsymbol{x}^{\mathrm{T}}\boldsymbol{A}\boldsymbol{x}$, and the rank of $\boldsymbol{A}$ is called the **rank** of $f=\boldsymbol{x}^{\mathrm{T}}\boldsymbol{A}\boldsymbol{x}$. Meanwhile, $f=\boldsymbol{x}^{\mathrm{T}}\boldsymbol{A}\boldsymbol{x}$ is called the **quadratic form of the symmetric matrix** $\boldsymbol{A}$.

注 $\boldsymbol{A}$ 的对角元素 a_{ii} 为 x_i^2 的系数, 非对角元素 $a_{ij}(i\neq j)$ 为 x_ix_j 系数的 1/2.

Note The diagonal element a_{ii} of $\boldsymbol{A}$ is the coefficient of x_i^2, and the off-diagonal elements $a_{ij}(i\neq j)$ are all the half of the corresponding coefficients of x_ix_j.

例 1 写出对应于二次型 $f = x_1x_2 + x_1x_3 + 2x_2^2 - 3x_2x_3$ 的矩阵及矩阵表示.

Example 1 Write out the matrix and the matrix representation associated with the quadratic form $f = x_1x_2 + x_1x_3 + 2x_2^2 - 3x_2x_3$.

解 二次型的对应矩阵为

Solution The corresponding matrix of the quadratic form is

$$\boldsymbol{A} = \begin{pmatrix} 0 & \frac{1}{2} & \frac{1}{2} \\ \frac{1}{2} & 2 & -\frac{3}{2} \\ \frac{1}{2} & -\frac{3}{2} & 0 \end{pmatrix}$$

相应的矩阵表示为

and the corresponding matrix representation is

$$f = x_1x_2 + x_1x_3 + 2x_2^2 - 3x_2x_3 = (x_1, x_2, x_3) \begin{pmatrix} 0 & \frac{1}{2} & \frac{1}{2} \\ \frac{1}{2} & 2 & -\frac{3}{2} \\ \frac{1}{2} & -\frac{3}{2} & 0 \end{pmatrix} \begin{pmatrix} x_1 \\ x_2 \\ x_3 \end{pmatrix} \quad \blacksquare$$

和几何中遇到的问题一样, 希望通过变量的线性变换来简化二次型. 为此, 我们引入如下定义.

Similar to the problems in Geometry, we hope to simplify quadratic forms by linear transformation of variables. Therefore, we introduce the following definition.

定义 2 设 $x_1, x_2, \cdots, x_n$ 和 $y_1, y_2, \cdots, y_n$ 是两组变量. 如下关系式

Definition 2 Let $x_1, x_2, \cdots, x_n$ and $y_1, y_2, \cdots, y_n$ be two sets of variables. The following relation, given by,

$$\begin{cases} x_1 = c_{11}y_1 + c_{12}y_2 + \cdots + c_{1n}y_n \\ x_2 = c_{21}y_2 + c_{22}y_2 + \cdots + c_{2n}y_n \\ \quad \cdots\cdots \\ x_n = c_{n1}y_n + c_{n2}y_n + \cdots + c_{nn}y_n \end{cases} \tag{6.5}$$

称为由 $x_1, x_2, \cdots, x_n$ 到 $y_1, y_2, \cdots, y_n$ 的一个**线性变换**.

is said to be a **linear transformation** from $x_1, x_2, \cdots, x_n$ to $y_1, y_2, \cdots, y_n$.

令

Let

$$\boldsymbol{C}=(c_{ij})_{n\times n},\quad \boldsymbol{x}=\begin{pmatrix}x_1\\x_2\\\vdots\\x_n\end{pmatrix},\quad \boldsymbol{y}=\begin{pmatrix}y_1\\y_2\\\vdots\\y_n\end{pmatrix}$$

则线性变换 (6.5) 可写为

then the linear transformation (6.5) can be written as

$$\boldsymbol{x}=\boldsymbol{C}\boldsymbol{y}$$

如果系数行列式不为零, 即 $|\boldsymbol{C}|\neq 0$, 那么线性变换 (6.5) 称为**非退化的**.

If the coefficient determinant is nonzero, i.e., $|\boldsymbol{C}|\neq 0$, then the linear transformation (6.5) is said to be **non-degenerate**.

如果 $\boldsymbol{C}=(c_{ij})$ 是正交矩阵, 则线性变换 (6.5) 称为**正交变换**.

If $\boldsymbol{C}=(c_{ij})$ is an orthogonal matrix, then the linear transformation (6.5) is said to be an **orthogonal transformation**.

例如, 在式 (6.2) 中,

For example, in Equation(6.2),

$$\begin{vmatrix}\cos\theta & -\sin\theta\\ \sin\theta & \cos\theta\end{vmatrix}=1\neq 0$$

因此线性变换 (6.2) 为非退化的, 且是正交变换.

it means that the linear transformation (6.2) is non-degenerate and is an orthogonal transformation.

设二次型为 $f(x_1,x_2,\cdots,x_n)=\boldsymbol{x}^{\mathrm{T}}\boldsymbol{A}\boldsymbol{x}$, 其中 $\boldsymbol{A}=\boldsymbol{A}^{\mathrm{T}}$. 经非退化线性变换 $\boldsymbol{x}=\boldsymbol{C}\boldsymbol{y}$, 得到

Suppose that the quadratic form is given by $f(x_1,x_2,\cdots,x_n)=\boldsymbol{x}^{\mathrm{T}}\boldsymbol{A}\boldsymbol{x}$, where $\boldsymbol{A}=\boldsymbol{A}^{\mathrm{T}}$. Using the non-degenerate linear transformation $\boldsymbol{x}=\boldsymbol{C}\boldsymbol{y}$, we get

$$f(x_1,x_2,\cdots,x_n)=(\boldsymbol{C}\boldsymbol{y})^{\mathrm{T}}\boldsymbol{A}(\boldsymbol{C}\boldsymbol{y})=\boldsymbol{y}^{\mathrm{T}}(\boldsymbol{C}^{\mathrm{T}}\boldsymbol{A}\boldsymbol{C})\boldsymbol{y}=\boldsymbol{y}^{\mathrm{T}}\boldsymbol{B}\boldsymbol{y}$$

显然, 它是关于变量 $y_1,y_2,\cdots,y_n$ 的新二次型, 其中 $\boldsymbol{B}=\boldsymbol{C}^{\mathrm{T}}\boldsymbol{A}\boldsymbol{C}$. 由 $\boldsymbol{B}^{\mathrm{T}}=\boldsymbol{C}^{\mathrm{T}}\boldsymbol{A}^{\mathrm{T}}\boldsymbol{C}=\boldsymbol{C}^{\mathrm{T}}\boldsymbol{A}\boldsymbol{C}=\boldsymbol{B}$ 知, $\boldsymbol{B}$ 为对称矩阵. 矩阵 $\boldsymbol{B}=\boldsymbol{C}^{\mathrm{T}}\boldsymbol{A}\boldsymbol{C}$ 反映了变换前后两个二次型的矩阵之间的关系. 与之相应, 我们引入如下定义.

Obviously, it is a new quadratic form with respect to the variables $y_1,y_2,\cdots,y_n$, where $\boldsymbol{B}=\boldsymbol{C}^{\mathrm{T}}\boldsymbol{A}\boldsymbol{C}$. Since $\boldsymbol{B}^{\mathrm{T}}=\boldsymbol{C}^{\mathrm{T}}\boldsymbol{A}^{\mathrm{T}}\boldsymbol{C}=\boldsymbol{C}^{\mathrm{T}}\boldsymbol{A}\boldsymbol{C}=\boldsymbol{B}$, we know that $\boldsymbol{B}$ is symmetric. The matrix $\boldsymbol{B}=\boldsymbol{C}^{\mathrm{T}}\boldsymbol{A}\boldsymbol{C}$ reflects the relation between the matrices associated with the original and the transformed quadratic forms. In correspondence, we introduce the following definition.

定义 3 如果有可逆的矩阵 $\boldsymbol{C}$, 使 $\boldsymbol{C}^{\mathrm{T}}\boldsymbol{AC}=\boldsymbol{B}$, 则称矩阵 $\boldsymbol{A}$ 与 $\boldsymbol{B}$ 合同, 记作 $\boldsymbol{A}\cong\boldsymbol{B}$.

Definition 3 If there exists an invertible matrix $\boldsymbol{C}$ such that $\boldsymbol{C}^{\mathrm{T}}\boldsymbol{AC}=\boldsymbol{B}$, then $\boldsymbol{A}$ is said to be congruent to $\boldsymbol{B}$, written as $\boldsymbol{A}\cong\boldsymbol{B}$.

矩阵的合同关系有以下性质:

The congruent relation of matrices has the following properties:

(1) **反身性** $\boldsymbol{A}\cong\boldsymbol{A}$. 事实上, $\boldsymbol{A}=\boldsymbol{E}^{\mathrm{T}}\boldsymbol{AE}$.

(1) **Reflexivity** $\boldsymbol{A}\cong\boldsymbol{A}$. In fact, $\boldsymbol{A}=\boldsymbol{E}^{\mathrm{T}}\boldsymbol{AE}$.

(2) **对称性** 若 $\boldsymbol{A}\cong\boldsymbol{B}$, 则 $\boldsymbol{B}\cong\boldsymbol{A}$. 事实上, 由 $\boldsymbol{B}=\boldsymbol{C}^{\mathrm{T}}\boldsymbol{AC}$, 得 $\boldsymbol{A}=(\boldsymbol{C}^{-1})^{\mathrm{T}}\boldsymbol{BC}^{-1}$.

(2) **Symmetry** If $\boldsymbol{A}\cong\boldsymbol{B}$ then $\boldsymbol{B}\cong\boldsymbol{A}$. In fact, from $\boldsymbol{B}=\boldsymbol{C}^{\mathrm{T}}\boldsymbol{AC}$ we get $\boldsymbol{A}=(\boldsymbol{C}^{-1})^{\mathrm{T}}\boldsymbol{BC}^{-1}$.

(3) **传递性** 若 $\boldsymbol{A}\cong\boldsymbol{B}$, $\boldsymbol{B}\cong\boldsymbol{C}$, 则 $\boldsymbol{A}\cong\boldsymbol{C}$. 事实上, 由 $\boldsymbol{B}=\boldsymbol{C}_1^{\mathrm{T}}\boldsymbol{AC}_1$, $\boldsymbol{C}=\boldsymbol{C}_2^{\mathrm{T}}\boldsymbol{BC}_2$, 即得 $\boldsymbol{C}=(\boldsymbol{C}_1\boldsymbol{C}_2)^{\mathrm{T}}\boldsymbol{A}(\boldsymbol{C}_1\boldsymbol{C}_2)$.

(3) **Transitivity** If $\boldsymbol{A}\cong\boldsymbol{B}$ and $\boldsymbol{B}\cong\boldsymbol{C}$, then $\boldsymbol{B}\cong\boldsymbol{C}$. In fact, from $\boldsymbol{B}=\boldsymbol{C}_1^{\mathrm{T}}\boldsymbol{AC}_1$ and $\boldsymbol{C}=\boldsymbol{C}_2^{\mathrm{T}}\boldsymbol{BC}_2$, we get $\boldsymbol{C}=(\boldsymbol{C}_1\boldsymbol{C}_2)^{\mathrm{T}}\boldsymbol{A}(\boldsymbol{C}_1\boldsymbol{C}_2)$.

思 考 题

Question

二次型和对称矩阵有什么关系?

What are the relations between quadratic forms and symmetric matrices?

6.2 二次型的标准形

6.2 Canonical Forms of Quadratic Forms

问题 与 $\boldsymbol{A}$ 合同的矩阵中最简单的形式是什么? 即二次型 $\boldsymbol{y}^{\mathrm{T}}(\boldsymbol{C}^{\mathrm{T}}\boldsymbol{AC})\boldsymbol{y}$ 的最简单形式如何?

Problem What is the simplest form in the matrices congruent to $\boldsymbol{A}$, namely, what is the simplest form of the quadratic form $\boldsymbol{y}^{\mathrm{T}}(\boldsymbol{C}^{\mathrm{T}}\boldsymbol{AC})\boldsymbol{y}$?

如果二次型 $\boldsymbol{y}^{\mathrm{T}}(\boldsymbol{C}^{\mathrm{T}}\boldsymbol{AC})\boldsymbol{y}$ 具有下面的形式

If the quadratic form $\boldsymbol{y}^{\mathrm{T}}(\boldsymbol{C}^{\mathrm{T}}\boldsymbol{AC})\boldsymbol{y}$ has the following form

$$d_1y_1^2+d_2y_2^2+\cdots+d_ry_r^2$$

其中 $d_i\neq 0(i=1,2,\cdots,r,r\leqslant n)$. 称这个形式为二次型 (6.3) 的一个**标准形**. 易知 $r=R(\boldsymbol{A})$. 矩阵

where $d_i\neq 0(i=1,2,\cdots,r,r\leqslant n)$. This form is said to be a **canonical form** of the quadratic form (6.3). Obviously, $r=R(\boldsymbol{A})$. The matrix

$$
\boldsymbol{D}=\begin{pmatrix} d_1 & & & & & & \\ & d_2 & & & & & \\ & & \ddots & & & & \\ & & & d_r & & & \\ & & & & 0 & & \\ & & & & & \ddots & \\ & & & & & & 0 \end{pmatrix}
$$

称为对称矩阵 $\boldsymbol{A}$ 的合同标准形矩阵, 简称为 $\boldsymbol{A}$ 的**标准形**.

is called the congruent canonical matrix of the symmetric matrix $\boldsymbol{A}$, for short, the **canonical form** of $\boldsymbol{A}$.

定理 1 任给二次型 $f=\boldsymbol{x}^{\mathrm{T}}\boldsymbol{A}\boldsymbol{x}$, 总有正交变换 $x=\boldsymbol{P}\boldsymbol{y}$, 使 f 化成标准型

Theorem 1 For any given quadratic form $f=\boldsymbol{x}^{\mathrm{T}}\boldsymbol{A}\boldsymbol{x}$, there always exists an orthogonal transformation $\boldsymbol{x}=\boldsymbol{P}\boldsymbol{y}$ such that f is reduced to the following canonical form

$$
f=\lambda_1 y_1^2+\lambda_2 y_2^2+\cdots+\lambda_n y_n^2
$$

其中 $\lambda_1,\lambda_2,\cdots,\lambda_n$ 是 f 的 $\boldsymbol{A}$ 的特征值.

where $\lambda_1,\lambda_2,\cdots,\lambda_n$ are eigenvalues of $\boldsymbol{A}$ associated with f.

这个定理是 5.4 节中定理 3 的等价说法.

This theorem is an equivalent statement of Theorem 3 in Section 5.4.

用正交变换将二次型化为标准型, 其特点是保持几何图形不变. 因此, 它在理论上和实际应用上都是非常重要的. 此方法的具体步骤就是 5.2 节所介绍的化实对称矩阵为对角矩阵的步骤.

Using orthogonal transformations to reduce a quadratic form to a canonical form can preserve the geometric graph invariant. Therefore, this approach is very important in theoretical and in practical applications. The practical procedure is to transform symmetric matrices into diagonal matrices, which has been mentioned in Section 5.2.

例 1 求一个正交变换 $\boldsymbol{x}=\boldsymbol{P}\boldsymbol{y}$, 将二次型

Example 1 Find an orthogonal transformation $\boldsymbol{x}=\boldsymbol{P}\boldsymbol{y}$ to reduce the following quadratic form

$$
f=2x_1x_2+2x_3x_4
$$

化为标准形.

to a canonical form.

解 f 对应的矩阵为

Solution The matrix associated with f is

$$\boldsymbol{A}=\begin{pmatrix}0&1&0&0\\1&0&0&0\\0&0&0&1\\0&0&1&0\end{pmatrix}$$

$\boldsymbol{A}$ 的特征方程为

The characteristic equation of $\boldsymbol{A}$ is

$$|\lambda\boldsymbol{E}-\boldsymbol{A}|=\begin{vmatrix}\lambda&-1&0&0\\-1&\lambda&0&0\\0&0&\lambda&-1\\0&0&-1&\lambda\end{vmatrix}=(\lambda+1)^2(\lambda-1)^2=0$$

$\boldsymbol{A}$ 的特征值为

The eigenvalues of $\boldsymbol{A}$ are given by

$$\lambda_1=\lambda_2=1,\quad \lambda_3=\lambda_4=-1$$

对于 $\lambda_1=\lambda_2=1$, 方程组 $(\boldsymbol{E}-\boldsymbol{A})\boldsymbol{x}=\boldsymbol{0}$ 的系数矩阵为

For $\lambda_1=\lambda_2=1$, the coefficient matrix associated with the system $(\boldsymbol{E}-\boldsymbol{A})\boldsymbol{x}=\boldsymbol{0}$ is

$$\boldsymbol{E}-\boldsymbol{A}=\begin{pmatrix}1&-1&0&0\\-1&1&0&0\\0&0&1&-1\\0&0&-1&1\end{pmatrix}\to\begin{pmatrix}1&-1&0&0\\0&0&0&0\\0&0&1&-1\\0&0&0&0\end{pmatrix}$$

相应的特征向量为

the corresponding eigenvectors are

$$\boldsymbol{\alpha}_1=\begin{pmatrix}1\\1\\0\\0\end{pmatrix},\quad \boldsymbol{\alpha}_2=\begin{pmatrix}0\\0\\1\\1\end{pmatrix}$$

对于 $\lambda_3=\lambda_4=-1$, 方程组 $(\boldsymbol{E}+\boldsymbol{A})\boldsymbol{x}=\boldsymbol{0}$ 的系数矩阵为

For $\lambda_3=\lambda_4=-1$, the coefficient matrix associated with the system $(\boldsymbol{E}+\boldsymbol{A})\boldsymbol{x}=\boldsymbol{0}$ is given by

$$\boldsymbol{E}-\boldsymbol{A}=\begin{pmatrix}1&1&0&0\\1&1&0&0\\0&0&1&1\\0&0&1&1\end{pmatrix}\to\begin{pmatrix}1&1&0&0\\0&0&0&0\\0&0&1&1\\0&0&0&0\end{pmatrix}$$

相应的特征向量为

the corresponding eigenvectors are

$$\boldsymbol{\alpha}_3 = \begin{pmatrix} 1 \\ -1 \\ 0 \\ 0 \end{pmatrix}, \quad \boldsymbol{\alpha}_4 = \begin{pmatrix} 0 \\ 0 \\ 1 \\ -1 \end{pmatrix}$$

由于 $\boldsymbol{\alpha}_1, \boldsymbol{\alpha}_2, \boldsymbol{\alpha}_3, \boldsymbol{\alpha}_4$ 彼此正交, 因此只需将其单位化, 即

Since $\boldsymbol{\alpha}_1, \boldsymbol{\alpha}_2, \boldsymbol{\alpha}_3, \boldsymbol{\alpha}_4$ are orthogonal to each other, it only needs to normalize them, namely,

$$\boldsymbol{\eta}_1 = \frac{1}{\sqrt{2}}\begin{pmatrix} 1 \\ 1 \\ 0 \\ 0 \end{pmatrix}, \quad \boldsymbol{\eta}_2 = \frac{1}{\sqrt{2}}\begin{pmatrix} 0 \\ 0 \\ 1 \\ 1 \end{pmatrix}, \quad \boldsymbol{\eta}_3 = \frac{1}{\sqrt{2}}\begin{pmatrix} 1 \\ -1 \\ 0 \\ 0 \end{pmatrix}, \quad \boldsymbol{\eta}_4 = \frac{1}{\sqrt{2}}\begin{pmatrix} 0 \\ 0 \\ 1 \\ -1 \end{pmatrix}$$

得到正交矩阵

we obtain the wanted orthogonal matrix

$$\boldsymbol{P} = (\boldsymbol{\eta}_1, \boldsymbol{\eta}_2, \boldsymbol{\eta}_3) = \frac{1}{\sqrt{2}}\begin{pmatrix} 1 & 0 & 1 & 0 \\ 1 & 0 & -1 & 0 \\ 0 & 1 & 0 & 1 \\ 0 & 1 & 0 & -1 \end{pmatrix}$$

令

Let

$$\boldsymbol{x} = \boldsymbol{P}\boldsymbol{y}$$

则得到标准形

we then get the corresponding canonical form

$$f = \boldsymbol{x}^{\mathrm{T}}\boldsymbol{A}\boldsymbol{x} = \boldsymbol{y}^{\mathrm{T}}(\boldsymbol{P}^{\mathrm{T}}\boldsymbol{A}\boldsymbol{P})\boldsymbol{y} = y_1^2 + y_2^2 - y_3^2 - y_4^2 \quad \blacksquare$$

例 2 设二次型 $f = 2x_1^2 + 3x_2^2 + 3x_3^2 + 2ax_2x_3 (a > 0)$ 可以通过正交变换可化为标准形 $f = y_1^2 + 2y_2^2 + 5y_3^2$, 求参数 a 及所用的正交变换.

Example 2 Suppose that the quadratic form, given by $f = 2x_1^2 + 3x_2^2 + 3x_3^2 + 2ax_2x_3 (a > 0)$, can be reduced to the canonical form $f = y_1^2 + 2y_2^2 + 5y_3^2$ by an orthogonal transformation. Determine the parameter a and the used orthogonal transformation.

分析 n 元二次型 $f=\boldsymbol{x}^{\mathrm{T}}\boldsymbol{A}\boldsymbol{x}$ 可以通过正交变换 $\boldsymbol{x}=\boldsymbol{P}\boldsymbol{y}$ 化成如下的标准型

Analysis Using an orthogonal transformation $\boldsymbol{x}=\boldsymbol{P}\boldsymbol{y}$, we can reduce the quadratic form $f=\boldsymbol{x}^{\mathrm{T}}\boldsymbol{A}\boldsymbol{x}$ in n variables to the following canonical form

$$f=\lambda_1 y_1^2+\lambda_2 y_2^2+\cdots+\lambda_n y_n^2$$

其中 $\lambda_1,\lambda_2,\cdots,\lambda_n$ 是 $\boldsymbol{A}$ 的特征值, 而且变换前后矩阵有下面的关系

where $\lambda_1,\lambda_2,\cdots,\lambda_n$ are eigenvalues of $\boldsymbol{A}$, moreover, the original and the transformed matrices have the following relation

$$\boldsymbol{P}^{\mathrm{T}}\boldsymbol{A}\boldsymbol{P}=\begin{pmatrix}\lambda_1 & & & \\ & \lambda_2 & & \\ & & \ddots & \\ & & & \lambda_n\end{pmatrix}$$

所以上式两边取行列式即可求得参数 a.

Hence, we can determine the parameter a by performing the operation of determinants on both sides.

解 变换前后二次型的矩阵分别为

Solution The original and the transformed matrices are respectively given by

$$\boldsymbol{A}=\begin{pmatrix}2&0&0\\0&3&a\\0&a&3\end{pmatrix},\quad \boldsymbol{\Lambda}=\begin{pmatrix}1&0&0\\0&2&0\\0&0&5\end{pmatrix}$$

设所求正交矩阵为 $\boldsymbol{P}$, 则有 $\boldsymbol{P}^{\mathrm{T}}\boldsymbol{A}\boldsymbol{P}=\boldsymbol{\Lambda}$, 此时两边取行列式, 并注意到 $|\boldsymbol{P}|=\pm1$, 得

Let $\boldsymbol{P}$ be the wanted orthogonal matrix. Then $\boldsymbol{P}^{\mathrm{T}}\boldsymbol{A}\boldsymbol{P}=\boldsymbol{\Lambda}$. Taking the operation of determinants on both sides and using the condition that $|\boldsymbol{P}|=\pm1$, we get

$$\left|\boldsymbol{P}^{\mathrm{T}}\right||\boldsymbol{A}||\boldsymbol{P}|=|\boldsymbol{P}|^2|\boldsymbol{A}|=|\boldsymbol{A}|=|\boldsymbol{\Lambda}|$$

即

that is,

$$2(9-a^2)=10$$

由 $a>0$, 得 $a=2$. $\boldsymbol{A}$ 的特征值为 $\lambda_1=1,\lambda_2=2,\lambda_3=5$.

Since $a>0$, we get $a=2$. As a result, the eigenvalues of $\boldsymbol{A}$ are $\lambda_1=1,\lambda_2=2,\lambda_3=5$.

当 $\lambda_1 = 1$ 时, 解 $(\boldsymbol{A}-\boldsymbol{E})\boldsymbol{x}=\boldsymbol{0}$, 求得特征向量为

For $\lambda_1 = 1$, solving $(\boldsymbol{A}-\boldsymbol{E})\boldsymbol{x}=\boldsymbol{0}$, we get the corresponding eigenvector, given by

$$\boldsymbol{\alpha}_1 = \begin{pmatrix} 0 \\ 1 \\ -1 \end{pmatrix}$$

同理, 可求得与 $\lambda_2 = 2, \lambda_3 = 5$ 对应的特征向量分别为

Similarly, we get that the eigenvectors associated with $\lambda_2 = 2$ and $\lambda_3 = 5$ are respectively given by

$$\boldsymbol{\alpha}_2 = \begin{pmatrix} 1 \\ 0 \\ 0 \end{pmatrix}, \quad \boldsymbol{\alpha}_3 = \begin{pmatrix} 0 \\ 1 \\ 1 \end{pmatrix}$$

因为对应于不同特征值的特征向量是相互正交的, 所以 $\boldsymbol{\alpha}_1, \boldsymbol{\alpha}_2, \boldsymbol{\alpha}_3$ 是正交向量组, 将它们单位化得

Since the eigenvectors associated with distinct eigenvalues are orthogonal to each other, we know that $\boldsymbol{\alpha}_1, \boldsymbol{\alpha}_2, \boldsymbol{\alpha}_3$ are orthogonal. Normalizing them, we get

$$\boldsymbol{\eta}_1 = \frac{1}{\|\boldsymbol{\alpha}_1\|}\boldsymbol{\alpha}_1 = \begin{pmatrix} 0 \\ \frac{1}{\sqrt{2}} \\ -\frac{1}{\sqrt{2}} \end{pmatrix}, \quad \boldsymbol{\eta}_2 = \frac{1}{\|\boldsymbol{\alpha}_2\|}\boldsymbol{\alpha}_2 = \begin{pmatrix} 1 \\ 0 \\ 0 \end{pmatrix}, \quad \boldsymbol{\eta}_3 = \frac{1}{\|\boldsymbol{\alpha}_3\|}\boldsymbol{\alpha}_3 = \begin{pmatrix} 0 \\ \frac{1}{\sqrt{2}} \\ \frac{1}{\sqrt{2}} \end{pmatrix}$$

以 $\boldsymbol{\eta}_1, \boldsymbol{\eta}_2, \boldsymbol{\eta}_3$ 为列即得所求的正交矩阵, 即

The wanted orthogonal matrix can be formed by taking $\boldsymbol{\eta}_1, \boldsymbol{\eta}_2, \boldsymbol{\eta}_3$ to be its columns, given by

$$\boldsymbol{P} = (\boldsymbol{\eta}_1, \boldsymbol{\eta}_2, \boldsymbol{\eta}_3) = \begin{pmatrix} 0 & 1 & 0 \\ \frac{1}{\sqrt{2}} & 0 & \frac{1}{\sqrt{2}} \\ -\frac{1}{\sqrt{2}} & 0 & \frac{1}{\sqrt{2}} \end{pmatrix}$$ ■

例 3 设

Example 3 Let

$$\boldsymbol{A} = \begin{pmatrix} 1 & -2 & 2 \\ -2 & 4 & -4 \\ 2 & -4 & 4 \end{pmatrix}$$

(1) 求一个正交矩阵 $\boldsymbol{P}$ 使 $\boldsymbol{P}^{-1}\boldsymbol{AP}$ 成为对角矩阵.

(1) Find an orthogonal matrix $\boldsymbol{P}$ such that $\boldsymbol{P}^{-1}\boldsymbol{AP}$ is a diagonal matrix.

(2) 求一个正交变换 $\boldsymbol{x}=\boldsymbol{Py}$, 将下面的二次型

(2) Find an orthogonal transformation $\boldsymbol{x}=\boldsymbol{Py}$ to reduce the following quadratic form

$$f=x_1^2-4x_1x_2+4x_1x_3+4x_2^2+4x_3^2-8x_2x_3$$

约化为标准形.

to a canonical form.

解 (1) $\boldsymbol{A}$ 的特征方程为

Solution (1) The characteristic polynomial of $\boldsymbol{A}$ is

$$|\boldsymbol{A}-\lambda\boldsymbol{E}|=\begin{vmatrix}1-\lambda & -2 & 2\\ -2 & 4-\lambda & -4\\ 2 & -4 & 4-\lambda\end{vmatrix}=-\lambda^2(\lambda-9)=0$$

易见, $\boldsymbol{A}$ 的特征值为 $\lambda_1=\lambda_2=0$, $\lambda_3=9$.

It is easy to see that the eigenvalues of $\boldsymbol{A}$ are given by $\lambda_1=\lambda_2=0$, $\lambda_3=9$.

对 $\lambda_1=\lambda_2=0$, 求解 $(\boldsymbol{A}-0\boldsymbol{E})\boldsymbol{x}=\boldsymbol{0}$, 可得如下线性无关的特征向量为

For $\lambda_1=\lambda_2=0$, solving $(\boldsymbol{A}-0\boldsymbol{E})\boldsymbol{x}=\boldsymbol{0}$, we get the following linearly independent eigenvectors, given by

$$\boldsymbol{\alpha}_1=\begin{pmatrix}2\\1\\0\end{pmatrix},\quad \boldsymbol{\alpha}_2=\begin{pmatrix}-2\\0\\1\end{pmatrix}$$

将它们正交化得

Orthogonalizing them, we get

$$\boldsymbol{\beta}_1=\boldsymbol{\alpha}_1=\begin{pmatrix}2\\1\\0\end{pmatrix},\quad \boldsymbol{\beta}_2=\boldsymbol{\alpha}_2-\frac{[\boldsymbol{\alpha}_2,\boldsymbol{\beta}_1]}{[\boldsymbol{\beta}_1,\boldsymbol{\beta}_1]}\boldsymbol{\beta}_1=\begin{pmatrix}-\frac{2}{5}\\ \frac{4}{5}\\ 1\end{pmatrix}$$

对于 $\lambda_3=9$, 求解 $(\boldsymbol{A}-9\boldsymbol{E})\boldsymbol{x}=\boldsymbol{0}$, 得到一个线性无关的特征向量, 即

For $\lambda_3=9$, solving $(\boldsymbol{A}-9\boldsymbol{E})\boldsymbol{x}=\boldsymbol{0}$, we get a linearly independent eigenvector, namely,

$$\boldsymbol{\alpha}_3=\begin{pmatrix}1\\-2\\2\end{pmatrix}$$

由于 $\boldsymbol{\alpha}_3$ 与 $\boldsymbol{\beta}_1, \boldsymbol{\beta}_2$ 正交, 故只需将 $\boldsymbol{\beta}_1$, $\boldsymbol{\beta}_2$, $\boldsymbol{\alpha}_3$ 单位化, 得

Since $\boldsymbol{\alpha}_3$ is orthogonal to $\boldsymbol{\beta}_1, \boldsymbol{\beta}_2$, it only needs to normalize $\boldsymbol{\beta}_1, \boldsymbol{\beta}_2, \boldsymbol{\alpha}_3$, as follows,

$$\boldsymbol{\eta}_1 = \begin{pmatrix} \dfrac{2}{\sqrt{5}} \\ \dfrac{1}{\sqrt{5}} \\ 0 \end{pmatrix}, \quad \boldsymbol{\eta}_2 = \begin{pmatrix} -\dfrac{2}{3\sqrt{5}} \\ \dfrac{4}{3\sqrt{5}} \\ \dfrac{5}{3\sqrt{5}} \end{pmatrix}, \quad \boldsymbol{\eta}_3 = \begin{pmatrix} \dfrac{1}{3} \\ -\dfrac{2}{3} \\ \dfrac{2}{3} \end{pmatrix}$$

则

Then

$$\boldsymbol{P} = (\boldsymbol{\eta}_1, \boldsymbol{\eta}_2, \boldsymbol{\eta}_3) = \begin{pmatrix} \dfrac{2}{\sqrt{5}} & -\dfrac{2}{3\sqrt{5}} & \dfrac{1}{3} \\ \dfrac{1}{\sqrt{5}} & \dfrac{4}{3\sqrt{5}} & -\dfrac{2}{3} \\ 0 & \dfrac{5}{3\sqrt{5}} & \dfrac{2}{3} \end{pmatrix}$$

为所求的正交矩阵, 且有

is the wanted orthogonal matrix, and

$$\boldsymbol{P}^{-1}\boldsymbol{A}\boldsymbol{P} = \boldsymbol{\Lambda}$$

其中

where

$$\boldsymbol{\Lambda} = \begin{pmatrix} 0 & & \\ & 0 & \\ & & 9 \end{pmatrix}$$

(2) 易见, f 对应的矩阵恰为 $\boldsymbol{A}$. 由 (1) 的计算可得正交变换 $\boldsymbol{x} = \boldsymbol{P}\boldsymbol{y}$, 则

(2) Obviously, the matrix associated with f is exactly given by $\boldsymbol{A}$. From the calculation of (1) we have the orthogonal transformation $\boldsymbol{x} = \boldsymbol{P}\boldsymbol{y}$, then

$$f = \boldsymbol{x}^{\mathrm{T}}\boldsymbol{A}\boldsymbol{x} = \boldsymbol{y}^{\mathrm{T}}\boldsymbol{\Lambda}\boldsymbol{y} = 9y_3^2$$ ■

注 (1) 实对称矩阵的属于不同特征值的特征向量是正交的, 所以本例中属于 $\lambda_3 = 9$ 的特征向量 $\boldsymbol{\alpha}_3$ 与 $\lambda_1 = \lambda_2 = 0$ 的特征向量正交, 只需将其标准化即可.

Note (1) For real symmetric matrices, the eigenvectors associated with distinct eigenvalues are orthogonal. Therefore, in this example, the eigenvector $\boldsymbol{\alpha}_3$ associated with $\lambda_3 = 9$ is orthogonal to the eigenvectors associated with $\lambda_1 = \lambda_2 = 0$, which means that it needs only to normalize it.

(2) 特征向量 (即齐次线性方程组的基础解系) 的取法是不唯一的, 所以正交矩阵 $\boldsymbol{P}$ 不唯一.

(2) The choice of the eigenvectors (namely, the system of fundamental solutions of the homonegeous linear system) is not unique, so is the orthogonal matrix $\boldsymbol{P}$.

(3) 在构成正交矩阵 $\boldsymbol{P}$ 时, 规范正交向量 $\boldsymbol{\eta}_1, \boldsymbol{\eta}_2, \boldsymbol{\eta}_3$ 的顺序是可变的, 但得到的对角阵中 $\lambda_1, \lambda_2, \lambda_3$ 的位置也要做相应的变化.

(3) During the course of constructing the orthogonal matrix $\boldsymbol{P}$, the order of the orthonormal vectors $\boldsymbol{\eta}_1, \boldsymbol{\eta}_2, \boldsymbol{\eta}_3$ can be changed, however, the positions of $\lambda_1, \lambda_2, \lambda_3$ in the diagonal matrix must be changed accordingly.

用正交变换约化二次型为标准形, 具有保持几何形状不变的优点. 除了正交变换, 还有多种方法可将二次型约化为标准形. 如配方法、初等变换法等. 下面通过实例来介绍**配方法**和**初等变换法**.

Using orthogonal transformations to reduce the quadratic form to a canonical form, the advantage is that the geometrical shape is invariable. Except for orthogonal transformations, there are several methods to reduce the quadratic form to a canonical form, such as the methods of completing square, elementary transformations, and so on. Next we introduce the methods of **completing square** and **elementary transformations** by some concrete example.

例 4 用配方法化二次型

Example 4 Use the method of completing square to reduce the quadratic form

$$f = x_1^2 + 2x_2^2 + 5x_3^2 + 2x_1x_2 + 2x_1x_3 + 6x_2x_3$$

成标准形, 并求对应的变换矩阵.

to a canonical form and find the corresponding transformation matrix.

解 由于 f 中含变量 x_1 的平方项, 故将含 x_1 的项归并起来配方可得

Solution Since f contains the squared term of x_1, collecting the terms containing x_1 and then completing the squares, we get

$$\begin{aligned} f &= (x_1^2 + 2x_1x_2 + 2x_1x_3) + 2x_2^2 + 5x_3^2 + 6x_2x_3 \\ &= (x_1 + x_2 + x_3)^2 - x_2^2 - x_3^2 - 2x_2x_3 + 2x_2^2 + 5x_3^2 + 6x_2x_3 \\ &= (x_1 + x_2 + x_3)^2 + x_2^2 + 4x_2x_3 + 4x_3^2 \end{aligned}$$

上式右端除第一项外已不再含 x_1. 继续配方, 得

Except for the first term, the right side of the above equation contains x_1 no more. Completing the squares sequentially, we get

$$f=(x_1+x_2+x_3)^2+(x_2+2x_3)^2$$

令

Let

$$\begin{cases} y_1=x_1+x_2+x_3 \\ y_2=x_2+2x_3 \\ y_3=x_3 \end{cases}$$

有

We obtain

$$\begin{pmatrix} y_1 \\ y_2 \\ y_3 \end{pmatrix}=\begin{pmatrix} 1 & 1 & 1 \\ 0 & 1 & 2 \\ 0 & 0 & 1 \end{pmatrix}\begin{pmatrix} x_1 \\ x_2 \\ x_3 \end{pmatrix}$$

所以

Hence,

$$\begin{pmatrix} x_1 \\ x_2 \\ x_3 \end{pmatrix}=\begin{pmatrix} 1 & 1 & 1 \\ 0 & 1 & 2 \\ 0 & 0 & 1 \end{pmatrix}^{-1}\begin{pmatrix} y_1 \\ y_2 \\ y_3 \end{pmatrix}=\begin{pmatrix} 1 & -1 & 1 \\ 0 & 1 & -2 \\ 0 & 0 & 1 \end{pmatrix}\begin{pmatrix} y_1 \\ y_2 \\ y_3 \end{pmatrix}$$

即

that is,

$$\begin{cases} x_1=y_1-y_2+y_3 \\ x_2=y_2-2y_3 \\ x_3=y_3 \end{cases}$$

经非退化线性变换 $\boldsymbol{x}=\boldsymbol{C}\boldsymbol{y}$ 后, f 被约化为如下的标准形

After performing the non-degenerate linear transformation $\boldsymbol{x}=\boldsymbol{C}\boldsymbol{y}$, we reduce f to the following canonical form

$$f=y_1^2+y_2^2$$

对应的变换矩阵为

and the corresponding transformation matrix is

$$\boldsymbol{C}=\begin{pmatrix} 1 & -1 & 1 \\ 0 & 1 & -2 \\ 0 & 0 & 1 \end{pmatrix} \quad (|\boldsymbol{C}|=1\neq 0)$$

■

例 5 用配方法化二次型

Example 5 Use the method of completing square to reduce the quadratic form

$$f = 2x_1x_2 + 2x_1x_3 - 6x_2x_3$$

成标准形, 并求所用的变换矩阵.

to a canonical form and find the corresponding transformation matrix.

解 显然, f 不含平方项. 由于含有 x_1x_2 乘积项, 取如下的非退化线性变换

Solution Obviously, f has no square term. Since f contains x_1x_2, perform the following non-degenetate linear transformation

$$\begin{cases} x_1 = y_1 + y_2 \\ x_2 = y_1 - y_2 \\ x_3 = \qquad y_3 \end{cases}$$

矩阵表示式为

The matrix representation is

$$\begin{pmatrix} x_1 \\ x_2 \\ x_3 \end{pmatrix} = \begin{pmatrix} 1 & 1 & 0 \\ 1 & -1 & 0 \\ 0 & 0 & 1 \end{pmatrix} \begin{pmatrix} y_1 \\ y_2 \\ y_3 \end{pmatrix}$$

即

that is,

$$\boldsymbol{x} = \boldsymbol{C}_1\boldsymbol{y}$$

代入 f 可得

Substituting it into f, we get

$$f = 2y_1^2 - 2y_2^2 - 4y_1y_3 + 8y_1y_3$$

再配方, 得

We then complete the squares and obtain

$$f = 2(y_1 - y_3)^2 - 2(y_2 - 2y_3)^2 + 6y_3^2$$

令

Let

$$\begin{cases} z_1 = y_1 - y_3 \\ z_2 = y_2 - 2y_3 \\ z_3 = y_3 \end{cases}$$

则

then

$$\begin{pmatrix} z_1 \\ z_2 \\ z_3 \end{pmatrix} = \begin{pmatrix} 1 & 0 & -1 \\ 0 & 1 & -2 \\ 0 & 0 & 1 \end{pmatrix} \begin{pmatrix} y_1 \\ y_2 \\ y_3 \end{pmatrix}$$

所以

Hence

$$\begin{pmatrix} y_1 \\ y_2 \\ y_3 \end{pmatrix} = \begin{pmatrix} 1 & 0 & -1 \\ 0 & 1 & -2 \\ 0 & 0 & 1 \end{pmatrix}^{-1} \begin{pmatrix} z_1 \\ z_2 \\ z_3 \end{pmatrix} = \begin{pmatrix} 1 & 0 & 1 \\ 0 & 1 & 2 \\ 0 & 0 & 1 \end{pmatrix} \begin{pmatrix} z_1 \\ z_2 \\ z_3 \end{pmatrix}$$

即

that is,

$$\begin{cases} y_1 = z_1 + z_3 \\ y_2 = z_2 + 2z_3 \\ y_3 = \qquad z_3 \end{cases}$$

$$\boldsymbol{y} = \boldsymbol{C}_2 \boldsymbol{z}$$

代入 f 可得标准形

Substituting it into f, we get the corresponding canonical form

$$f = 2z_1^2 - 2z_2^2 + 6z_3^2$$

对应的变换矩阵为

the corresponding transformation matrix used is

$$\boldsymbol{C} = \boldsymbol{C}_1 \boldsymbol{C}_2 = \begin{pmatrix} 1 & 1 & 0 \\ 1 & -1 & 0 \\ 0 & 0 & 1 \end{pmatrix} \begin{pmatrix} 1 & 0 & 1 \\ 0 & 1 & 2 \\ 0 & 0 & 1 \end{pmatrix}$$

$$= \begin{pmatrix} 1 & 1 & 3 \\ 1 & -1 & -1 \\ 0 & 0 & 1 \end{pmatrix} \quad (|\boldsymbol{C}| = -2 \neq 0)$$ ■

事实上, 约化二次型为标准形就是寻求可逆矩阵 $\boldsymbol{C}$, 使 $\boldsymbol{C}^{\mathrm{T}}\boldsymbol{A}\boldsymbol{C}$ 成为对角矩阵, 这里 $\boldsymbol{A}$ 为二次型的矩阵. 下面介绍一种利用矩阵初等变换将 $\boldsymbol{A}$ 约化为对角矩阵的方法.

In fact, reducing the quadratic form to a canonical form is to finding an invertible matrix $\boldsymbol{C}$ such that $\boldsymbol{C}^{\mathrm{T}}\boldsymbol{A}\boldsymbol{C}$ is a diagonal matrix, where $\boldsymbol{A}$ is the matrix associated with a quadratic form. In the following part we introduce a method for reducing $\boldsymbol{A}$ to a diagonal form by elementary operations on matrices.

定理 2 对实对称矩阵 $\boldsymbol{A}$, 一定存在一系列初等矩阵 $\boldsymbol{E}_1, \boldsymbol{E}_2, \cdots, \boldsymbol{E}_s$ 使得

Theorem 2 For a real symmetric matrix $\boldsymbol{A}$, there must exist a sequence of elementary matrices $\boldsymbol{E}_1, \boldsymbol{E}_2, \cdots, \boldsymbol{E}_s$ such that

$$\boldsymbol{E}_s^{\mathrm{T}} \cdots \boldsymbol{E}_2^{\mathrm{T}} \boldsymbol{E}_1^{\mathrm{T}} \boldsymbol{A} \boldsymbol{E}_1 \boldsymbol{E}_2 \cdots \boldsymbol{E}_s = \mathrm{diag}(d_1, d_2, \cdots, d_n)$$

关于初等矩阵, 易见

For elementary matrices, it is easy to see that

$$\boldsymbol{E}^{\mathrm{T}}(i,j)=\boldsymbol{E}(i,j),\quad \boldsymbol{E}^{\mathrm{T}}(i(k))=\boldsymbol{E}(i(k)),\quad \boldsymbol{E}^{\mathrm{T}}(ij(k))=\boldsymbol{E}(ji(k))$$

记 $\boldsymbol{C}=\boldsymbol{E}_1\boldsymbol{E}_2\cdots\boldsymbol{E}_s$. 则定理 2 还表明:

Denoted by $\boldsymbol{C}=\boldsymbol{E}_1\boldsymbol{E}_2\cdots\boldsymbol{E}_s$. Theorem 2 also shows that

对 $\boldsymbol{A}$ 同时施行一系列同类的初等行、列变换, 得到对角矩阵, 而相应地将这一系列的初等列变换施加于单位矩阵, 就得到变换矩阵 $\boldsymbol{C}$. 具体做法如下:

We can diagonalize $\boldsymbol{A}$ by performing a sequence of elementary row and column operations of the same type on $\boldsymbol{A}$ simultaneously, and then we can obtain $\boldsymbol{C}$ by performing the corresponding elementary operations on a unit matrix. The concrete implementation is as follows.

(1) 将 n 阶单位阵 $\boldsymbol{E}$ 放在二次型的矩阵 $\boldsymbol{A}$ 的下面, 形成一个 $2n\times n$ 矩阵.

(1) Put the unit matrix $\boldsymbol{E}$ of order n under $\boldsymbol{A}$ and form a $2n\times n$ matrix.

(2) 对此矩阵作相同的行、列初等变换, 将 $\boldsymbol{A}$ 化成对角形的同时, 将单位阵变换成了可逆的变换矩阵 $\boldsymbol{C}$.

(2) Perform elementary row and column operations of the same type on it. As long as $\boldsymbol{A}$ is reduced to a diagonal form, $\boldsymbol{E}$ is transformed into the wanted invertible transformation matrix $\boldsymbol{C}$.

这就是所谓的**初等变换法**.

This is the so called **method of elementary operations**.

例 6 用初等变换法将例 4 中二次型 f 化为标准形.

Example 6 Use the method of elementary operations to reduce the quadratic form in Example 4 to a canonical form.

解 对下面矩阵作相同的行、列初等变换

Solution Perform the row and column operations of the same type on the following matrix

$$\begin{pmatrix}\boldsymbol{A}\\ \boldsymbol{E}\end{pmatrix}=\begin{pmatrix}1&1&1\\1&2&3\\1&3&5\\1&0&0\\0&1&0\\0&0&1\end{pmatrix}\xrightarrow[r_2-r_1,r_3-r_1]{c_2-c_1,c_3-c_1}\begin{pmatrix}1&0&0\\0&1&2\\0&2&4\\1&-1&-1\\0&1&0\\0&0&1\end{pmatrix}\xrightarrow[r_3-2r_2]{c_3-2c_2}\begin{pmatrix}1&0&0\\0&1&0\\0&0&0\\1&-1&1\\0&1&-2\\0&0&1\end{pmatrix}$$

则经非退化线性变换 $\boldsymbol{x}=\boldsymbol{C}\boldsymbol{y}$ 就将 f 化成标准形

then f is reduced to the following canonical form by the non-degenerate linear transformation $\boldsymbol{x}=\boldsymbol{C}\boldsymbol{y}$

$$f=y_1^2+y_2^2$$

其中

where

$$\boldsymbol{C}=\begin{pmatrix}1&-1&1\\0&1&-2\\0&0&1\end{pmatrix}$$

■

例 7 用初等变换法将例 5 中二次型 f 化为标准形.

Example 7 Use the method of elementary operations to reduce the quadratic form in Example 5 to a canonical form.

解 对下面矩阵作相同的行、列初等变换

Solution Perform the row and column operations of the same type on the following matrix

$$\begin{pmatrix}\boldsymbol{A}\\\boldsymbol{E}\end{pmatrix}=\begin{pmatrix}0&1&1\\1&0&-3\\1&-3&0\\1&0&0\\0&1&0\\0&0&1\end{pmatrix}\xrightarrow[r_1+r_2]{c_1+c_2}\begin{pmatrix}2&1&-2\\1&0&-3\\-2&-3&0\\1&0&0\\1&1&0\\0&0&1\end{pmatrix}\xrightarrow[r_2-1/2r_1,r_3+r_1]{c_2-1/2c_1,c_3+c_1}$$

$$\begin{pmatrix}2&0&0\\0&-\frac{1}{2}&-2\\0&-2&-2\\1&-\frac{1}{2}&1\\1&\frac{1}{2}&1\\0&0&1\end{pmatrix}\xrightarrow[r_3-4r_2]{c_3-4c_2}\begin{pmatrix}2&0&0\\0&-\frac{1}{2}&0\\0&0&6\\1&-\frac{1}{2}&3\\1&\frac{1}{2}&-1\\0&0&1\end{pmatrix}$$

则经非退化线性变换 $\boldsymbol{x}=\boldsymbol{C}\boldsymbol{y}$ 将 f 约化为如下的标准形

then f is reduced to the following canonical form by performing the non-degenerate linear transformation $\boldsymbol{x}=\boldsymbol{C}\boldsymbol{y}$

$$f=2y_1^2-\frac{1}{2}y_2^2+6y_3^2$$

其中

where

$$C = \begin{pmatrix} 1 & \frac{-1}{2} & 3 \\ 1 & \frac{1}{2} & -1 \\ 0 & 0 & 1 \end{pmatrix}$$

■

思 考 题

Question

二次型的标准形是否唯一?

Is the canonical form of a quadratic form unique?

6.3 正定二次型

6.3 Positive Definite Quadratic Forms

本节讨论一种特殊的实二次型.

This section discusses a special kind of real quadratic forms.

定义 1 如果对于任意的 n 维非零实向量 $\boldsymbol{x} = (c_1, c_2, \cdots, c_n)^{\mathrm{T}}$, 都有 $f(c_1, c_2, \cdots, c_n) > 0$ 则称实二次型 $f(x_1, x_2, \cdots, x_n) = \boldsymbol{x}^{\mathrm{T}}\boldsymbol{A}\boldsymbol{x}$ 为**正定二次型**.

Definition 1 The real quadratic form $f(x_1, x_2, \cdots, x_n) = \boldsymbol{x}^{\mathrm{T}}\boldsymbol{A}\boldsymbol{x}$ is called a **positive definite quadratic form** if we always have $f(c_1, c_2, \cdots, c_n) > 0$ for any nonzero real vector $\boldsymbol{x} = (c_1, c_2, \cdots, c_n)^{\mathrm{T}}$.

如果二次型 $f(x_1, x_2, \cdots, x_n) = \boldsymbol{x}^{\mathrm{T}}\boldsymbol{A}\boldsymbol{x}$ 是正定的, 则称实对称矩阵 $\boldsymbol{A}$ 为**正定矩阵**.

If the quadratic form $f(x_1, x_2, \cdots, x_n) = \boldsymbol{x}^{\mathrm{T}}\boldsymbol{A}\boldsymbol{x}$ is positive definite, then the real symmetric matrix $\boldsymbol{A}$ is called a **positive definite matrix**.

与正定二次型的定义类似, 还有下面的概念.

Similar to the definition of positive definite quadratic forms, we have the following concepts.

定义 2 设 $f(x_1, x_2, \cdots, x_n)$ 是一个实二次型, 对任意的 n 维实向量 $\boldsymbol{x} = (c_1, c_2, \cdots, c_n)^{\mathrm{T}} \neq \mathbf{0}$, 如果都有 $f(c_1, c_2, \cdots, c_n) < 0$, 那么 $f(x_1, x_2, \cdots, x_n)$ 称为**负定**的; 如果都有 $f(c_1, c_2, \cdots, c_n) \geqslant 0$, 那么 $f(x_1, x_2, \cdots, x_n)$ 称为**半正定**的; 如果都有 $f(c_1, c_2, \cdots, c_n) \leqslant 0$, 那么 $f(x_1, x_2, \cdots, x_n)$ 称为**半负定**的; 如果它既不是半正定的, 又不是半负定的, 那么 $f(x_1, x_2, \cdots, x_n)$ 就称为**不定**的.

Definition 2 Let $f(x_1, x_2, \cdots, x_n)$ be a real quadratic form and let $\boldsymbol{x} = (c_1, c_2, \cdots, c_n)^{\mathrm{T}} \neq \mathbf{0}$ be an arbitrary n dimensional real vector. $f(x_1, x_2, \cdots, x_n)$ is said to be **negative definite** if $f(c_1, c_2, \cdots, c_n) < 0$, is said to be **positive semidefinite** if $f(c_1, c_2, \cdots, c_n) \geqslant 0$, and is said to be **negative semidefinite** if $f(c_1, c_2, \cdots, c_n) \leqslant 0$. That the quadratic form is neither positive semidefinite nor negative semidefinite is said to be **indefinite**.

例如, $f = x_1^2 + x_2^2 + \cdots + x_n^2$ 是正定二次型, $f = x_2^2 + \cdots + x_n^2$ 是半正定二次型.

For example, $f = x_1^2 + x_2^2 + \cdots + x_n^2$ is a positive definite quadratic form and $f = x_2^2 + \cdots + x_n^2$ is a positive semidefinite quadratic form.

设 $f(x_1, x_2, \cdots, x_n) = \boldsymbol{x}^{\mathrm{T}} \boldsymbol{A} \boldsymbol{x}$ 正定. 令 $\boldsymbol{x} = \boldsymbol{C}\boldsymbol{y}$, $\boldsymbol{C}$ 可逆, 则有

Suppose that $f(x_1, x_2, \cdots, x_n) = \boldsymbol{x}^{\mathrm{T}} \boldsymbol{A} \boldsymbol{x}$ is positive definite. Let $\boldsymbol{x} = \boldsymbol{C}\boldsymbol{y}$ and let $\boldsymbol{C}$ be invertible. Then we have

$$f(x_1, x_2, \cdots, x_n) = \boldsymbol{x}^{\mathrm{T}} \boldsymbol{A} \boldsymbol{x} = \boldsymbol{y}^{\mathrm{T}} (\boldsymbol{C}^{\mathrm{T}} \boldsymbol{A} \boldsymbol{C}) \boldsymbol{y}$$

即二次型 $\boldsymbol{y}^{\mathrm{T}} (\boldsymbol{C}^{\mathrm{T}} \boldsymbol{A} \boldsymbol{C}) \boldsymbol{y}$ 也正定.

that is to say, the quadratic form $\boldsymbol{y}^{\mathrm{T}} (\boldsymbol{C}^{\mathrm{T}} \boldsymbol{A} \boldsymbol{C}) \boldsymbol{y}$ is also positive definite.

事实上, 对任意非零实向量

In fact, for any nonzero real vector

$$\boldsymbol{y} = \begin{pmatrix} c_1 \\ c_2 \\ \vdots \\ c_n \end{pmatrix}$$

因为 $\boldsymbol{C}$ 可逆, 所以

since $\boldsymbol{C}$ is invertible, then

$$\boldsymbol{x} = \boldsymbol{C} \begin{pmatrix} c_1 \\ c_2 \\ \vdots \\ c_n \end{pmatrix} \neq \mathbf{0}$$

由于 $f(x_1, x_2, \cdots, x_n)$ 正定, 所以

Since $f(x_1, x_2, \cdots, x_n)$ is positive definite, then

$$(c_1, c_2, \cdots, c_n)(\boldsymbol{C}^{\mathrm{T}}\boldsymbol{A}\boldsymbol{C})\begin{pmatrix} c_1 \\ c_2 \\ \vdots \\ c_n \end{pmatrix} = \boldsymbol{x}^{\mathrm{T}}\boldsymbol{A}\boldsymbol{x} > 0$$

因此二次型 $\boldsymbol{y}^{\mathrm{T}}(\boldsymbol{C}^{\mathrm{T}}\boldsymbol{A}\boldsymbol{C})\boldsymbol{y}$ 也正定.

that is to say, $\boldsymbol{y}^{\mathrm{T}}(\boldsymbol{C}^{\mathrm{T}}\boldsymbol{A}\boldsymbol{C})\boldsymbol{y}$ is also positive definite.

以上说明: **正定二次型经过非退化线性替换时不改变二次型的正定性**. 于是有

In summary, **performing a non-degenerate linear transformation on a positive definite quadratic form does not change the positive definiteness of the quadratic form.** Consequently,

定理 1 实二次型 $f(x_1, x_2, \cdots, x_n) = \boldsymbol{x}^{\mathrm{T}}\boldsymbol{A}\boldsymbol{x}$ 正定的充分必要条件是它的标准形中, 诸 $d_i > 0 \quad (i = 1, 2, \cdots, n)$.

Theorem 1 The real quadratic form $f(x_1, x_2, \cdots, x_n) = \boldsymbol{x}^{\mathrm{T}}\boldsymbol{A}\boldsymbol{x}$ is positive definite if and only if $d_i > 0 (i = 1, 2, \cdots, n)$ in its canonical form.

所以, 若 $f(x_1, x_2, \cdots, x_n)$ 正定, 当且仅当 $\boldsymbol{A}$ 合同于

That is to say, $f(x_1, x_2, \cdots, x_n)$ is positive definite if and only if $\boldsymbol{A}$ is congruent to

$$\boldsymbol{D} = \begin{pmatrix} d_1 & & & \\ & d_2 & & \\ & & \ddots & \\ & & & d_n \end{pmatrix}$$

其中 $d_i > 0 \quad (i = 1, 2, \cdots, n)$.

where $d_i > 0 \quad (i = 1, 2, \cdots, n)$.

显然, $\boldsymbol{D}$ 合同于单位矩阵 $\boldsymbol{E}$. 因而, 有如下定理.

Obviously, $\boldsymbol{D}$ is congruent to the unit matrix $\boldsymbol{E}$. Then we have the following theorem.

定理 2 实二次型 $f(x_1, x_2, \cdots, x_n) = \boldsymbol{x}^{\mathrm{T}}\boldsymbol{A}\boldsymbol{x}$ 正定的充分必要条件是它的矩阵 $\boldsymbol{A}$ 与单位矩阵合同.

Theorem 2 The real quadratic form, given by $f(x_1, x_2, \cdots, x_n) = \boldsymbol{x}^{\mathrm{T}}\boldsymbol{A}\boldsymbol{x}$, is positive definite if and only if $\boldsymbol{A}$ is congruent to the unit matrix $\boldsymbol{E}$.

若 $\boldsymbol{A}$ 合同于 $\boldsymbol{E}$, 即 $\boldsymbol{A}=\boldsymbol{C}^{\mathrm{T}}\boldsymbol{C}$, 则 $|\boldsymbol{A}|>0$. 因此有如下推论.

If $\boldsymbol{A}$ is congruent to $\boldsymbol{E}$, that is, $\boldsymbol{A}=\boldsymbol{C}^{\mathrm{T}}\boldsymbol{C}$, then $|\boldsymbol{A}|>0$. Therefore, we have the follow corollary.

推论 1 如果 $\boldsymbol{A}$ 是正定矩阵, 则 $|\boldsymbol{A}|>0$.

Corollary 1 If $\boldsymbol{A}$ is a positive definite matrix, then $|\boldsymbol{A}|>0$.

我们还知道, 对实对称矩阵 $\boldsymbol{A}$ 而言, 存在正交矩阵 $\boldsymbol{P}$, 使得

We also know that, for a real symmetric matrix $\boldsymbol{A}$, there exists an orthogonal matrix $\boldsymbol{P}$ such that

$$\boldsymbol{P}^{\mathrm{T}}\boldsymbol{A}\boldsymbol{P}=\boldsymbol{P}^{-1}\boldsymbol{A}\boldsymbol{P}=\operatorname{diag}(\lambda_1,\lambda_2,\cdots,\lambda_n)$$

所以, $\boldsymbol{A}$ 正定 $\Leftrightarrow \lambda_i>0(i=1,2,\cdots,n)$. 因此有

Hence $\boldsymbol{A}$ is positive definite$\Leftrightarrow \lambda_i>0(i=1,2,\cdots,n)$. Thus we have

定理 3 实二次型 $f(x_1,x_2,\cdots,x_n)=\boldsymbol{x}^{\mathrm{T}}\boldsymbol{A}\boldsymbol{x}$(或实对称矩阵 $\boldsymbol{A}$) 正定的充分必要条件是 $\boldsymbol{A}$ 的特征值均大于零.

Theorem 3 A real quadratic form $f(x_1,x_2,\cdots,x_n)=\boldsymbol{x}^{\mathrm{T}}\boldsymbol{A}\boldsymbol{x}$ (or a real symmetric matrix $\boldsymbol{A}$) is positive definite if and only if all eigenvalues of $\boldsymbol{A}$ are positive.

有时, 我们希望从二次型的矩阵较为直接地判定它的正定性, 为此引入如下定义.

Sometimes we hope to determine the positive definiteness of a quadratic form by its matrix directly. Therefore, we introduce the following definition.

定义 3 子式

Definition 3 The subdeterminants, given by

$$\Delta_i=\begin{vmatrix} a_{11} & a_{12} & \cdots & a_{1i}\\ a_{21} & a_{22} & \cdots & a_{2i}\\ \vdots & \vdots & & \vdots\\ a_{i1} & a_{i2} & \cdots & a_{ii}\end{vmatrix}\quad (i=1,2,\cdots,n)$$

称为 $\boldsymbol{A}=(\boldsymbol{a}_{ij})_{n\times n}$ 的顺序主子式.

are called the sequence principal subdeterminants of $\boldsymbol{A}=(\boldsymbol{a}_{ij})_{n\times n}$.

定理 4 实对称矩阵 $\boldsymbol{A}$(或实二次型 $f(x_1,x_2,\cdots,x_n)=\boldsymbol{X}^{\mathrm{T}}\boldsymbol{A}\boldsymbol{X}$) 正定, 当且仅当 $\boldsymbol{A}$ 的各阶顺序主子式全为正, 即

Theorem 4 The real symmetric matrix $\boldsymbol{A}$ (or the real quadratic form $f(x_1,x_2,\cdots,x_n)=\boldsymbol{X}^{\mathrm{T}}\boldsymbol{A}\boldsymbol{X}$) is positive definite if and only if all the sequence principal subdeterminants of $\boldsymbol{A}$ are positive, that is,

$$\Delta_r = \begin{vmatrix} a_{11} & a_{12} & \cdots & a_{1r} \\ a_{21} & a_{22} & \cdots & a_{2r} \\ \vdots & \vdots & & \vdots \\ a_{r1} & a_{r2} & \cdots & a_{rr} \end{vmatrix} > 0 \quad (r = 1, 2, \cdots, n)$$

对称矩阵 $\boldsymbol{A}$ 为负定当且仅当 $\boldsymbol{A}$ 的奇数阶顺序主子式为负, 偶数阶顺序主子式为正, 即

The real symmetric matrix $\boldsymbol{A}$ is negative definite if and only if its principal subdeterminants of odd order are negative and those of even order are positive, that is,

$$(-1)^r \begin{vmatrix} a_{11} & a_{12} & \cdots & a_{1r} \\ a_{21} & a_{22} & \cdots & a_{2r} \\ \vdots & \vdots & & \vdots \\ a_{r1} & a_{r2} & \cdots & a_{rr} \end{vmatrix} > 0 \quad (r = 1, 2, \cdots, n)$$

这个定理称为赫尔维茨定理.

This theorem is called the Hurwitz theorem.

例 1 判定下面二次型的正定性.

Example 1 Determine the positive definiteness of the following quadratic form.

$$f = (x_1, x_2, x_3) \begin{pmatrix} 3 & 2 & 0 \\ 2 & 3 & 0 \\ 0 & 0 & 1 \end{pmatrix} \begin{pmatrix} x_1 \\ x_2 \\ x_3 \end{pmatrix}$$

解 由于 $|\boldsymbol{A} - \lambda\boldsymbol{E}| = (1-\lambda)^2(5-\lambda)$, 得 $\boldsymbol{A}$ 的特征值为 1, 1, 5. 根据定理 3, $\boldsymbol{A}$ 为正定矩阵, 从而 f 为正定二次型. ■

Solution Since $|\boldsymbol{A} - \lambda\boldsymbol{E}| = (1-\lambda)^2(5-\lambda)$, we get the eigenvalues of $\boldsymbol{A}$ are 1, 1, 5. From Theorem 3 we see that $\boldsymbol{A}$ is positive definite, hence f is positive definite. ■

例 2 判别下面二次型的正定性.

Example 2 Determine the positive definiteness of the following quadratic form

$$f = -5x_1^2 - 6x_2^2 - 4x_3^2 + 4x_1x_2 + 4x_1x_3$$

解 f 的对应矩阵为

Solution The matrix associated with f is

$$\boldsymbol{A} = \begin{pmatrix} -5 & 2 & 2 \\ 2 & -6 & 0 \\ 2 & 0 & -4 \end{pmatrix}$$

因为

Since

$$a_{11}=-5<0,\quad \begin{vmatrix} a_{11} & a_{12} \\ a_{21} & a_{22} \end{vmatrix} = \begin{vmatrix} -5 & 2 \\ 2 & -6 \end{vmatrix} = 26>0,\quad |\boldsymbol{A}|=-80<0$$

则根据定理 4 知, f 为负定二次型. ■

From Theorem 4 we know that f is a negative definite quadratic form. ■

例 3 设 $f=x_1^2+4x_2^2+4x_3^2+2\lambda x_1x_2-2x_1x_3+4x_2x_3$, 问 λ 取何值时, 使得 f 为正定二次型.

Example 3 Suppose that $f=x_1^2+4x_2^2+4x_3^2+2\lambda x_1x_2-2x_1x_3+4x_2x_3$. What values do λ take on such that f is a positive definite quadratic form.

解 对应于 f 的矩阵为

Solution The matrix associated with f is

$$\boldsymbol{A}=\begin{pmatrix} 1 & \lambda & -1 \\ \lambda & 4 & 2 \\ -1 & 2 & 4 \end{pmatrix}$$

由

Since

$$a_{11}=1>0,\quad \begin{vmatrix} a_{11} & a_{12} \\ a_{21} & a_{22} \end{vmatrix} = 4-\lambda^2,\quad |\boldsymbol{A}|=-4(\lambda-1)(\lambda+2),$$

根据定理 4 知, 当

From Theorem 4 we know that as

$$\begin{cases} 4-\lambda^2>0 \\ -4(\lambda-1)(\lambda+2)>0 \end{cases}$$

即当 $-2<\lambda<1$ 时, 所给二次型 f 正定.■

that is, as $-2<\lambda<1$, f is a positive definite quadratic form. ■

例 4 已知 $\boldsymbol{A}$ 是 n 阶正定矩阵. 证明: $|\boldsymbol{A}+\boldsymbol{E}|>1$.

Example 4 Known that $\boldsymbol{A}$ is an positive definite matrix of order n. Show that $|\boldsymbol{A}+\boldsymbol{E}|>1$.

证 因为 $\boldsymbol{A}$ 正定, 所以 $\boldsymbol{A}$ 的所有特征值 λ_i 都大于零, 那么 $\boldsymbol{A}+\boldsymbol{E}$ 的特征值 λ_i+1 均大于 1, 再由 $|\boldsymbol{A}+\boldsymbol{E}|=\prod\limits_{i=1}^{n}(\lambda_i+1)$ 得到 $|\boldsymbol{A}+\boldsymbol{E}|>1$. ■

Proof Since $\boldsymbol{A}$ is positive definite, the eigenvalues λ_i of $\boldsymbol{A}$ are all positive. Then the eigenvalues λ_i+1 of $\boldsymbol{A}+\boldsymbol{E}$ are all greater than 1. From $|\boldsymbol{A}+\boldsymbol{E}|=\prod\limits_{i=1}^{n}(\lambda_i+1)$ we get $|\boldsymbol{A}+\boldsymbol{E}|>1$. ■

思 考 题

判定二次型的正定性有几种常用的方法?

Questions

How many commonly used methods are there for determining the positive definiteness of a quadratic form?

习 题 6

Exercise 6

1. 写出下列二次型对应的矩阵.

1. Write out the matrices associated with the following quadratic forms.

(1) $f = x_1^2 - 2x_1x_2 + 3x_1x_3 - 2x_2^2 + 8x_2x_3 + 3x_3^2$;

(2) $f = x_1x_2 - x_1x_3 + 2x_2x_3 + x_4^2$.

2. 已知下列二次型的矩阵, 写出对应的二次型:

2. For the following given matrices of the quadratic forms, write out the corresponding quadratic forms:

$$(1)\ \boldsymbol{A} = \begin{pmatrix} 1 & -1 & -3 & 1 \\ -1 & 0 & -2 & \frac{1}{2} \\ -3 & -2 & \frac{1}{3} & -\frac{3}{2} \\ 1 & \frac{1}{2} & -\frac{3}{2} & 0 \end{pmatrix};\quad (2)\ \boldsymbol{A} = \begin{pmatrix} 0 & 1 & \frac{1}{2} & -\frac{3}{2} \\ 1 & 0 & -1 & -1 \\ \frac{1}{2} & -1 & 0 & 3 \\ -\frac{3}{2} & -1 & 3 & 0 \end{pmatrix}.$$

3. 用配方法化下列二次型为标准形, 并求所用的变换矩阵:

3. Use the method of completing square to transform the following quadratic forms into canonical forms and find the used transformation matrices:

(1) $f = x_1^2 + 2x_1x_2 + 2x_2^2 + 4x_2x_3 + 4x_3^2$;

(2) $f = x_1^2 - 3x_2^2 - 2x_1x_2 + 2x_1x_3 - 6x_2x_3$.

4. 对如下给定的矩阵 $\boldsymbol{A}$, 用初等变换的求一非奇异矩阵 $\boldsymbol{C}$, 使 $\boldsymbol{C}^{\mathrm{T}}\boldsymbol{A}\boldsymbol{C}$ 为对角矩阵.

4. For each matrix $\boldsymbol{A}$, use elementary operations to find a nonsingular matrix $\boldsymbol{C}$ such that $\boldsymbol{C}^{\mathrm{T}}\boldsymbol{A}\boldsymbol{C}$ is a diagonal matrix.

$$(1)\ \boldsymbol{A} = \begin{pmatrix} 1 & 2 & 0 \\ 2 & 0 & 1 \\ 0 & 1 & 3 \end{pmatrix};\quad (2)\ \boldsymbol{A} = \begin{pmatrix} 0 & 1 & -2 \\ 1 & 0 & -1 \\ -2 & -1 & 0 \end{pmatrix}.$$

5. 用正交线性变换化下列二次型为标准形:

5. Use the orthogonal linear transformation to transform the following quadratic forms into canonical forms:

(1) $f = x_1^2 + 2x_2^2 + 3x_3^2 - 4x_1x_2 - 4x_2x_3$; (2) $f = 2x_1x_2 + 2x_3x_4$.

6. 求 a 的值, 使下列二次型正定:

6. Find the values of a such that the following quadratic forms are positive definite:

(1) $f = x_1^2 + x_2^2 + 5x_3^2 + 2ax_1x_2 - 2x_1x_3 + 4x_2x_3$;

(2) $f = 5x_1^2 + x_2^2 + ax_3^2 + 4x_1x_2 - 2x_1x_3 - 2x_2x_3$.

7. 设 $\boldsymbol{A}$ 与 $\boldsymbol{B}$ 是 n 阶正定矩阵. 证明: $\boldsymbol{A}+\boldsymbol{B}$ 为正定矩阵.

7. Let $\boldsymbol{A}$ and $\boldsymbol{B}$ be positive definite matrices of order n. Prove that $\boldsymbol{A}+\boldsymbol{B}$ is a positive definite matrix.

8. 若 $\boldsymbol{A}$ 是 n 阶正定矩阵, 证明: 伴随矩阵 $\boldsymbol{A}^*$ 也是正定矩阵.

8. Let $\boldsymbol{A}$ be a positive definite matrix of order n. Prove that the adjoint matrix $\boldsymbol{A}^*$ of $\boldsymbol{A}$ is also a positive definite matrix.

9. 设对称矩阵 $\boldsymbol{A}$ 为正定矩阵. 证明: 存在可逆矩阵 $\boldsymbol{U}$, 使 $\boldsymbol{A}=\boldsymbol{U}^{\mathrm{T}}\boldsymbol{U}$.

9. Let $\boldsymbol{A}$ be a positive definite matrix. Prove that there exists an invertible matrix $\boldsymbol{U}$ such that $\boldsymbol{A}=\boldsymbol{U}^{\mathrm{T}}\boldsymbol{U}$.

补充题 6

Supplement Exercise 6

1. 若 $\boldsymbol{A}$ 与 $\boldsymbol{A}-\boldsymbol{E}$ 是正定矩阵, 证明 $\boldsymbol{E}-\boldsymbol{A}^{-1}$ 也是正定矩阵.

1. If $\boldsymbol{A}$ and $\boldsymbol{A}-\boldsymbol{E}$ are positive definite matrices, prove that $\boldsymbol{E}-\boldsymbol{A}^{-1}$ is also a positive definite matrix.

2. 若 $\boldsymbol{A}$ 与 $\boldsymbol{B}$ 是两个 n 阶实对称矩阵, 并且 $\boldsymbol{A}$ 是正定的. 证明: 存在一个实可逆矩阵 $\boldsymbol{U}$, 使得 $\boldsymbol{U}^{\mathrm{T}}\boldsymbol{A}\boldsymbol{U}$ 和 $\boldsymbol{U}^{\mathrm{T}}\boldsymbol{B}\boldsymbol{U}$ 都是对角矩阵.

2. If $\boldsymbol{A}$ and $\boldsymbol{B}$ are real symmetric matrices of order n and if $\boldsymbol{A}$ is positive definite, prove that there exists a real invertible matrix $\boldsymbol{U}$ such that $\boldsymbol{U}^{\mathrm{T}}\boldsymbol{A}\boldsymbol{U}$ and $\boldsymbol{U}^{\mathrm{T}}\boldsymbol{B}\boldsymbol{U}$ are diagonal matrices simultaneously.

3. 设 $\boldsymbol{A}$ 是一个实对称矩阵. 求证: 对充分大的 t, $t\boldsymbol{E}+\boldsymbol{A}$ 是对称正定矩阵.

3. Let $\boldsymbol{A}$ be a real symmetric matrix. Prove that $t\boldsymbol{E}+\boldsymbol{A}$ is a symmetric positive definite matrix for sufficiently large t.

习题参考答案

习题 1

1. (1) $1-a$; (2) 0; (3) 0; (4) $a_{11}a_{22}-a_{21}a_{12}$;
   (5) -3; (6) 48; (7) 90; (8) 0.

2. (1) $(-1)^{\frac{(n-1)(n-2)}{2}}n!$; (2) $a_1a_2\cdots a_n\left(1+\sum\limits_{i=1}^{n}\dfrac{1}{a_i}\right)$;

   (3) $\left(\sum\limits_{i=1}^{n}x_i-m\right)(-m)^{n-1}$; (4) $a(x+a)^n$.

3. 略.
4. 4; 0.
5. $x_1=0,x_2=1,\cdots,x_n=n-1$.
6. -12.
7. (1) $x=2,y=0,z=-2$; (2) $x_1=3,x_2=-4,x_3=-1,x_4=1$.
8. $\lambda=2$, 或 5 或 8.
9. $a\neq 1$ 且 $b\neq 0$.

补充题 1

1–4. 略.

5. 0, 1.

6. $f(x)=7-5x^2+2x^3$.

习题 2

1. (1) $\begin{pmatrix}-2 & 13 & 22\\ -2 & -17 & 20\\ 4 & 29 & -2\end{pmatrix}$; (2) $\begin{pmatrix}6 & 1 & 6\\ 0 & -7 & 4\\ 2 & 5 & -4\end{pmatrix}$.

2. (1) $\begin{pmatrix}6 & -7 & 9\\ 20 & -5 & -7\end{pmatrix}$;

   (2) $a_{11}x_1^2+a_{22}x_2^2+a_{33}x_3^2+2a_{12}x_1x_2+2a_{13}x_1x_3+2a_{23}x_2x_3$;

(3) $\begin{pmatrix} \lambda^n & n\lambda^{n-1} \\ 0 & \lambda^n \end{pmatrix}$; (4) $\begin{pmatrix} 35 \\ 6 \\ 49 \end{pmatrix}$;

(5) 10; (6) $\begin{pmatrix} 3 & 6 & 9 \\ 2 & 4 & 6 \\ 1 & 2 & 3 \end{pmatrix}$.

3. (1) $\begin{pmatrix} 1 & 1 \\ 0 & 0 \end{pmatrix}$; (2) $\begin{pmatrix} 1 & 0 \\ 0 & 1 \end{pmatrix}$; (3) $\begin{pmatrix} -8 & -12 \\ 6 & 9 \end{pmatrix}$.

4. (1) $f(\boldsymbol{A}) = \begin{pmatrix} 0 & 0 & 0 \\ 0 & 0 & 0 \\ 0 & 1 & 0 \end{pmatrix}$; (2) $\boldsymbol{B}^{-1} = -2\boldsymbol{E} - \boldsymbol{B}$.

5–6. 略.

7. (1) $\begin{pmatrix} 5 & -2 \\ -2 & 1 \end{pmatrix}$; (2) $\begin{pmatrix} \cos\theta & \sin\theta \\ -\sin\theta & \cos\theta \end{pmatrix}$; (3) $\begin{pmatrix} 1 & 3 & -2 \\ -\frac{3}{2} & -3 & \frac{5}{2} \\ 1 & 1 & -1 \end{pmatrix}$.

8. (1) $\boldsymbol{X} = \begin{pmatrix} 2 \\ 0 \end{pmatrix}$; (2) $\boldsymbol{X} = \begin{pmatrix} -2 & 2 & 1 \\ -\frac{8}{3} & 5 & -\frac{2}{3} \end{pmatrix}$;

(3) $\boldsymbol{X} = \begin{pmatrix} 1 & 1 \\ \frac{1}{4} & 0 \end{pmatrix}$; (4) $\boldsymbol{X} = \begin{pmatrix} 3 & -8 & -6 \\ 2 & -9 & -6 \\ -2 & 12 & 9 \end{pmatrix}$.

9. (1) $\boldsymbol{X} = \begin{pmatrix} 1 \\ 0 \\ 0 \end{pmatrix}$; (2) $\boldsymbol{X} = \begin{pmatrix} 5 \\ 0 \\ 3 \end{pmatrix}$.

10. 略.

11. $\boldsymbol{A}^{-1} = \frac{1}{2}(\boldsymbol{A} - \boldsymbol{E}), (\boldsymbol{A} + 2\boldsymbol{E})^{-1} = \frac{1}{4}(3\boldsymbol{E} - \boldsymbol{A})$.

12. $|2\boldsymbol{AB}| = -160, \left|\boldsymbol{AB}^{\mathrm{T}}\right| = -20, |\boldsymbol{A}^{-1}\boldsymbol{B}^{-1}| = -\frac{1}{20}, |-\boldsymbol{A}^{-1}\boldsymbol{B}| = \frac{5}{4}$.

13. $(-1)^n 2^{2n-1}$.

14. $\frac{1}{2}\boldsymbol{A}$.

15. (1) $|\boldsymbol{A}^8| = 10^{16}$; (2) $\boldsymbol{A}^4 = \begin{pmatrix} 5^4 & 0 & 0 & 0 \\ 0 & 5^4 & 0 & 0 \\ 0 & 0 & 2^4 & 0 \\ 0 & 0 & 2^6 & 2^4 \end{pmatrix}$;

(3) $\boldsymbol{A}^{-1} = \begin{pmatrix} \frac{3}{25} & \frac{4}{25} & 0 & 0 \\ \frac{4}{25} & -\frac{3}{25} & 0 & 0 \\ 0 & 0 & \frac{1}{2} & 0 \\ 0 & 0 & -\frac{1}{2} & \frac{1}{2} \end{pmatrix}$.

16–17. 略.

18. $\begin{pmatrix} 0 & \boldsymbol{B}^{-1} \\ \boldsymbol{A}^{-1} & 0 \end{pmatrix}$.

19. $\begin{pmatrix} 1 & 0 & -1 & 0 \\ -1 & -3 & 0 & -1 \\ 4 & 3 & 1 & 2 \\ -3 & -4 & 3 & 0 \end{pmatrix}$.

补充题 2

1. 略.
2. 3.
3. $7^{n-1}\boldsymbol{A}$.
4. $\begin{pmatrix} 3 & 0 & 0 \\ 0 & 3 & 0 \\ 0 & 0 & -1 \end{pmatrix}$.
5. $\frac{125}{36}$.
6. $\frac{1}{2}$.
7. $|\boldsymbol{AB}| = 2, |\boldsymbol{BA}| = 0$.
8. 40.
9. $-4\boldsymbol{A}$.
10. 略.
11. $\begin{pmatrix} 2 & 0 & 0 \\ 0 & -4 & 0 \\ 0 & 0 & 2 \end{pmatrix}$.

12. (1) 略; (2) $\begin{pmatrix} \frac{1}{2} & 0 & 0 \\ 0 & \frac{7}{2} & -\frac{1}{2} \\ 0 & 9 & -4 \end{pmatrix}$.

13. $|\boldsymbol{A}|^{(n-1)^2}$.

习题 3

1. (1) 行最简形 $\begin{pmatrix} 1 & 0 & 0 \\ 0 & 1 & 0 \\ 0 & 0 & 1 \end{pmatrix}$, $R(\boldsymbol{A}) = 3$;

(2) 行最简形 $\begin{pmatrix} 1 & 0 & -1 & -1 \\ 0 & 1 & -4 & -1 \\ 0 & 0 & 0 & 0 \end{pmatrix}$, $R(\boldsymbol{B}) = 2$;

(3) 行最简形 $\begin{pmatrix} 1 & 3 & 0 & 0 & 0 \\ 0 & 0 & 1 & 0 & 0 \\ 0 & 0 & 0 & 1 & 0 \\ 0 & 0 & 0 & 0 & 1 \end{pmatrix}$, $R(\boldsymbol{C}) = 4$.

2. $a = 2$.

3. $\lambda = 3, R(\boldsymbol{A}) = 2; \lambda \neq 3, R(\boldsymbol{A}) = 3$.

4. 可能有等于零的 $r-1$ 阶子式和 r 阶子式, 但没有不等于零的 $r+1$ 阶子式.

5. (1) $\begin{pmatrix} -2 & 1 & 0 \\ \frac{13}{6} & -1 & \frac{1}{6} \\ \frac{4}{3} & -1 & \frac{1}{3} \end{pmatrix}$; (2) $\frac{1}{24}\begin{pmatrix} 24 & 0 & 0 & 0 \\ -12 & 12 & 0 & 0 \\ -12 & -4 & 8 & 0 \\ -9 & -5 & -2 & 6 \end{pmatrix}$;

(3) $\begin{pmatrix} 1 & 1 & -2 & -4 \\ 0 & 1 & 0 & -1 \\ -1 & -1 & 3 & 6 \\ 2 & 1 & -6 & -10 \end{pmatrix}$.

6. (1), (3) 对应的非齐次线性方程组有唯一解;

(2), (4) 对应的非齐次线性方程组无解;

(5) 对应的非齐次线性方程组有无穷多解.

7. $\begin{pmatrix} x_1 \\ x_2 \\ x_3 \\ x_4 \end{pmatrix} = k_1 \begin{pmatrix} -2 \\ 1 \\ 0 \\ 0 \end{pmatrix} + k_2 \begin{pmatrix} 1 \\ 0 \\ 0 \\ 1 \end{pmatrix}$ (k_1, k_2 为任意实数).

8. $\begin{pmatrix} x_1 \\ x_2 \\ x_3 \\ x_4 \end{pmatrix} = k_1 \begin{pmatrix} 4 \\ -2 \\ 1 \\ 0 \end{pmatrix} + k_2 \begin{pmatrix} -1 \\ -2 \\ 0 \\ 1 \end{pmatrix} + \begin{pmatrix} -1 \\ 1 \\ 0 \\ 0 \end{pmatrix}$ (k_1, k_2 为任意实数).

9. (1) 当 $\lambda \neq 1, -2$ 时, 有唯一解

$$x_1 = \frac{-\lambda - 1}{\lambda + 2}, \quad x_2 = \frac{1}{\lambda + 2}, \quad x_3 = \frac{(\lambda + 1)^2}{(\lambda + 2)}$$

(2) 当 $\lambda = -2$ 时, 无解;

(3) 当 $\lambda = 1$ 时, 有无穷多个解, 其解为

$$\begin{pmatrix} x_1 \\ x_2 \\ x_3 \end{pmatrix} = k_1 \begin{pmatrix} -1 \\ 1 \\ 0 \end{pmatrix} + k_2 \begin{pmatrix} -1 \\ 0 \\ 1 \end{pmatrix} + \begin{pmatrix} 1 \\ 0 \\ 0 \end{pmatrix} \quad (k_1, k_2\text{为任意实数})$$

10. (1) $a = 2$ 且 $b \neq -2$ 时无解;

(2) $a \neq 2$ 时有唯一解;

(3) $a = 2$ 且 $b = 2$ 时, 有无穷解, 此时 $\begin{pmatrix} x_1 \\ x_2 \\ x_3 \\ x_4 \end{pmatrix} = \begin{pmatrix} 0 \\ \frac{3}{7} \\ 0 \\ \frac{1}{7} \end{pmatrix} + k \begin{pmatrix} 0 \\ -2 \\ 1 \\ 0 \end{pmatrix}$ (k 为任意实数).

11–12. 略.

补充题 3

1. $\begin{pmatrix} 3 & 2 & 1 \\ 4 & 3 & 2 \\ 16005 & 12004 & 8003 \end{pmatrix}$.

2. $\begin{pmatrix} 0 & 1 & 1 \\ 1 & 0 & 0 \\ 0 & 0 & 1 \end{pmatrix}$.

3. $\begin{pmatrix} x_1 \\ x_2 \\ x_3 \end{pmatrix} = \begin{pmatrix} 10 \\ -15 \\ 12 \end{pmatrix}$, $\begin{pmatrix} x_1 \\ x_2 \\ x_3 \end{pmatrix} = \begin{pmatrix} 2 \\ -3 \\ 4 \end{pmatrix}$.

4. $a = -1$.

5. $t = 3$.

习题 4

1. $\boldsymbol{\alpha}=\begin{pmatrix}11\\8\\7\\14\end{pmatrix}$.

2. $\boldsymbol{\beta}=2\boldsymbol{\alpha}_1-\boldsymbol{\alpha}_2-3\boldsymbol{\alpha}_3$.

3. $x\neq 5$.

4. 是.

5. 略.

6. (1) 错; (2) 正确; (3) 错; (4) 错; (5) 错.

7–8. 略.

9. (1) $R(\boldsymbol{\alpha}_1,\boldsymbol{\alpha}_2,\boldsymbol{\alpha}_3)=3$, 向量组本身为一最大线性无关组;

(2) $R(\boldsymbol{\alpha}_1,\boldsymbol{\alpha}_2,\boldsymbol{\alpha}_3)=2$, $\boldsymbol{\alpha}_1,\boldsymbol{\alpha}_2$ 为一最大线性无关组;

(3) $R(\boldsymbol{\alpha}_1,\boldsymbol{\alpha}_2,\boldsymbol{\alpha}_3)=3$, $\boldsymbol{\alpha}_1,\boldsymbol{\alpha}_2,\boldsymbol{\alpha}_3$ 为一最大线性无关组.

10. $a\neq 6$ 且 $b\neq -2$ 时, $\boldsymbol{\alpha}_1,\boldsymbol{\alpha}_2,\boldsymbol{\alpha}_3,\boldsymbol{\alpha}_4$ 线性无关;

$a=6$ 或 $b=-2$ 时, $\boldsymbol{\alpha}_1,\boldsymbol{\alpha}_2,\boldsymbol{\alpha}_3,\boldsymbol{\alpha}_4$ 线性相关.

11. $\boldsymbol{\alpha}_1,\boldsymbol{\alpha}_2,\boldsymbol{\alpha}_3$ 为一个最大线性无关组, 并且 $\boldsymbol{\alpha}_4=\boldsymbol{\alpha}_1+3\boldsymbol{\alpha}_2-\boldsymbol{\alpha}_3$.

12. $a=15$, $b=5$.

13. (1) 基础解系为 $\boldsymbol{\xi}_1=\begin{pmatrix}-2\\1\\0\\0\\0\end{pmatrix}$, $\boldsymbol{\xi}_2=\begin{pmatrix}-3\\0\\1\\0\\0\end{pmatrix}$, $\boldsymbol{\xi}_3=\begin{pmatrix}-4\\0\\0\\1\\0\end{pmatrix}$, $\boldsymbol{\xi}_4=\begin{pmatrix}-5\\0\\0\\0\\1\end{pmatrix}$,

通解为 $\boldsymbol{X}=k_1\boldsymbol{\xi}_1+k_2\boldsymbol{\xi}_2+k_3\boldsymbol{\xi}_3+k_4\boldsymbol{\xi}_4$($k_1,k_2,k_3,k_4$ 为任何实数);

(2) 基础解系为 $\boldsymbol{\xi}_1=\begin{pmatrix}\frac{2}{7}\\\frac{5}{7}\\1\\0\end{pmatrix}$, $\boldsymbol{\xi}_2=\begin{pmatrix}\frac{3}{7}\\\frac{4}{7}\\0\\1\end{pmatrix}$, 通解为 $\boldsymbol{X}=k_1\boldsymbol{\xi}_1+k_2\boldsymbol{\xi}_2$($k_1,k_2$ 为任何实数);

(3) 基础解系为 $\boldsymbol{\xi}_1=\begin{pmatrix}0\\1\\0\\4\end{pmatrix}$, $\boldsymbol{\xi}_2=\begin{pmatrix}-4\\0\\1\\-3\end{pmatrix}$; 通解为 $\boldsymbol{X}=k_1\boldsymbol{\xi}_1+k_2\boldsymbol{\xi}_2$($k_1,k_2$ 为任何实数);

(4) 基础解系为 $\boldsymbol{\xi}_1=\begin{pmatrix}-1\\1\\0\\0\\0\end{pmatrix}$, $\boldsymbol{\xi}_2=\begin{pmatrix}-1\\0\\-1\\0\\1\end{pmatrix}$, 通解为 $\boldsymbol{X}=k_1\boldsymbol{\xi}_1+k_2\boldsymbol{\xi}_2$($k_1,k_2$ 为任意实数).

14. (1) 通解为 $\boldsymbol{X}=\begin{pmatrix}x_1\\x_2\\x_3\end{pmatrix}=\begin{pmatrix}1\\0\\0\end{pmatrix}+k_1\begin{pmatrix}-2\\1\\0\end{pmatrix}+k_2\begin{pmatrix}-3\\0\\1\end{pmatrix}$($k_1,k_2$ 为任意实数);

(2) 通解为 $\boldsymbol{X}=\begin{pmatrix}x_1\\x_2\\x_3\\x_4\end{pmatrix}=\begin{pmatrix}-2\\5\\0\\0\end{pmatrix}+k_1\begin{pmatrix}-1\\2\\1\\0\end{pmatrix}+k_2\begin{pmatrix}5\\-7\\0\\1\end{pmatrix}$($k_1,k_2$ 为任意实数);

(3) 通解为 $\boldsymbol{X}=\begin{pmatrix}x_1\\x_2\\x_3\end{pmatrix}=\begin{pmatrix}-1\\2\\0\end{pmatrix}+k\begin{pmatrix}-2\\1\\1\end{pmatrix}$($k$ 为任意实数);

(4) 通解为 $\boldsymbol{X}=\begin{pmatrix}x_1\\x_2\\x_3\\x_4\\x_5\end{pmatrix}=\begin{pmatrix}-1\\8\\0\\0\\0\end{pmatrix}+k_1\begin{pmatrix}-1\\-1\\1\\0\\0\end{pmatrix}+k_2\begin{pmatrix}2\\-3\\0\\1\\0\end{pmatrix}+k_2\begin{pmatrix}1\\-3\\0\\0\\1\end{pmatrix}$($k_1,k_2,k_3$ 为任意实数).

15. 通解为 $\boldsymbol{X}=\begin{pmatrix}\frac{1}{2}\\\frac{1}{2}\\0\\1\end{pmatrix}+k\begin{pmatrix}0\\1\\-1\\-1\end{pmatrix}$($k$ 为任意实数).

16. (1) $a\neq 1$ 且 $b\neq 0$ 时, 方程组有唯一解, 其唯一解由克拉默法则求出, 为

$$x_1=\frac{3-4b}{b(1-a)},\quad x_2=\frac{3}{b},\quad x_3=\frac{4b-3}{b(1-a)}.$$

(2) 当 $b=0$ 时, 方程组无解.

(3) 当 $b\neq 0, a=1$ 时

当 $b\neq\frac{3}{4}$ 时, 秩 $(\boldsymbol{A})=2<$ 秩 $(\overline{\boldsymbol{A}})=3$, 方程组无解.

当 $b=\dfrac{3}{4}$ 时, 通解为 $\boldsymbol{X}=\begin{pmatrix}x_1\\x_2\\x_3\end{pmatrix}=\begin{pmatrix}0\\4\\0\end{pmatrix}+k\begin{pmatrix}-1\\0\\1\end{pmatrix}$, 其中 k 为任意常数.

17. (1) 当 $k_1\neq 2$ 时, 秩 $(\boldsymbol{A})=$ 秩 $(\bar{\boldsymbol{A}})=4$, 方程组有唯一解;

(2) 当 $k_1=2$ 时, 若 $k_2\neq 1$, 则秩 $(\boldsymbol{A})=3<$ 秩 $(\bar{\boldsymbol{A}})=4$, 方程组无解;

(3) 当 $k_1=2$ 时, 若 $k_2=1$, 则秩 $(\boldsymbol{A})=$ 秩 $(\bar{\boldsymbol{A}})=3$, 方程组有无穷多解, 且通解为

$$X=\begin{pmatrix}x_1\\x_2\\x_3\\x_4\end{pmatrix}=\begin{pmatrix}-8\\3\\0\\2\end{pmatrix}+k\begin{pmatrix}0\\-2\\1\\0\end{pmatrix}\text{(其中}k\text{为任意常数)}$$

18. $\begin{pmatrix}1&2&0\\1&0&0\\-1&-1&0\end{pmatrix}$.

19–23. 略

补充题 4

1. $\lambda=-2$.

2. (1)$a=-1,b\neq 0$; (2) $a\neq -1$; (3) $a=-1$ 且 $b=0$.

3. 可以; 不可以.

4. $t=-1$.

5. $t=\dfrac{-1}{4}$.

6–8. 略

9. 通解为 $\boldsymbol{X}=\begin{pmatrix}x_1\\x_2\\x_3\end{pmatrix}=k\begin{pmatrix}13\\-5\\-1\end{pmatrix}+\begin{pmatrix}6\\-1\\1\end{pmatrix}$ (k 为任意实数).

10. 通解为 $\boldsymbol{X}=k\begin{pmatrix}1\\-2\\1\\0\end{pmatrix}+\begin{pmatrix}1\\1\\1\\1\end{pmatrix}$ (k 为任意实数).

习题 5

1. (1) 特征值: $\lambda_1=-1$(三重),

相应的特征向量: $k\begin{pmatrix}1\\1\\-1\end{pmatrix}$ (k 为任意实数);

(2) 特征值: $\lambda_1=0,\lambda_2=-1,\lambda_3=9$,

相应特征向量: $k\begin{pmatrix}-1\\-1\\1\end{pmatrix}, k\begin{pmatrix}-1\\1\\0\end{pmatrix}, k\begin{pmatrix}1/2\\1/2\\1\end{pmatrix}$ (k 为任意实数);

(3) 特征值: $\lambda_1=\lambda_2=-1, \lambda_3=\lambda_4=1$,

相应特征向量: $k_1\begin{pmatrix}1\\0\\0\\-1\end{pmatrix}+k_2\begin{pmatrix}0\\1\\-1\\0\end{pmatrix}, k_1\begin{pmatrix}1\\0\\0\\1\end{pmatrix}+k_2\begin{pmatrix}0\\1\\1\\0\end{pmatrix}$ (k_1, k_2 为任意实数);

(4) 特征值: $\lambda_1=\lambda_2=7, \lambda_3-2$,

相应特征向量: $k_1\begin{pmatrix}-1/2\\1\\0\end{pmatrix}+k_2\begin{pmatrix}-1\\0\\1\end{pmatrix}, k\begin{pmatrix}1\\1/2\\1\end{pmatrix}$ (k_1, k_2, k 为任意实数).

2. $\boldsymbol{A}$ 看成实数域上的矩阵, 没有特征值;

$\boldsymbol{A}$ 看成复数域上的矩阵, 特征值为: $1+\sqrt{3}\mathrm{i}$ 与 $1-\sqrt{3}\mathrm{i}$;

相应的特征向量为: $k\begin{pmatrix}\mathrm{i}\\1\end{pmatrix}$ 与 $k\begin{pmatrix}-\mathrm{i}\\1\end{pmatrix}$ (k 为任意实数).

3–6. 略.

7. (1) 不可以; (2), (3), (4) 可以.

8. (1) 特征值为: $\lambda_1=1, \lambda_2=2, \lambda_3=4$, 相应的特征向量为: $\begin{pmatrix}1\\-1\\-1\end{pmatrix}, \begin{pmatrix}0\\1\\-1\end{pmatrix}, \begin{pmatrix}2\\1\\1\end{pmatrix}$;

(2) $\boldsymbol{P}=\begin{pmatrix}1&0&2\\-1&1&1\\1&-1&1\end{pmatrix}, \boldsymbol{P}^{-1}\boldsymbol{AP}=\begin{pmatrix}1&&\\&2&\\&&4\end{pmatrix}$.

9. (1) 特征值: $\lambda_1=-4, \lambda_2=-6, \lambda_3=-12$; $\begin{pmatrix}-4&&\\&-6&\\&&-12\end{pmatrix}$.

(2) $|\boldsymbol{B}|=-288$; $|\boldsymbol{A}-5\boldsymbol{E}|=-72$.

10. $\begin{pmatrix}1&2^{100}-1&1-2^{100}\\0&2-2^{100}&2^{100}-1\\0&2-2^{101}&2^{101}-1\end{pmatrix}$.

11. $\boldsymbol{\beta}_1=\boldsymbol{\alpha}_1=\begin{pmatrix}1\\0\\-1\\1\end{pmatrix}$,

$$\boldsymbol{\beta}_2=\boldsymbol{\alpha}_2-\frac{[\boldsymbol{\alpha}_2,\boldsymbol{\beta}_1]}{[\boldsymbol{\beta}_1,\boldsymbol{\beta}_1]}\boldsymbol{\beta}_1=\begin{pmatrix}1\\-1\\0\\1\end{pmatrix}-\frac{2}{3}\begin{pmatrix}1\\0\\-1\\1\end{pmatrix}=\begin{pmatrix}\frac{1}{3}\\-1\\\frac{2}{3}\\\frac{1}{3}\end{pmatrix},$$

$$\boldsymbol{\beta}_3=\boldsymbol{\alpha}_3-\frac{[\boldsymbol{\alpha}_3,\boldsymbol{\beta}_1]}{[\boldsymbol{\beta}_1,\boldsymbol{\beta}_1]}\boldsymbol{\beta}_1-\frac{[\boldsymbol{\alpha}_3,\boldsymbol{\beta}_2]}{[\boldsymbol{\beta}_2,\boldsymbol{\beta}_2]}\boldsymbol{\beta}_2=\begin{pmatrix}-\frac{1}{5}\\\frac{3}{5}\\\frac{3}{5}\\\frac{4}{5}\end{pmatrix}.$$

12. (1) $\sqrt{6}$, (2) $3\sqrt{2}$.

13. (1) 是, (2) 不是.

14–17. 略.

18. (1) $\boldsymbol{P}=\begin{pmatrix}\frac{2\sqrt{5}}{5}&\frac{2\sqrt{5}}{15}&\frac{1}{3}\\\frac{-\sqrt{5}}{5}&\frac{4\sqrt{5}}{15}&\frac{2}{3}\\0&\frac{\sqrt{5}}{3}&\frac{-2}{3}\end{pmatrix}$, $\boldsymbol{P}^{-1}\boldsymbol{AP}=\begin{pmatrix}1&0&0\\0&1&0\\0&0&-8\end{pmatrix}$;

(2) $\boldsymbol{P}=\begin{pmatrix}\frac{1}{\sqrt{3}}&-\frac{1}{\sqrt{2}}&-\frac{1}{\sqrt{6}}\\\frac{1}{\sqrt{3}}&\frac{1}{\sqrt{2}}&-\frac{1}{\sqrt{6}}\\\frac{1}{\sqrt{3}}&0&\frac{2}{\sqrt{6}}\end{pmatrix}$, $\boldsymbol{P}^{-1}\boldsymbol{AP}=\begin{pmatrix}0&0&0\\0&3&0\\0&0&3\end{pmatrix}$;

(3) $\boldsymbol{P}=\begin{pmatrix}\frac{\sqrt{2}}{2}&0&\frac{1}{2}&\frac{1}{2}\\0&\frac{\sqrt{2}}{2}&\frac{-1}{2}&\frac{1}{2}\\\frac{\sqrt{2}}{2}&0&\frac{-1}{2}&\frac{-1}{2}\\0&\frac{\sqrt{2}}{2}&\frac{1}{2}&\frac{-1}{2}\end{pmatrix}$, $\boldsymbol{P}^{-1}\boldsymbol{AP}=\begin{pmatrix}4&0&0&0\\0&4&0&0\\0&0&2&0\\0&0&0&6\end{pmatrix}$;

(4) $\boldsymbol{P}=\begin{pmatrix}\frac{2}{3} & \frac{2}{3} & \frac{1}{3}\\ \frac{1}{3} & \frac{-2}{3} & \frac{2}{3}\\ \frac{-2}{3} & \frac{1}{3} & \frac{2}{3}\end{pmatrix}$, $\boldsymbol{P}^{-1}\boldsymbol{A}\boldsymbol{P}=\begin{pmatrix}2 & 0 & 0\\ 0 & 5 & 0\\ 0 & 0 & -1\end{pmatrix}$.

补充题 5

1. $a=1$.

2. 2, 2, 2.

3. 1, 7, 7.

4. 1, 2.

5. 18.

6. $a\neq -9$.

7. $\boldsymbol{A}=\frac{1}{3}\begin{pmatrix}7 & 0 & -2\\ 0 & 5 & -2\\ -2 & -2 & 6\end{pmatrix}$.

8. $\boldsymbol{P}^{-1}\boldsymbol{A}\boldsymbol{P}=\begin{pmatrix}2 & 0 & 0\\ 0 & 2 & 0\\ 0 & 0 & 6\end{pmatrix}, \boldsymbol{P}=\begin{pmatrix}1 & 1 & 1\\ -1 & 0 & -2\\ 0 & 1 & 3\end{pmatrix}$.

9. (1) $x=0, y=-2$; (2) $\boldsymbol{P}=\begin{pmatrix}0 & 0 & 1\\ 2 & 1 & 0\\ -1 & 1 & -1\end{pmatrix}$.

10. 略.

习题 6

1. (1) $\boldsymbol{A}=\begin{pmatrix}1 & -1 & \frac{3}{2}\\ -1 & -2 & 4\\ \frac{3}{2} & 4 & 3\end{pmatrix}$; (2) $\boldsymbol{A}=\begin{pmatrix}0 & \frac{1}{2} & \frac{-1}{2} & 0\\ \frac{1}{2} & 0 & 1 & 0\\ \frac{-1}{2} & 1 & 0 & 1\end{pmatrix}$.

2. (1) $f=x_1^2-2x_1x_2-6x_1x_3+2x_1x_4-4x_2x_3+x_2x_4+\frac{1}{3}x_3^2-3x_3x_4$;

(2) $f=2x_1x_2+x_1x_3-3x_1x_4-2x_2x_3-2x_2x_4+6x_3x_4$.

3. (1) 标准形为 $f=y_1^2+y_2^2$, $\boldsymbol{X}=\boldsymbol{C}\boldsymbol{Y}, \boldsymbol{C}=\begin{pmatrix}1 & -1 & 2\\ 0 & 1 & -2\\ 0 & 0 & 1\end{pmatrix}$;

(2) 标准形为 $f=y_1^2-y_2^2$, $\boldsymbol{X}=\boldsymbol{C}\boldsymbol{Y}$, $\boldsymbol{C}=\begin{pmatrix} 1 & \frac{1}{2} & \frac{-3}{2} \\ 0 & \frac{1}{2} & \frac{-1}{2} \\ 0 & 0 & 1 \end{pmatrix}$.

4. (1) $\begin{pmatrix} 1 & -2 & \frac{-1}{2} \\ 0 & 1 & \frac{1}{4} \\ 0 & 0 & 1 \end{pmatrix}$; (2) $\begin{pmatrix} 1 & \frac{-1}{2} & 1 \\ 1 & \frac{1}{2} & 2 \\ 0 & 0 & 1 \end{pmatrix}$.

5. (1) $\boldsymbol{P}=\frac{1}{3}\begin{pmatrix} 2 & 2 & 1 \\ -1 & 2 & -2 \\ -2 & 1 & 2 \end{pmatrix}$, $\boldsymbol{P}^{\mathrm{T}}\boldsymbol{A}\boldsymbol{P}=2y_1^2-y_2^2+5y_3^2$;

(2) $\boldsymbol{P}=\frac{1}{\sqrt{2}}\begin{pmatrix} 1 & 0 & 1 & 0 \\ 1 & 0 & -1 & 0 \\ 0 & 1 & 0 & 1 \\ 0 & 1 & 0 & -1 \end{pmatrix}$, $\boldsymbol{P}^{\mathrm{T}}\boldsymbol{A}\boldsymbol{P}=y_1^2+y_2^2-y_3^2-y_4^2$.

6. (1) $-0.8<a<0$; (2) $a>2$.

7–9. 略.

补充题 6

1-3. 略.

中一英名词索引

B

伴随矩阵 (adjoint matrix) 2.2, 2.3, 3.1
半负定的 (negative semidefinite) 6.3
半正定的 (positive semidefinite) 6.3
保秩变换 (rank-preserving operation) 3.2
倍 (multiple) 3.2
倍乘性 (multiple-multiplication) 1.2
倍加 (multiple-adding) 3.1
倍加性 (multiple-additivity) 1.2
必要性 (necessity) 2.1, 3.1, 4.2, 5.2
标准形 (canonical form, standard form) 3.1, 3.2, 6.2, 6.3
标准正交向量组 (orthonormal vector set) 5.3, 5.4
不定的 (indefinite) 6.3

C

差 (difference) 1.1, 4.1
长度 (length) 5.3
乘积 (multiplication, product) 1.1, 2.1
充分必要条件 (充要条件、当且仅当) (necessary and sufficient condition, if and only if) 1.4, 2.2, 2.4, 3.1, 4.1, 4.2, 4.3, 5.1, 5.2, 5.3, 6.3
充分性 (sufficiency) 2.2, 3.1, 4.2, 5.2
重根 (repeated root, multiple root) 5.1, 5.4
重数 (multiplicity, multiple numbers) 5.1, 5.2, 5.4
重特征值 (multiple eigenvalue) 5.1
初等变换 (elementary operation) 3.1, 3.2, 4.2, 4.3, 4.4, 6.2
初等变换法 (method of elementary operations) 6.2
初等行变换 (elementary row operation) 3.1, 3.2, 3.3, 4.3, 4.4
初等列变换 (elementary column operation) 3.1
初等矩阵 (elementary matrix) 3.1, 3.2, 4.3, 6.2
传递性 (transitivity) 3.1, 4.1, 5.2, 6.1

D

代数和 (algebraic sum) 1.1
代数余子式 (algebraic cofactor) 1.2, 1.3, 1.4
单根 (simple root) 5.1
单位矩阵 (unit (identity) matrix) 2.1
单位向量 (unit (identity) vector) 4.1, 4.2, 5.3, 5.4
导出组 (derived system) 4.4
等价 (equivalence) 4.1, 4.3
等价关系 (equivalence relation) 3.1, 5.2
等式 (方程)(equation) 1.1
递推公式 (recursive formula) 1.1, 1.3
第一 (二、三) 类初等矩阵 (elementary matrix of the first (second, third) kind) 3.1
定理 (theorem) 1.2
定义 (definition) 1.1
对称 (symmetry) 3.1, 4.1, 5.2, 6.1
对称矩阵 (symmetric matrix) 2.1, 5.4, 6.2, 6.3
对角化 (diagonalization) 5.2, 5.4
对角行列式 (diagonal determinant) 1.1
对角线法则 (diagonal rule) 1.1
对角矩阵 (diagonal matrix) 2.1, 5.2, 5.4, 6.2
多项式 (polynomial) 1.2, 1.4, 5.1

E

二次型 (quadratic form) 6.1, 6.2, 6.3
二阶行列式 (second order determinant) 1.1

二元线性方程组 (system of linear equations in two unknowns) 1.1

F

反对称矩阵 (skew-symmetric matrix) 2.1
反身性 (reflexivity) 3.1, 4.1, 5.2, 6.1
范德蒙德行列式 (Vandermonde determinant) 1.3, 5.1
方阵 (square matrix) 2.1, 2.2, 2.3, 2.4, 3.1, 3.2, 3.3, 5.1, 5.2, 5.3
方阵的对角线 (diagonal of matrix) 2.1
方阵的行列式 (determinant of square matrix) 2.2
方阵的迹 (trace of matrix) 5.1
非负性 (nonnegativity) 5.3
非零解 (nonzero solution) 1.4, 2.2, 3.3, 4.2, 4.3, 4.4, 5.1
非齐次线性方程组 (system of non-homogeneous linear equations) 1.4, 3.3, 4.1, 4.4
非奇异 (满秩) 矩阵 (nonsingular matrix) 2.2, 3.2
非退化的 (non-degenerate) 6.1
分块矩阵 (block matrix) 2.4
分块对角矩阵 (block diagonal matrix) 2.4
分量 (component) 4.1, 4.2, 4.3
分配律 (distributive law) 2.1
复矩阵 (complex matrix) 2.1, 5.4
负矩阵 (negative matrix) 2.1
负向量 (negative vector) 4.1

G

概念 (concept) 2.1, 3.2, 4.1, 5.3, 6.3
格拉姆-施密特正交化 (Gram-Schmidt orthogonalization) 5.3
过渡矩阵 (transition matrix) 4.3

H

行阶梯形矩阵 (row echelon matrix) 3.1, 3.2, 4.2, 4.3
行矩阵 (row matrix) 2.1, 4.1
行列式 (determinant) 1.1
行列式的展开定理 (expansion theorem of determinant) 1.2
行向量 (row vector) 2.1, 4.1, 4.2, 4.3, 5.3
行最简形矩阵 (row-reduced matrix) 3.1, 3.3, 4.3, 4.4
和 (sum) 1.1
合同的 (congruent) 6.1
换行 (row-interchanging) 3.1
霍尔维茨定理 (Hurwitz theorem) 6.3

J

基底 (basis) 4.3
基础解系 (system of fundamental solutions) 4.4, 5.1, 5.2, 5.4, 6.2
极大线性无关向量组 (maximal linearly independent vector subset) 4.3, 4.4, 5.1
极大无关组 (maximal independent subset) 4.3
计算 (calculation, computation, evaluation) 1.1
加法 (addition) 2.1, 4.1
交换律 (commutative law) 2.1
结合律 (associative law) 2.1
解 (solution) 1.1
解的结构 (structure of solutions) 4.4
解向量 (solution vector) 4.1, 4.4
距离 (distance) 5.3
矩阵 (matrix) 2.1
矩阵表示 (matrix representation) 6.1, 6.2
矩阵乘法 (matrix multiplication) 2.1
矩阵的初等变换 (elementary operation on matrix) 3.1
矩阵的加法 (addition of matrices) 2.1
矩阵的相似 (similarity of matrices) 5.2
矩阵的转置 (transpose of matrix) 2.1
矩阵的秩 (rank of matrix) 3.2, 4.3
矩阵方程 (matrix equation) 2.3

K

k 阶子式 (k-th order subdeterminant) 3.2
可对角化的 (diagonalizable) 5.2
可加性 (additivity) 1.2

可交换的 (commutable) 2.1
可逆矩阵 (invertible matrix) 2.2, 2.3, 2.4, 3.1, 3.2, 4.3, 5.2, 6.1
克拉默法则 (Cramer's rule) 1.4, 3.3
柯西-施瓦茨不等式 (Cauchy-Schwarz inequality) 5.3

L

列矩阵 (column matrix) 2.1
列向量 (column vector) 2.1, 4.1, 4.1, 4.2, 4.3
零矩阵 (zero matrix) 2.1
零向量 (zero vector) 4.1
零解 (zero solution) 1.4, 2.2, 3.3, 4.2, 4.3, 4.4

M

m 次多项式 (polynomial of degree m) 2.1
幂 (power) 2.1
命题 (proposition) 习题 2.6, 习题 3.6

N

n 阶方阵 (square matrix of order n) 2.1
n 阶行列式 (n-th order determinant) 1.1
n 维行向量 (n dimensional row vector) 4.1
n 维列向量 (n dimensional column vector) 4.1
n 维向量 (n dimensional vector) 4.1
n 元二次型 (quadratic form in n variables) 6.1
n 元线性方程组 (system of linear equations in n unknowns) 1.1, 1.4, 3.3, 4.3, 4.4
内积 (inner product) 5.3
逆否定理 (converse-negative theorem) 1.4, 2.2
逆矩阵 (inverse matrix) 2.2, 2.3, 3.1
可逆矩阵 (invertible matrix) 2.2, 3.1, 5.2, 5.4, 6.2

P

配方法 (method of completing square) 6.2

Q

齐次线性方程组 (system of homogeneous linear equations) 1.4, 3.3, 4.2, 4.3, 4.4
齐次性 (homogeneity) 5.3
奇异 (降秩) 矩阵 (singular matrix) 2.2, 3.2

S

三角形不等式 (triangle inequality) 5.3
三角形行列式 (triangular determinant) 1.1, 1.2, 1.3
三角形矩阵 (triangular matrix) 2.1
三阶行列式 (third order determinant) 1.1
三元线性方程组 (system of linear equations in three unknowns) 1.1
上三角形行列式 (upper triangular matrix determinant) 1.1, 1.3
上三角形矩阵 (upper triangular matrix) 2.1
实对称矩阵 (real symmetric matrix) 5.4
实矩阵 (real matrix) 2.1
实数 (real number) 2.1
数乘 (scalar multiplication) 2.1, 2.2, 4.1
数量矩阵 (scalar matrix) 2.1
数学归纳法 (mathematical induction) 1.3
数域 (number field) 5.1, 5.4, 6.1
顺序主子式 (sequence principal subdeterminant) 6.3

T

特解 (particular solution) 4.4
特征多项式 (characteristic polynomial, eigenpolynomial) 5.1, 5.2, 6.2
特征方程 (characteristic equation) 5.1, 5.2, 6.2
特征向量 (eigenvector, characteristic vector) 5.1, 5.2, 5.4, 6.2
特征值 (eigenvalue, characteristic value) 5.1, 5.2, 5.4, 6.2
通 (一般) 解 (general solution) 4.4
同解变换 (equivalent operation) 3.1
同型矩阵 (matrices of the same size) 2.1
推论 (corollary) 1.2
退化的 (degenerate) 6.1

W

维数 (dimension) 4.3
唯一解 (unique solution) 1.1, 1.4, 2.2, 4.1, 4.4

X

系数行列式 (coefficient determinant) 1.4, 3.3, 4.1, 4.2, 4.3
系数矩阵 (coefficient matrix) 2.1, 2.2, 4.2, 4.3, 4.4, 6.2
下三角形矩阵 (lower triangular matrix) 2.1
下三角形行列式 (lower triangular matrix determinant) 1.1, 1.2, 1.3
线性变换 (linear transformation, operator) 6.1
线性表示 (linear representation) 4.1
线性代数 (linear algebra)
线性等价 (linearly equivalent) 3.3
线性方程组 (system of linear equations, linear system) 1.1
线性关系 (linear relation) 4.3
线性无关 (linearly independent) 4.1
线性相关 (linearly dependent) 4.1
线性运算 (linear operation) 2.1, 4.1
线性组合 (linear combination) 4.1
相似的 (similar) 5.2
向量 (vector) 4.1
向量的内积 (inner product of vectors) 5.3
向量空间 (vector space) 4.3
向量组 (vector set) 4.1
向量组的秩 (rank of vector set) 4.3
消元法 (elimination) 1.1, 3.1
性质 (property) 1.2

Y

右乘 (postmultiply) 2.1
余子式 (cofactor) 1.1
元素 (element) 1.1, 2.1
运算 (operation) 1.2

Z

增广矩阵 (augmented matrix) 2.1, 3.3, 4.4
整数 (integer number) 2.1
正定 (positive definite) 6.3
正定二次型 (positive definite quadratic form) 6.3
正定矩阵 (positive definite matrix) 6.3
正交的 (orthogonal) 5.3
正交变换 (orthogonal transformation) 6.1
正交矩阵 (orthogonal matrix) 5.3, 5.4, 6.2
正交化 (orthogonalizaton) 5.3, 5.4
正交向量组 (orthogonal vector set) 5.3
证明 (proof) 1.3
秩 (rank) 3.2
主变量 (principal variable) 3.3, 4.4
主子式 (principal subdeterminant) 6.3
注 (note) 1.1
转置 (transpose) 1.2, 2.1
转置行列式 (transpose determinant) 1.2
转置矩阵 (transpose matrix) 2.1, 5.1
子矩阵 (submatrix) 2.4
子块 (subblock) 2.4
子式 (subdeterminant) 3.2, 6.3
自由未知量 (free unknows) 3.3, 4.4
左乘 (premultiply) 2.1
坐标 (coordinate) 4.1, 4.3